# Routledge Handbook of Interdisciplinary Research Methods

The landscape of contemporary research is characterized by a renewed interest in the potential of interdisciplinarity. Yet there are few discussions of the development of interdisciplinary methods and their 'behaviour' in the field.

The *Routledge Handbook of Interdisciplinary Research* presents a bold intervention by showcasing a diversity of stimulating approaches. Over 50 experienced researchers illustrate the challenges, but also the rewards of doing interdisciplinary research through discussions of their own practice and that of others. Each section is dedicated to an aspect of interdisciplinary methodology, including collection, classification, validation and communication to research audiences. Featured projects cover a variety of scales and topics, from small art-science collaborations to the 'big data' of mass observations.

Most importantly, the *Handbook* presents a *distinctive* approach, defamiliarizing and reworking established practices such as experimenting, archiving, observing, prototyping and translating. The focus is on knowledge as process, the compounding of methods, and the role of interdisciplinary methods in activating the present.

**Celia Lury** is Professor and Director of the Centre for Interdisciplinary Methodologies at the University of Warwick.

**Rachel Fensham** is Professor of Dance and Theatre Studies and Assistant Dean of the Digital Studio, Faculty of Arts, University of Melbourne.

**Alexandra Heller-Nicholas** is a writer on contemporary cinema and a Research Associate at the Victorian College of the Arts, University of Melbourne.

**Sybille Lammes** is Professor of New Media and Digital Culture at the University of Leiden.

**Angela Last** is Lecturer in Environmental Humanities at the University of Leicester.

**Mike Michael** is Professor of Sociology in the Department of Sociology, Philosophy and Anthropology at the University of Exeter.

**Emma Uprichard** is Reader at the Centre for Interdisciplinary Methodologies at the University of Warwick.

# Routledge Handbook of Interdisciplinary Research Methods

*Edited by Celia Lury, Rachel Fensham,*
*Alexandra Heller-Nicholas, Sybille Lammes,*
*Angela Last, Mike Michael and Emma Uprichard*

LONDON AND NEW YORK

First published 2018 by Routledge

2 Park Square, Milton Park, Abingdon, Oxfordshire OX14 4RN

52 Vanderbilt Avenue, New York, NY 10017

*Routledge is an imprint of the Taylor & Francis Group, an informa business*

First issued in paperback 2020

*British Library Cataloguing-in-Publication Data*
A catalogue record for this book is available from the British Library

*Library of Congress Cataloging-in-Publication Data*
Names: Lury, Celia, editor.
Title: Routledge handbook of interdisciplinary research methods / edited by Celia Lury [and six others].
Other titles: Handbook of interdisciplinary research methods
Description: Abingdon, Oxon ; New York, NY : Routledge, 2018. | Includes bibliographical references and index.
Identifiers: LCCN 2018011178 | ISBN 9781138886872 (hardback) | ISBN 9781315714523 (ebook)
Subjects: LCSH: Interdisciplinary research.
Classification: LCC Q180.55.148 R68 2018 | DDC 001.4—dc23
LC record available at https://lccn.loc.gov/2018011178

ISBN: 978-1-138-88687-2 (hbk)
ISBN: 978-0-367-65988-2 (pbk)

Typeset in Bembo
by Keystroke, Neville Lodge, Tettenhall, Wolverhampton

# Contents

Contents

Contents

# Figures

Figures

# Contributors

**Barbara Adam,** Professor Emerita (Cardiff University), is a social theorist who has applied the focus on time to the breadth of social science concerns – from education and environmental matters to transport and work – and has published extensively on the subject. The future featured mainly as the unresolved part of this agenda-setting work until an ESRC Professorial Fellowship enabled her to concentrate explicitly on this elusive subject. She is founder editor of the journal *Time & Society*.

**Yoko Akama** is Associate Professor in Communication Design, School of Media and Communication, RMIT University, Australia. Her research focuses on participatory design practices, and design artefacts, language and processes, and their role in enabling people to live in changing economic, cultural and environmental circumstances.

**Catherine Ayres** is a PhD candidate in the School of Sociology at the Australian National University. Her research focuses on the sometimes conflicting ways we conceptualize and experience 'Nature', specifically in the realm of national parks and other protected areas.

**Harmony Bench** is Assistant Professor in the Department of Dance at Ohio State University. Her writing can be found in *Dance Research Journal*, *The International Journal of Performance Arts and Digital Media*, *Participations* and *The International Journal of Screendance*, for which she currently serves as co-editor with Simon Ellis. She is working on a book manuscript tentatively entitled *Dance as Common: Movement as Belonging in Digital Cultures*, as well as *Mapping Touring*, a digital humanities and database project focused on the performance engagements of early twentieth-century dance companies.

**David Bissell** is Senior Lecturer in the Research School of Social Sciences at the Australian National University. He combines qualitative research on embodied practices with social theory to explore the social, political and ethical consequences of mobile lives.

**Monika Büscher** is Professor of Sociology at Lancaster University, Director of the Centre for Mobilities Research and Associate Director at the Institute for Social Futures. Her research explores the digital dimension of contemporary 'mobile lives' with a focus on IT ethics and risk governance. She edits the book series *Changing Mobilities* (Routledge) with Peter Adey.

**Jane Calvert** is a Reader in Science, Technology and Innovation Studies at the University of Edinburgh. Her current research focuses on attempts to engineer living things in the emerging field of synthetic biology. She is interested in interdisciplinary collaborations of all sorts.

**Rebecca Coleman** is Senior Lecturer in the Sociology Department, Goldsmiths, University of London, where she researches and teaches on visual and inventive methodologies, futures and presents, bodies and images. She is currently working on a multi-media book project titled *Engaging Futures: Methods, Materials, Media* (in preparation, Goldsmiths Press). She has recently published *Deleuze and Research Methodologies* (co-edited with Jessica Ringrose, 2013, Edinburgh University Press).

**Alberto Corsín Jiménez** is Reader in Social Anthropology in the Department of the History of Science at the Spanish National Research Council in Madrid. He is the author of *An Anthropological Trompe l'Oeil for a Common World* (Berghahn, 2013) and editor of *Prototyping Cultures: Art, Science and Politics in Beta* (Routledge, 2016), *Culture and Well-Being: Anthropological Approaches to Freedom and Political Ethics* (Pluto, 2008) and *The Anthropology of Organisations* (Ashgate, 2007). His current work examines the rise of an urban commons movement and the development of open-source urban hardware projects by architects, artists and engineers.

**Claire Craig** worked as project administrator on the ERC-funded project, 'The Cultural Politics of Dirt in Africa', from 2013 to 2015. Her work included liaising between the teams in Nairobi and Lagos, ensuring the smooth running of reporting-back procedures, and managing the project blog. She holds an undergraduate degree in Anthropology and a Masters degree in Social Research Methods, both from the University of Sussex.

**Gail Davies** is Professor in Human Geography at the University of Exeter. Her work incorporates insights from geography, science and technology studies and anthropology to explore the spatiality of knowledge practices, biotechnology and emerging animal geographies. She is currently working on a Wellcome Trust Collaborative Award, which is charting changing relationships across the 'Animal Research Nexus'.

**Leila Dawney** is a Senior Lecturer in Human Geography at the University of Brighton, UK. As a theorist of power, affect and embodiment, her research concerns the forms of experience and subjectivity that are produced in and through spaces of late capitalism, and the development of conceptual and methodological tools for thinking about the politics of experience. She is a member of the Authority Research Network.

**Carl DiSalvo** is an Associate Professor in the School of Literature, Media, and Communication at the Georgia Institute of Technology. At Georgia Tech he directs the Public Design Workshop: a design research studio that explores socially engaged design and civic media. DiSalvo's scholarship draws together theories and methods from design research and design studies, the social sciences and the humanities, to analyse the social and political qualities of design, and to prototype experimental systems and services. His first book, *Adversarial Design* (2012), is part of the Design Thinking, Design Theory series at MIT Press. He is also a co-editor of the MIT Press journal *Design Issues*. DiSalvo's experimental design work has been exhibited and supported by the ZKM, Grey Area Foundation for the Arts, Times Square Arts Alliance, Science Gallery Dublin and the Walker Art Center.

**Tuur Driesser** is a PhD candidate at the Centre for Interdisciplinary Methodologies, University of Warwick, UK. His research examines how smart cities are made through digital maps and data visualizations, exploring different methods and methodologies for the study of, with and through maps.

**Luciana Duranti** is Professor of Archival Theory, Diplomatics, and the Management of Digital Records in the master's and doctoral archival programmes of the School of Library, Archival and Information Studies of the University of British Columbia (UBC). She is Director of the Centre for the International Study of Contemporary Records and Archives (CISCRA – www.ciscra. org) and of InterPARES, the largest and longest living publicly funded research project on the long-term preservation of authentic electronic records (1998–2019).

**Catriona Elder** is Associate Professor in the Department of Sociology and Social Policy at the University of Sydney. The focus of her scholarship is race relations and national identity, making contributions to the development of theory emerging in Critical Whiteness Studies in Australia. She is co-coordinator of a WUN *International Indigenous Research Network* where she works with Indigenous and non-Indigenous scholars on issues related to Indigenous knowledges, intercultural research and ethics. She is also part of a team working on an Australian Research Council project (*Telling it like it is*) exploring the contemporary experiences of Aboriginal people in northern Australia.

**Sasha Engelmann** is a creative geographer exploring the poetics and politics of air. Over the past three years she has carried out site-based ethnographic fieldwork at Studio Tomás Saraceno in Berlin, especially related to Saraceno's *Aerocene* project. She is currently Lecturer in GeoHumanities at Royal Holloway University, London, and Director of Artists in Residence at the Centre for GeoHumanities. Sasha holds a DPhil in Geography and the Environment from the University of Oxford.

**Rachel Fensham** is a dance and theatre scholar, and is Professor and Assistant Dean of the Digital Studio, Faculty of Arts, University of Melbourne. In her field of dance and performance studies, her funded research includes a study of regional theatre and young adult audiences and a digital mapping project of performance venues. Recent publications include the edited volume *Dancing Naturally: Nature, Neo-Classicism and Modernity in Early Twentieth Century Dance* (Palgrave, 2011) and digital humanities case studies in *Transmission in Motion* (Routledge, 2016) and *Digital Movement* (Palgrave, 2015). With Peter M. Boenisch she co-edits the book series *New World Choreographies* (Palgrave).

**Masato Fukushima** is Professor of Social Anthropology and STS at the University of Tokyo, Japan. He has extensively published on issues such as religion and politics in Southeast Asia, situated cognition in modern institutions and the dynamics of experimentability in contemporary culture, science and design.

**Anne Galloway** leads the More-Than-Human Lab (http://morethanhumanlab.org/) and teaches in the Culture+Context Design programme at Victoria University of Wellington, New Zealand. Trained as an anthropologist, her ethnographic research critically examines entanglements of people, animals, spaces and technologies, and explores creative methods for public engagement around related matters of concern. Anne also spends as much time as possible at her rural home, where she shepherds small flocks of rare-breed sheep and ducks.

**Carolin Gerlitz** is Professor of Digital Media and Methods at the University of Siegen, Germany, and member of the Digital Methods Initiative Amsterdam. Her research focuses on platform studies, software cultures, apps, digital research methods, issue mapping, quantification in digital media and the value of social media data.

**Priska Gisler** is a sociologist and Professor and Head of Research Unit 'Intermediality' at University of the Arts, Berne. Her research fields are: social studies of science, ArtsSciences, human–animal relations, and educational policy.

**Connor Graham** is a Senior Lecturer at Tembusu College and a Research Fellow at the Science, Technology, and Society Research Cluster at the Asia Research Institute at NUS. He has conducted ethnographic studies of technology in Australia, China, England, Northern Ireland and Singapore and has worked and published with anthropologists, computer scientists, historians, sociologists and scholars of information systems and science, technology and society (STS). His research centres on living and dying in the time of the Internet, with a particular focus on human–technology relations. Specifically, he researches perspectives on and approaches to the design of new information and communication technologies. Recently he has been situating his research in Asia.

**Jennifer Green** is a Postdoctoral Fellow in the School of Languages and Linguistics at the University of Melbourne. Green has worked for over 30 years in Central Australia on projects documenting Indigenous languages, cultural history and visual arts. Her doctoral research on women's sand stories pioneered methods for the recording and analysis of multimodal narrative practices. Recent collaborative work on Indigenous sign languages from Central Australia has resulted in an online sign language dictionary.

**Gay Hawkins** is a Research Professor at the Institute for Culture and Society Western Sydney University, Australia. Trained in sociology she researches in the areas of materiality, political theory, environments and markets. Her research is interdisciplinary and informed by current debates in political philosophy, science and technology studies and social theory. Her most recent book, co-authored with Kane Race and Emily Potter, is *Plastic Water: The Social and Material Life of Bottled Water* (MIT Press, 2015).

**Alexandra Heller-Nicholas** has written five books on cult, horror and exploitation cinema, including *Rape-Revenge Films: A Critical Study* (McFarland, 2011), *Suspiria* (Auteur, 2015) and *Ms. 45* (Columbia University Press/Wallflower, 2017). She is an award-winning film critic, editor of the film journal *Senses of Cinema*, and a recipient of the 2017 Australian Film Institute Research Centre Fellowship. Alexandra is currently a research assistant at the Victorian College of the Arts (University of Melbourne).

**Maja Horst** is Professor and Head of the Department of Media, Cognition and Communication at the University of Copenhagen. Her research focuses on public communication about science and technology and on issues of research organization and management.

**Alan Irwin** is a Professor in the Department of Organization at Copenhagen Business School. He has published over a number of years on issues of science and technology policy, environmental sociology and science–public relations.

**Thomas Jellis** is a British Academy Postdoctoral Fellow at the School of Geography at the University of Oxford and a Research Fellow at Keble College. His research interests include experimental spaces, social theory, questions of disciplinarity, and research techniques. Thomas is currently working on a new project that examines exhaustion.

**Ann Kirori** has worked extensively in the NGO sector in Kenya. She joined the ERC-funded project, 'The Cultural Politics of Dirt in Africa', as a researcher in education and schools around Nairobi. Her work involved collecting qualitative data on the perceptions of dirt in various neighbourhoods of Nairobi through in-depth interviews, focus group discussions and observations. Other responsibilities in the project included transcribing and translating the data collected.

**Sybille Lammes** is Professor of New Media and Digital Culture at the University of Leiden. Her background is in media studies which she has always approached from an interdisciplinary angle, including cultural studies, game studies, science and technology studies and gender studies, and through working together with researchers with an interest in human–computer interaction, critical geography, social sciences and philosophy. Among others, she is co-editor of *Playful Identities: The Ludification of Digital Media Cultures* (AUP, 2015), *Playful Mapping in the Digital Age* (INT, 2016), *Time Travellers: Temporality and Digital Mapping* (Manchester University Press, 2018 fc.), and *The Playful Citizen: Power, Creativity, Knowledge* (Amsterdam University Press, 2019 fc.).

**Angela Last** is Lecturer in Human Geography at the University of Leicester. She is interested in how people make sense of the world, whether this is through research practices or the adoption of geopolitical imaginaries. She has published on methods and 'experimental geographies' in academic journals, as well as her blog *Mutable Matter*.

**Joanna Latimer** is Professor of Sociology, Science & Technology at the University of York. She is interested in the worlds people make together and the biopolitics in which they are entangled and has written widely on the cultural, social and existential effects and affects of how science & medicine is done. Her many articles and books, include *The Conduct of Care*, shortlisted for the BSA Philip Abram's Memorial Prize, and *The Gene, The Clinic and The Family: Diagnosing Dysmorphology, Reviving Medical Dominance*, winner of the 2014 Foundation for the Sociology of Health & Illness annual book prize.

**Ramon Lobato** is Senior Research Fellow in the School of Media and Communication at RMIT University, Melbourne. His research interests include media markets, piracy and media globalization. His books include *Shadow Economies of Cinema: Mapping Informal Film Distribution* (British Film Institute, 2012), *The Informal Media Economy* (Polity, 2012, with J. Thomas), *Geoblocking and Global Video Culture* (Institute of Network Cultures, 2016, ed. with J. Meese) and *Netflix Nations: The Geography of Digital Distribution* (NYU Press, 2018).

**Celia Lury** is Professor and Director of the Centre for Interdisciplinary Methodologies at the University of Warwick. Recent publications on methodology include: *Inventive Methods*, co-edited with Nina Wakeford (Routledge, 2012), and 'Measure and Value', co-edited with Lisa Adkins (*Sociological Review* Special Issue, 2012).

**Nina Lykke**, Professor Emerita, Gender Studies, Linköping University, Sweden, is co-director of GEXcel International Collegium for Advanced Transdisciplinary Gender Studies as well as scientific leader of the Swedish-International Research School, InterGender. She has published extensively within the areas of feminist theory, intersectionality studies, feminist cultural studies and feminist technoscience studies, including *Cosmodolphins* (Zed Books, 2000, with Mette Bryld), *Bits of Life* (University of Washington Press, 2008, with Anneke Smelik), *Feminist Studies*

(Routledge, 2010), *Writing Academic Texts Differently* (Routledge, 2014) and *Assisted Reproduction Across Borders* (Routledge, 2016, with Merete Lie). Her current research is a queerfeminist, autophenomenographic and poetic exploration of cancer cultures, death and mourning.

**Laura U. Marks** works on media art and philosophy. Her most recent books are *Hanan al-Cinema: Affections for the Moving Image* (MIT Press, 2015) and *Enfoldment and Infinity: An Islamic Genealogy of New Media Art* (MIT Press, 2010). She curates programmes of experimental media for festivals and art spaces worldwide. Marks teaches in the School for the Contemporary Arts at Simon Fraser University, Vancouver, Canada, where she is the Grant Strate University Professor.

**Derek McCormack** is an Associate Professor at the School of Geography and the Environment, University of Oxford. He has written about affective spaces and nonrepresentational theory and is currently writing about atmospheres and the elemental. His book *Atmospheric Things: On the Allure of Elemental Envelopment* is forthcoming with Duke University Press.

**Greg McInerny** is an Assistant Professor at the Centre for Interdisciplinary Methodologies at the University of Warwick. Originally trained as an ecologist, Greg now works among the broader topics of information visualization, scientific software and statistical modelling.

**Axel Meunier** is an independent researcher, cartographer and workshop facilitator. He is interested in how we live and work in a network-based material culture with multiple layers of reality. In the academy, he has worked at Sciences Po's médialab, where his research concerns digital mapping and the participation of mixed communities of humans and non-humans in the making of environmental issues. He believes everyone should be able to make their own research tools. He also practises performance in the art field, where the artwork is not a finished product but a prototype to experiment new social situations.

**Mike Michael** is Professor of Sociology in the Department of Sociology, Philosophy and Anthropology at the University of Exeter. His research interests have touched on the relation of everyday life to technoscience, the role of culture in biomedicine, and the interplay of design and social scientific perspectives. Recent major publications include (co-authored with Marsha Rosengarten) *Innovation and Biomedicine: Ethics, Evidence and Expectation in HIV* (Palgrave, 2013) and *Actor-Network Theory: Trials, Trails and Translations* (Sage, 2017).

**Anders Munk** is Associate Professor in Techno-Anthropology and Director of the Techno-Anthropology Lab at the University of Aalborg in Copenhagen. His research interests include controversy mapping, science and technology studies (STS), public engagement with science (PES), pragmatism, actor-network theory (ANT), and new digital methods for the social sciences and humanities. Anders holds a PhD in Human Geography from the University of Oxford and an MA in European Ethnology from the University of Copenhagen. He has worked as a senior visiting researcher at the SciencesPo Médialab and received research funding from the ESRC and the Carlsberg Foundation.

**Rolland Munro** has written widely on key issues in social theory, including affect, class, identity and motility. He is Professor of Philosophy of Organisation, University of Leicester, and Honorary Professor, Department of Sociology, University of York.

**Job Mwaura** is a Doctoral Fellow in the Department of Media Studies at the University of Witwatersrand, South Africa. His PhD research focuses on digital activism in Kenya. His research interests include social media studies in Africa, citizen journalism and political communication. In 2016, he won a fellowship with Next Generation of Social Science Research in Africa (SSRC) as well as a Post-Graduate Merit Award at Wits University. He recently won a research grant with the French Institute in Nairobi (IFRA). Before joining the ERC-funded 'Cultural Politics of Dirt in Africa' project as a researcher in media and communication, he worked as a researcher and a lecturer in Media Studies at Moi University, Kenya.

**Tahani Nadim** is a sociologist of science and runs the Bureau for Troubles at the Natural History Museum in Berlin (Museum für Naturkunde). Her research examines, among other things, data practices and imaginaries in the biosciences, constructions of biodiversity in museums and the technics and politics of global taxonomy.

**Jane Nebe** is a PhD candidate at the Graduate School of Education, University of Bristol. She was the researcher in education and schools for 'The Cultural Politics of Dirt in Africa' project in Lagos, Nigeria. Her research interests revolve around poor academic achievement and educational aspirations, using qualitative research methods and mixed-methods designs.

**Stephanie Newell** is Professor of English and Senior Research Fellow in International and Area Studies at Yale University and Professor Extraordinaire in the English Department at Stellenbosch University. Her research focuses on the public sphere in colonial West Africa, particularly newspapers and pamphlets. She was Principal Investigator of 'The Cultural Politics of Dirt in Africa' project between 2013 and 2015, a project that transferred to Yale University from 2015–2016. Her most recent book, *The Power to Name: A History of Anonymity in Colonial West Africa* (Ohio, 2013), was shortlisted for the Herskovitz Prize in African Studies. Her current book, *Histories of Dirt in West Africa: Media and Urban Life in Colonial and Postcolonial Lagos*, is under contract with Duke University Press for publication in 2019.

**Patrick Oloko** (PhD) is a Senior Lecturer in the Department of English at the University of Lagos in Nigeria. As Regional Coordinator of the Lagos team on the ERC-funded project, 'The Cultural History of Dirt in West Africa', he supervised field work and data gathering by the project researchers, interfaced with government authorities and the public on matters relating to the project, and worked closely with the Principal Investigator to ensure the smooth running of the project. His research and publications focus on contemporary Nigerian writing, particularly fiction.

**Rebeccah Onwong'a** has a Master of Biology (Human Ecology) degree from the Vrije Universiteit Brussel, and a Bachelor of Science (Botany/Zoology) from the University of Nairobi. Alongside this training, she has training in social science research. She worked on the ERC-funded 'The Cultural History of Dirt in Africa' project as a researcher in health and the environment. Her work involved collecting qualitative data on the perceptions of dirt in various neighbourhoods in Nairobi Kenya. She used in-depth interviews, focus group discussions and observations to collect the data, and was responsible for transcribing and translating the data collected. Currently, she works as a tutorial fellow at the Department of Ecology and Conservation, Technical University of Kenya.

**Jussi Parikka** is a professor at the Winchester School of Art, University of Southampton. He has worked on media archaeology and is also the author of the books *Digital Contagions* (2nd edition, Peter Lang, 2016), *Insect Media* (University of Minnesota Press, 2010) and *A Geology of Media* (University of Minnesota Press, 2015).

**Sarah Pink** is Distinguished Professor in the School of Media and Communication at RMIT University, Australia and Visiting Professor at Halmstad University, Sweden, and Loughborough University, UK. Her current research focuses on emerging technologies and design for well-being. Her recent books include *Uncertainty and Possibility* (with Yoko Akama and Shanti Sumartojo, Bloomsbury, 2018) and *Anthropologies and Futures* (with Juan Salazar, Andrew Irving and Johannes Sjoberg, Bloomsbury, 2017).

**Jonathon Potskin** is a doctoral student in the School of Social and Political Science, University of Sydney. His main research is on Indigenous youth and rap music in Canada and Australia: using music to continue cultural knowledge systems. Jonathon is a Nehiyaw (Cree) and Apeetogason (Mètis) and is a Member of the Sawridge First Nation in Alberta, Canada, who has a passion for sharing his cultural knowledge of the Cree and the Mètis people of Western Canada. Jonathon's topic for his Master's degree in Anthropology was Indigenous Research Methodologies and Decolonization of Anthropology.

**Holger Pötzsch** (PhD) is Associate Professor in Media and Documentation Studies at UiT Tromsø, Norway. He has published on themes such as war memories in films and games, borders and cultural production, as well as the role of technology in processes of bordering. Pötzsch has been involved as a researcher in the NRC project 'Border Aesthetics' (2010–2012) and the FP 7 project 'EUBORDERSCAPES' (2012–2016). He currently leads the development of the WAR/GAME-project at UiT Tromsø. His work has been published in journals such as *Environment & Planning D: Society & Space*, *New Media & Society*, *Games & Culture*, *Nordicom Review*, *Journal of Borderlands Studies*, *Memory Studies* and *Media, War & Conflict*.

**Charles C. Ragin** is Chancellor's Professor of Sociology and Political Science at the University of California, Irvine. His main interests are methodology, political sociology and comparative-historical research. His books include *Intersectional Inequality* (University of Chicago, 2016, with Peer Fiss), *Handbook of Case-Based Methods* (Sage, 2009, with David Byrne), *Configurational Comparative Methods* (Sage, 2009, with Benoit Rihoux), *Redesigning Social Inquiry: Fuzzy Sets and Beyond* (University of Chicago, 2008), *Fuzzy-Set Social Science* (University of Chicago, 2000), *The Comparative Method: Moving Beyond Qualitative and Quantitative Strategies* (University of California, 1987) and *What is a Case?* (Cambridge University, 1992, with Howard S. Becker).

**Matthew Reason** is Professor of Theatre and Performance at York St John University (UK). Publications include *Documentation, Disappearance and the Representation of Live Performance* (Palgrave, 2006), *The Young Audience: Exploring and Enhancing Children's Experiences of Theatre* (Trentham, 2010) and, co-edited with Dee Reynolds, *Kinesthetic Empathy in Creative and Cultural Contexts* (Intellect, 2012). He has recently published a new edited collection, with Anja Lindelof, *Experiencing Liveness in Contemporary Performance* (Routledge, 2017).

**Richard Rogers** is Professor of New Media & Digital Culture, Media Studies, University of Amsterdam. He is director of the Govcom.org Foundation as well as the Digital Methods Initiative, and author of *Information Politics on the Web* (MIT Press, 2004) and *Digital Methods* (MIT Press, 2013). Rogers has received research grants from, among other institutions, the Open Society Foundations, Ford Foundation, MacArthur Foundation and Gates Foundation.

**Helen Scalway** received her training (MA Fine Art) at Chelsea College of Art, London University of the Arts. Her practice is concerned with the representation of complex contemporary spaces, which are still coming into being. She works through the creation of diagrams, drawings, collages and models. She is currently an Honorary Research Associate in the Geography Department at Royal Holloway, University of London.

**Miguel Angel Sicart** is an Associate Professor at the IT University, where he teaches Game Design and Play Design. He is the author of *The Ethics of Computer Games* (MIT Press, 2009), *Beyond Choices: The Design of Ethical Gameplay* (MIT Press, 2013) and *Play Matters* (MIT Press, 2014).

**Ana Teixeira de Melo** has a PhD in Clinical Psychology and is a postdoctoral research fellow in the Centre for Social Studies and the Faculty of Psychology and Education Sciences of the University of Coimbra, in Portugal. She is a fellow of the Foundation for Science and Technology (SFRH/BPD/77781/2011). She has been conducting both basic and applied research with a focus on family and parenting, and, more recently, love as a complex system. She is also interested in issues related to interdisciplinarity, complexity and complex thinking, research methods, epistemology and the philosophy of science.

**Manuel Tironi** is Associate Professor at the Department of Sociology, P. Universidad Católica de Chile, where he convenes the Critical Studies on the Anthropocene group. His latest research focuses on toxicity, ecological politics and geological modes of knowing.

**Olutoyosi Tokun** worked as a researcher in Lagos, Nigeria, on the health and environment aspect of the ERC-funded project 'The Cultural Politics of Dirt in Africa'. Her work involved engaging with managers and users and in the fields of public health, the environment and waste management. She holds a Master of Public Health (MPH) degree from Monash University, Australia, and a Bachelor of Science degree from the Department of Biochemistry, University of Lagos. Her current research interests include the social and environmental determinants of health, particularly among Africans.

**Emma Uprichard** is Reader at the Centre for Interdisciplinary Methodologies at the University of Warwick. She is co-editor with David Byrne of four volumes on 'Cluster Analysis' (Sage, 2012). Her disciplinary background is Sociology (with Education), but she turns to complexity science, which spreads across the disciplinary boundaries to rethink social science research and to research socio-spatial and temporal change relating to a range of multi-disciplinary, global substantive topics, including: time, childhood, food and cities.

**John Uwa** was a Researcher in Media and Communications for the ERC-funded project, 'The Cultural Politics of Dirt in Africa'. His brief was to interact with media producers and consumers in order to collect data, through interviews and various media platforms, on the varying perceptions of 'dirt' in urban Lagos. He is currently a PhD student in the Department of English, University of Lagos. His research focuses on dramatic theatre (literary and popular), particularly the transformation of Nigerian popular theatre through the emergence of Nigerian Stand-Up comedy. He has published in *Okike* and other literary journals.

**Tommaso Venturini** is Lecturer at King's College London in the Digital Humanities Department. He is also Associate Researcher at the Médialab of Sciences Po, Paris, which he founded with Bruno Latour and coordinated for six years. He has been the leading scientist of the projects EMAPS (climaps.eu – EU FP7) and MEDEA (projetmedea.hypotheses.org – ANR). His research focuses on Digital Methods, STS and Social Modernization. He teaches Controversy

Mapping, Data Journalism and Information Design at graduate and undergraduate level. He trained in sociology and media studies at the University of Bologna, completed a PhD in Society of Information at the University of Milano Bicocca and a post-doc on social modernization in the Department of Philosophy of the University of Bologna. He has been a visiting student at UCLA and was a visiting researcher at the CETCOPRA of Paris 1 Pantheon Sorbonne. During his studies, he founded a web design agency and led several online communication projects.

**Moritz Wedell** holds a PhD from the Humboldt-University, Berlin. He taught German Medieval Literature and Culture at the Universities of Berlin and Zurich, and at the University of California at Berkeley. In his research, he addresses medieval poetics and historical anthropology, in particular the history of numerical knowledge (*Zählen*, Göttingen, 2011; *Was zählt*, Wien, 2012), and more recently the idea of human creativity.

**Margaret Wertheim** is a writer, artist and curator whose work focuses on the intersections of science and the wider cultural landscape. Her books include *The Pearly Gates of Cyberspace* (W.W. Norton, 2000), a history of Western concepts of space, and *Physics on the Fringe* (Walker and Co., 2011), about the scientific equivalent of 'outsider art'. Based in Los Angeles, Wertheim is founder and director of the Institute for Figuring (IFF), an organization devoted to the poetic and aesthetic dimensions of science and mathematics. Through the IFF, she and her twin sister Christine Wertheim (a faculty member at the California Institute of the Arts), have created exhibitions for the Hayward Gallery in London, the Science Gallery in Dublin, the Andy Warhol Museum in Pittsburgh and the Smithsonian in Washington DC. The sisters' *Crochet Coral Reef* project is among the largest science and art endeavours in the world and has been conducted in over 35 cities and countries, including Australia, USA, Germany and the UAE. In 2015 Wertheim was a Vice-Chancellor's Fellow at the University of Melbourne.

**Alex Wilkie** is a sociologist of science and technology and a Senior Lecturer in Design at Goldsmiths, University of London. His research interests combine aspects of social theory, science and technology studies with experimental design research that bear on theoretical, methodological and substantive areas including: aesthetics, constructivist and speculative thought, situated design practice, healthcare and computational technologies, human–computer interaction design, as well as involvement, engagement and participation with science and technology. Alex is a Co-Director of the Centre for Invention and Social Process (Department of Sociology, Goldsmiths) and the Director of Research in Design. He has recently edited *Studio Studies* with Ignacio Farías (Routledge, 2015) and *Speculative Research* with Martin Savransky and Marsha Rosengarten (Routledge, 2017) and is currently preparing the edited volume *Inventing the Social* with Noortje Marres and Michael Guggenheim (Mattering Press).

# Preface

**Homo Pontifex**
Method or hyphen, those are soft bridges;
viaduct or bridge, those are hard unions or methods.
Watch: I am constructing a new footbridge;
moving from matter to the sign and from the abstract
to the concrete, I am bridging the hard and the soft. Whether of one
or the other kind, I find bridges everywhere.
Examples: the method of translation mobilises two grammars
and a bilingual dictionary, it bridges languages;
the method
for producing
living mutation
moves through
genetic
manipulations;
it bridges
organisms
and soon species;
the method
for transmuting
elements passes
through radioactive
decay;
it bridges inert bodies.
Bridging, respectively, languages,
living beings and elements, we bridge, transversely,
the soft empire of signs with the hard realms of physics and biology . . .
First labour, to build bridges in the hard;
second work, to think of soft bridges. To launch oneself between
the second and the first, the final enterprise. Bridging, in general, becomes
an activity so large that it coincides perhaps with the whole human project, in that
our very body bridges flesh and word.
*Michel Serres, translation by Steven Connor*

# Introduction

## Activating the present of interdisciplinary methods

*Celia Lury*

> Has not philosophy restricted itself to exploring – inadequately – the 'on' with respect to transcendence, the 'under', with respect to substance and the subject and the 'in' with respect to the immanence of the world and the self? Does this not leave room for expansion, in following out the 'with' of communication and contract, the 'across' of translation, the 'among' and 'between' of interferences, the 'through' of the channels through which Hermes and the Angels pass, the 'alongside' of the parasite, the 'beyond' of detachment . . . all the spatio-temporal variations preposed by all the prepositions, declensions and inflections?
>
> *Michel Serres (1994: 83) in Steven Connor, www.stevenconnor.com/milieux/*

Methods are fundamental to the paradigms that structure the production of knowledge: they contribute to the history of disciplines and inform lines of enquiry. In some cases, methods are part of core disciplinary knowledge; often, the methods used in a discipline contribute to its unity and continuity. This handbook seeks to reflect on and contribute to the ways in which methods might shape the future development of *interdisciplinary* research.

Of course, what counts as interdisciplinarity is widely contested. On the one hand, 'the history of intellectual disciplines is longer, more differentiated and more "indisciplined" than has conventionally been presented in the stories that disciplines have told about themselves' (Osborne 2013: 4). On the other, interdisciplinarity itself has a long history, a variety of definitions, and shifting relations to the multidisciplinary and transdisciplinary (Apostel, Berger, Briggs and Merchaud 1972; Barry and Born 2013; Lykke 2010; Nowotny, Scott and Gibbons 2001; Stenner 2014) while recent years have seen the rise of anti-disciplines (Pickering 1993), non-disciplines and post-disciplinary practices as well as a variety of re-disciplinarizing dynamics (Osborne 2013).

Acknowledging this complex, entangled history, this essay nevertheless proposes some simplified definitions with which to begin. Multidisciplinarity is understood to be an additive approach – using knowledge from more than one discipline which are not themselves transformed by being used in conjunction with one another. Transdisciplinarity aspires to be a more holistic approach, and aims to displace disciplinary formations. Interdisciplinarity is characterized as interaction *across* and *between* disciplines. Importantly, this interaction is not oriented toward either a synthesis or a disappearance of disciplines. Instead, interdisciplinarity emerges through *interferences* between disciplines and between disciplines and other forms of knowledge.

1

In their study of interdisciplinarity, Barry, Born and Weszkalnys (2008) locate the current interest in interdisciplinary research in terms of its (supposed) ability to make science *accountable* to society. Understanding this accountability in terms of a curtailment of the autonomy of academic knowledge production linked to an instrumental emphasis on problem solving, they insist that it is both possible and necessary to identify the *autonomy* of interdisciplinarity:

> [Interdisciplinarity] can be associated with the development of fields and initiatives in which new kinds of autonomy are defended against a reduction of research to questions of account-ability or innovation. It can generate knowledge practices and forms, and may have effects, that cannot be understood merely as instrumental, or as responses to broader political demands or social and economic transformations. In short, autonomy may be associated as much with interdisciplinary as with disciplinary research.
>
> *2008: 23–24*

It certainly seems as if a lot is expected from interdisciplinarity today: as Boix Mansilla, Lamont and Sato observe, interdisciplinary research 'is increasingly viewed by . . . scientific funding agencies and policy makers as the philosopher's stone, capable of turning vulgar metals into gold' (2016: 572). This handbook aims to demonstrate the value of interdisciplinary methods, not for their ability to turn what they touch into gold, but for the potential they hold to develop the autonomy *and* accountability of interdisciplinary research. As will be developed in detail below, this potential is understood in terms of *the activation of the present*.

## -Ings!

The working shorthand for the approach to interdisciplinary methods collected together in this handbook was -ing! Our collective aim as editors was to focus discussion on a critical exploration of how the relations between questions and answers, practice, process and outcomes, epistemology and ontology, validity and value, are made anew in the practice of interdisciplinary enquiry. Contributors were asked to describe the do-*ing* of their chosen methods.

An inspiration for this approach was the artist Richard Serra's claim that 'Drawing is a verb'. His art work *Verb List* (1967–68) serves as a kind of manifesto for this pronouncement. In pencil, on two sheets of paper, in four columns of scripts, the artist lists the infinitives of 84 verbs – *to roll, to crease, to fold, to store*, for example – and 24 possible states or conditions – *of gravity, of entropy, of photosynthesis, of nature*. In interview, he says, 'The problem I was trying to resolve . . . was: How do you apply an activity or a process to a material and arrive at a form that refers back to its own making?' He continues, 'That reference was mostly established by line. In a sense you can't form anything without drawing' (Garrels 2011: 61). The art critic Rosalind Krauss suggests that the list describes Serra's practice in terms of action that 'simply acts, and acts, and acts' (1985: 101). Serra himself draws attention to the relations in which the action that 'simply acts' takes place: he describes the list as a series of 'actions *to relate to oneself, material, place, and process*' (Buchloh 2000: 7; emphasis added).

The form of the verbs in Serra's list is infinitive, that is, they are the basic form of a verb, without an inflection binding the verb to either a specific subject or a specific tense. The word 'infinitive' is derived from Late Latin *infinitus* meaning 'unlimited', 'not finite' or 'unfinished'. In using this form of the verb, Serra calls up the possibility that each action might be conducted in an infinite variety of ways. Additionally, the verbs in the list are transitive, that is, in linguistic terms, they can imply or express an object (to roll pastry, to crease paper, to fold metal, to store data, for example). This formulation – of implication or expression – is important; the verbs – methods or doings – are not 'applied' to materials – they imply or implicate an object in a process, 'something/happening' as Thomas Jellis says in his discussion of Experimenting. This, perhaps, is why 24 examples 'of' a variety of objects, conditions or states are also included in the list by Serra: verbs are expressions *of* objects, conditions or states, and objects, states and conditions are the implication or expression *of* verbs.

Perhaps most significantly for this handbook, *how* the verbs Serra lists imply or express an object is a problem, indeed it is 'the' problem. As Serra puts it, the problem is how to accomplish a form by or in doing. Drawing on this insight, the proposal advanced here is that the determination or individuation of a problem is an accomplishment of the doing of a method or methods, that is, it is an accomplishment of a practice that involves a referral back to person, place, matter and process. For Serra, this accomplishment cannot be assumed: similarly, the methodological activity of composition, as Emma Uprichard describes it, may or may not arrive at a form, have purchase on a question, or individuate a problem. In short, forming a problem is the always contingent outcome of actions that fold a referral back to person, place, material and process to the referral forward of the doing or practice of methods.

In contrast to Serra's use of the infinitive form of verbs in *Verb List* the approach adopted here places emphasis on a specific verb form: what are, in the English language,[1] known as *gerunds*, that is, active present tense forms that function as nouns. This verb form – ending in –ing – is typically the object of prepositions, the variety of which, Serres suggests, philosophy has neglected.[2] The suggestion put forward here then is that by pre-posing problems – that is, by 'following out the "with" of communication and contract, the "across" of translation, the "among" and "between" of interferences, the "through" of the channels through which Hermes and the Angels pass, the "alongside" of the parasite, the "beyond" of detachment' (Serres 1994: 83) interdisciplinary methods can activate the present. That is, pre-posing problems in the doing of interdisciplinary methods allows for their individuation as 'figures of suspension and expectation', able to function as 'traps for the emergence of compossibility' (Corsín Jiménez 2014a: 383; see also Prototyping and Dissenting, both this volume and Lury and Wakeford 2012).

Put rather grandly, the handbook's concern with -ings is intended to identify the potential of interdisciplinary methods to compose problems as interruptions of the (historical) present. That is, the aim is to emphasize the role of interdisciplinary methods in the activation of the present: the determination of a situation as a problem, that is, 'a state of things in which something that will perhaps matter is unfolding amidst the usual activity of life' (Berlant 2008: 4). More prosaically, the aim is to consider how interdisciplinary methods might constitute some aspect of what is given, the present – in all its geo-political complexity – as a problem, which is to say, as a situation that may be methodologically activated in specific, precise ways. This might involve, as Mike Michael suggests, identifying a 'pattern of pasts and prospects', or require, as Manuel Tironi puts it, empowering 'a situation with the capacity to provoke new relations . . . [crafting] a space for being in the presence of [the values of others] and their consequences'.

Approaching interdisciplinary methods as ways of giving a problem the form of the active present necessarily obliges the researcher to be attentive to the methodological potential of complex (spatio-)temporalities. As Matthew Reason says in his discussion of Drawing, 'Drawing is at once immediate, and yet takes time'. He continues:

> When I ask a participant to draw me a picture I am inviting a different dynamic than if I had simply asked them to talk. I do not expect them to respond instantly. Instead drawing imposes a slowing down, a pause for reflection in the returning to memories.

Catherine Ayres and David Bissel use the term Suspending to describe the analytical potential of acknowledging the multiple durations present in an interview. They say:

> Different durations resonate at different times, sometimes immediately, and sometimes years after the initial encounter. Following Ingold's (1993) observations about the multiple co-existent temporalities of landscapes, we want to show how the interview 'landscape' is steeped in the pasts and possible futures of researcher and researched alike, a site in which trajectories converge and transform. We want to revisit the interview event between Catherine and John to draw out 'suspending' as a methodological intervention filled with theoretical, practical and ethical possibilities for thinking empirical encounters.

Jussi Parikka says of Digging:

> As a method, digging opens up historically constructed material reality. It does not merely expose 'ruins' but the multiple historical realities where material infrastructures have been layered, revealing different 'distinctive temporalities and evolutionary paths' (Mattern 2015: 14). In this sense, digging opens the different temporalities that are all the time layered in infrastructures of cities, in media technological objects and in everyday situations.

While Alex Wilkie observes of Speculating:

> Speculation, however, requires a shift in approach from analysing how probabilistic futures are manifested, managed and contested in the present – how actors imagine, model, predict, coordinate and in turn configure the future to the present – to the construction of adequate concepts and devices for exploring possible latent futures that matter. A word of caution is in order here, however: speculation is both prospective and retrospective. It applies as much to the politics of explaining past events (what might have been) as it does to the capturing of future possibilities (what might be).

And Gail Davies says of the diagram, 'it hasn't got a beginning, it hasn't got an end but nonetheless the incommensurable meanings are there, written in, but it hasn't got to have that linear structure of time'.

In summary, approaching methods as -ings focuses attention on the pre-posing of problems, that is, it understands the role of methods as ways to activate the spatio-temporal variations, the 'declensions and inflections' of the present. To push this argument a bit further, the presentation of methods as -ings highlights their methodological potential to not only *take* but also *make* (space and) time (Back and Puwar 2012; also see Timing by Barbara Adam, and the discussion of tensing methods by Emma Uprichard).

## -Ings and things

Lucy Kimbell's discussion of design(-ing) gives some more indication of what this approach might involve (2015a, 2015b). Kimbell starts by stressing the importance of understanding design as doing or practice, building on Reckwitz's understanding of practice as

> a routinized type of behavior which consists of several elements, interconnected to one another: forms of bodily activities, forms of mental activities, 'things' and their use, a

background knowledge in the form of understanding, know-how, states of emotion and motivational knowledge.

*Reckwitz 2002: 249 in Kimbell 2015b: 132; see also Rachel Fensham and Alexandra Heller-Nicholas's discussion of practice in this volume*

To supplement Reckwitz's understanding, however, Kimbell is concerned to show how relations between designing and design(s) may happen in a variety of ways. To do this, she describes the relations between designing and design(s) as having two forms: design-as-practice, and designs-in-practice. In the terms of approach being developed here, the two forms make visible the methodological potential of exploring a problem through the spatio-temporal variations that emerge when the referral forward of practice is combined with a referral back to relations to person, place, material and process.

To elaborate: Kimbell's first term – design-as-practice – 'mobilizes a way of thinking about the work of designing that acknowledges that design practices are habitual, possibly rule-governed, often routinized, conscious or unconscious, and that they are embodied and situated'. In addition, however, Kimbell says, 'Design-as-practice cannot conceive of designing (the verb) without the artifacts that are created and used by the bodies and minds of people doing designing' (2015b: 135). The second term – designs-in-practice – 'acknowledges the emergent nature of design outcomes as they are enacted in practice' (2015b: 135). Together, these two terms draw attention to how designs emerge from designing, and in doing so provides a way to think about how scale emerges from scaling, a map from mapping, a sample from sampling and a translation from translating.

Such outcomes might appear so inevitable as to need no acknowledgement – what else would emerge from the activity of scaling but a scale? But if we ask what emerges from the method of reading, we can see that as well as a reader, a text and perhaps even a writer ('the author') might also be produced in the organization of a referral back to person, material, place and process. When we also remember that the term 'computers' historically referred to people (often women) who computed, rather than to machines, we might wonder whether the (alleged) crisis in the humanities stems in part from the fact that reading as a practice is no longer routinely organized to refer back to a person but, increasingly, to machines (Hayles 2005). And we might think again about what relations to person, place, material and process are involved when a document is formatted as 'read-only'.

Continuing in these terms, we can further reflect on the significance of the privileging of referral back to place (alongside the referral back to person, material and process) in methods of indigeneity. In their contribution to this handbook, Catriona Elder and Jonathon Potskin discuss this privileging in terms of the importance attached in Indigenous methodologies to the inalienable connection of Indigenous peoples to specific geographic spaces. In Australia, they say, it is the custodianship of Indigenous peoples to land – and the responsibility built into that relationship –– that informs Indigenous methodology. Relatedly, they draw on Margaret Kovach to describe the referral that takes place within a Nêhiyaw epistemology: "'so while I speak of knowledges (e.g., values, language), it should be assumed that they are nested, created, and re-created within the context of relationships with other living beings'".

Such accounts compel us to reflect on the significance of the spatio-temporal relations in which problems are pre-posed (and subjects and objects predicated), and acknowledge the important role that feminist, environmental, post-colonial and Indigenous scholars have played in highlighting this significance. In the practices of such scholars the autonomy of interdisciplinarity is explicitly put into diverse relations of accountability (even if this word is avoided because of its association with a requirement to problem-solve in instrumental ways; see Elder and

Potskin). This work allows us to see that how method implies or expresses its objects allows the knowledge that is produced to move in particular ways: as evidence that is context-independent, for example, or as local or situated (see Newell *et al.*, this volume); to be protected by intellectual property laws or held in trust collectively (see Corsín Jiménez, this volume); to contribute to the consolidation or dissolution of hierarchies of expertise both between disciplines and between academic and non-academic modes of knowing; and to the definition of ethical and non-ethical ways of conducting research (see Newell *et al.*).

If we consider the relations between measuring and value in Western epistemologies in this respect, we can see that they do not exist in a simple linear relation to each other, with one, value, as the outcome of the other, measuring; rather, as Brighenti puts it: 'the relation between measure and value is necessarily circular – better, entangled. In this light, value exists *before* as well as *after* measure, and precisely in such "circumnavigation of measure" lies a transformation and concretion of the nature of value' (2017: 3).

As part of this circumnavigation, Mike Michael suggests, the activity of valuing relates to both value and valuation, while validating relates to validity and validation. He observes:

> At base, the former term of each pair implies an 'external' or 'objective' relation by which some thing or some activity is judged against pre-given standards: value against market forces, validity against scientific or epistemological criteria. Conversely, the latter term of each pair – valuation and validation – points to an intersubjective or interactional relation in which some thing or some activity is assessed by means of standards and criteria. Thus, validation, say of a person's work within an organization, might entail shared negotiation of criteria as to what is of worth; and valuation of a property might involve the situated weighting and juggling of criteria in the process of moving through a house, say, to come to a proposed money figure.
>
> *Michael, this volume*

Brighenti summarizes: 'A whole social imaginary may evolve from, and concrete around, the gap between the cold side of measure (which Dewey called "estimate") and the hot side of it ("esteem"). Understanding this gap calls for a wide interpretive framework' (2017: 5) as Manuel Tironi also indicates in his discussion of Dissenting, a method to explore 'the political capacities that might be unleashed when value mismatches in interdisciplinary projects are not worked through but enhanced as moments of democratic expansion'.

The usefulness of Kimbell's two terms for the approach to interdisciplinary methods being proposed here, then, is that they draw attention to the importance of a methodologically informed referral back to person, place, material and process in the practice or doing of method. It is in this folding of multiple kinds of relations into the doing of methods – in the work of pre-positioning – that problems are composed, their form making possible the drawing of lines of enquiry involving specific kinds of analysis – whether these analyses involve predictions, the establishing of cause and effect, the refining of concepts, the revision of classifications, or the devising of speculations. Put another way, it is in this folding that methods acquire the capacity to form a problem that has the ability to activate the present, to pause, to de- and re-compose the latency of what is *given* in relations between past, present and future (see Tahani Nadim, this volume). And such an emphasis is vital if we are to understand and take responsibility for how (disciplinary and interdisciplinary) methods enact the world (see, for example, Annelise Riles' (2006) discussion of 'the agency of legal form'). As Ramon Lobato shows in his discussion of Rescaling as a methodological tactic, the activity of scaling always makes a particular kind of scale (this scale rather than that scale), and then, this scale (rather than that scale) contributes, or not, to – for example – economies of scale that require and exploit specific calibrations in time and space.

## Compound-ing the problem

These points will be developed further below, but for now let me make just one further observation about -ings. It will be obvious from a cursory look at the contents page that most of the methods collected here are not those conventionally described in methods textbooks. This is not simply because they are interdisciplinary, but because they are interdisciplinary in a specific way: they are *compound* methods, that is, they are a combination of methods (see Emma Uprichard, this volume, for further discussion). So, for example, Visualizing, as described by Greg McInerny, routinely involves the combination of statistical, visual, coding and diagrammatic practices. Experimenting, as described by Thomas Jellis, involves both participating and relaying, with each of these itself composed of a variety of practices.

One of the reasons to focus on compound methods as we do here is to detach techniques from specific disciplinary uses (and related proprietary claims), and describe them instead in terms that can be recognized *across* disciplines. So, for example, the discussion of sampling below concerns the use of this method in music and film. Nevertheless, it is of relevance to techniques of sampling in social research insofar as it calls attention to the links between representativeness and representation, and to the creative possibilities opened up by activating the relation between parts and wholes. Another reason to describe the methods collected here as compound, however, is to recognize the ways in which methods have deeply affective and political dimensions as well – or as part of – their onto-epistemological agency. This is recognized by Gail Davies and Helen Scalway (this volume) in a discussion of the term 'recognizing': '[recognizing] implied to me a relationship to the other, a process of knowing, and the two together, of recognizing the implications of this knowledge in the context of the other'.

Describing the methods included here as compound is also designed to demonstrate inter- rather than meta- or trans-disciplinarity: that is, it is to show how methods emerge from within a necessarily contingent, more-or-less enduring interaction between disciplines. Of course, the contributions to the handbook – and this essay itself – describe -ings in terms of the interaction between only some, and not all, disciplines. It could not be otherwise. And while most of the methods discussed here will probably be primarily of interest to scholars in the humanities and the social sciences, the hope is that each -ing is described in such a way that it could be adopted and adapted in movements across (other) disciplines. Indeed, making visible the possibility of what Gay Hawkins (this volume) describes as the Qualifying of methods is one of the ways in which the handbook aims to contribute to the autonomy as well as the accountability of interdisciplinarity.

To consider the value of considering interdisciplinary methods as -ings further, I turn now to four of the most salient vectors of the contemporary configuration of interdisciplinary research. These are: the global formulation of problems, understood in terms of scale, position and depth or perspective; the emphasis on collaboration and the affiliated term participation; the changing infrastructures of research; and futures thinking. While all these aspects of the contemporary formation of interdisciplinary research can be understood in terms of a push to make science accountable to society in instrumental terms, this essay aims to show how they can (also) inform its autonomy.

## Vectors of interdisciplinary research

### Global

Recent accounts of interdisciplinary research suggest that it is of special value in relation to what are described as global problems, but the basis of this claim is not always clear.

First, consider some of the ways in which interdisciplinarity and the global are currently being brought into relation with one another (see Angela Last's chapter in this volume for further discussion of these relations). One is to do with size or magnitude, that is, some problems – such as climate change or disease – are understood to be too 'big' to be successfully addressed by single disciplines. Similarly, it is sometimes suggested that a 'world' literature cannot be understood by literary studies alone, but requires an engagement with geo-political disciplines. A second way in which the global and interdisciplinarity are brought together concerns the heterogeneity of the actors said to be required to address global problems, and the necessity of including both human and non-human participants. Moreover, as Sasha Engelmann and Derek McCormack say in a discussion of Sensing atmospheres, 'The atmospheric is not a domain circumscribed by phenomenological modes of conscious sensing . . . Indeed, much of the data and processes that can now be sensed operate below and before thresholds of human awareness' (see also Clough 2008; Fraser 2009). A further way in which interdisciplinarity is brought into relation with the global is that it is commonly seen to be necessary to address problems insofar as they might be described as complex or wicked problems (Rittel and Webber 1973), or problems that 'resist telling' (Crenshaw 1995; and see the discussion by Nina Lykke and Angela Last in this volume) especially insofar as such problems are not seen to be amenable to being fixed once and for all.[3]

In all these ways, interdisciplinarity and the global might seem to stand in relation to each other in mutually productive ways. Yet, the term global and the related term globalization remain the source of considerable debate in at least some disciplines. It is by no means accepted that they can capture the intensities and unevenness of the variety of mobilities that cross-cut the contemporary world (Büscher, Urry and Witchger 2010; and see Monika Büscher's discussion of Moving methods). For this reason, some interdisciplinary scholars prefer to explore the unevenness that might be introduced into an understanding of such mobilities by focusing on 'inter'-relations. In this respect, we might learn from the example of those scholars who drew on the idea of 'Asia as a method' (Chen 2010; Yoshimi 2005[1960]) to develop both an intellectual movement – Inter-Asia Cultural Studies (IACS) – and a methodology – Inter-Asia methodology.

In these developments, a strong contrast is drawn with area studies (see also Chow 2006): the aim is to emphasize the location from which research questions are articulated, not to compile similar data from different countries. Tejaswini Niranjana (2013) writes, 'We don't see [Asia as a method] as taking up a "regional" study, which often means "applying" Western concepts to Asian material, or demonstrating complete and authoritative knowledge about a place in a historical or ethnographic mode'.

She describes how Asia as method was put into circulation in IACS: 'an intellectual movement, conjoined with social movements in diverse ways in each locale, entailing physical travel and exchanges so as the better to speak *across* places and to each other' (2013). So, for example, the Inter-Asia journal on which the project was originally premised was envisaged as a platform for movements traversing all the above, across different planes and points of intersection.

Many members of IACS conduct comparative studies, but the inter- of inter-Asia complicates the practice of comparison (see also Connor Graham on Projecting, this volume), since it involves the exploration of similarities and differences within and between not one but multiple Asias: 'The name Inter-Asia which has gained so much currency today was proposed as a grammatical impossibility – the term suggests the improbable existence of multiple Asias. But it manages to capture what we had in mind' (Chua Beng-Huat, quoted in Niranjana 2013).

Niranjana develops this approach as 'inter-Asia methodology', a term she develops to describe 'knowledge production about Asian locations premised on the multiplication of frames of reference', and on comparative research addressing the conditions of emergence of specific phenomena. Inter-Asia as methodology involves 'visibilising the normative frame which often has the "West" as its key reference point' (2013), comparative work (2000), and the identification of terms to become part of a broader conceptual framework. For Niranjana this includes what she calls the 'pressing' of concepts, that is a mutual or lateral interrogation of concepts with each other. She gives the examples of Partha Chatterjee's *political society* and Kuan-Hsing Chen's *minjian*, saying that each involves 'the analysis of deep historical processes to reveal their interconnections and how they implicate simultaneously different Asian locations, the production of genealogies of the Asian present' (2013).

## Collaboration

Perhaps not surprisingly, interdisciplinary research is routinely understood to require collaboration (although as Angela Last points out it can be conducted by an individual researcher). The dimensions of this collaboration are many and various. So, for example, Boix Mansilla *et al.* (2016) identify the importance of what they call the cognitive-emotional-interactional platform of interdisciplinary research. They use the term platform to describe what emerges when

> researchers practically engage one another to work on a common problem . . . In this shared space, researchers define problems to study, exchange expertise, build personal relations, project and maintain academic self-concepts, and yoke for status; what they create together constitutes a basis that shapes how they collaborate with each other – such as shared language, key concepts, tacit rules of interaction, group culture and identity, and collective mission.
>
> *2016: 573*

The platform is 'both a site of and springboard for collaborative activities – a dynamically co-constructed space with a set of rules and objectives that members develop – and both resultant of and contributing to collaboration' (2016: 574).

Their study highlights the importance of acknowledging the lived reality of interdisciplinary collaboration, along with an acknowledgement of the friction that is often – perhaps always – involved in interdisciplinary research. As Mike Michael observes in Compromising, 'It goes without saying that interdisciplinary collaboration can be a fraught business'. Friction in collaboration can have many sources, including the difficulty of reconciling different epistemic cultures, styles of thinking and modes of interaction as made clear in the influential formulation of both the potential and the limits of a specific kind of platform for interdisciplinary research – the collaboratory – by the anthropologist Paul Rabinow (2003). These challenges are further multiplied when collaboration extends outside the academy, to involve not only members of different (academic) disciplines but also representatives of business, government and the third sector, social movements and the public.

One currently influential form of collaboration of this kind are what is described as experiments in participation (Lezaun, Marres and Tironi 2017), as experimental participation (Whatmore and Landström 2011), or as a shift in emphasis from the experimental as a knowledge-site to the experimental as a social process (Corsín Jiménez 2014a; see also http://limn.it/prototyping-relationships-on-techno-political-hospitality/; and Engelmann and McCormack this volume). Examples of experimental practices that take the form of social process can be found, Corsín Jiménez suggests, in open access publishing, or in open collaborative scientific exchanges, 'where sociality and social exchange often become the limit-tests of experimentation itself, such as in debates about interdisciplinary exchanges' (2014a: 382; see also Shah, Sneya and Chattapadhyay 2016). Other examples he gives include the use of social media to enable 'new para-sites of collaboration, where researchers and informants mutually co-design and modulate an epistemic space, or, simply, occasions where researchers . . . are drawn into a research problem at the request of their informants' (Corsín Jiménez 2014a: 382). As he points out, 'Where researchers once entered the field as outsiders (academics), they are now suddenly and unexpectedly being turned into insiders (colleagues, advisors). The traditional entry and exit points of knowledge-creation now face a permanent threat of abduction and destabilisation' (2014a: 382).

Another approach to understanding the possibilities for enquiry of collaboration between academics and those outside the academy is described by Noortje Marres (2012) as 'material participation'. This term draws attention to the fact that while participation traditionally refers to human activity (securing the involvement of people in forms of enquiry), it may also refer to the deployment of seemingly mundane artefacts in settings of public deliberation. Drawing on John Dewey's understanding of 'ontological trouble', Marres describes material participation as the design of objects, devices, or more generally material settings or environments in such a way that publics can form and act in relation to a problem. Given that participation can be more-or-less active, more-or-less deliberative, such devices, she argues, have the capacity to turn everyday activities 'into an index of public participation', conscripting subjects into an 'ecological public'. At the same time, such methods raise issues of accountability (to whom is the research accountable, and how can that accountability be demonstrated?), not least because the sociality of the society to which researchers are required to be accountable is itself being reconfigured in these methods as part of the process of collaboration. As Jussi Parikka observes in his discussion of Digging (this volume):

> The collective workshops that are starting to define a methodological – even if most often still outside academic settings – attitude to digital culture indicate an important trend: cryptoparties, hackathons, game jams and other sorts of activities that run over one night or multiple days (and sometimes nights) define a fan-styled enthusiasm which attaches curiosity, dedication and often a critical attitude to working with machines whether in terms of coding, hardware hacking or the social and legal issues around digital culture from surveillance to economy (for example, copyright). Instead of mere Do-It-Yourself (DIY)-ethos, there are suggestions of more socially-oriented hack and other activities of DIWO – Do-It-With-Others (Garrett and Catlow 2012).

## Infrastructure

Marres' emphasis on material participation leads nicely into a consideration of the ways in which current changes in the research infrastructure afford new possibilities for interdisciplinary research. Let me give two examples to illustrate some of these changes and their relation to interdisciplinary research. The first example concerns the development of a software program for scraping the web for prices (Gross and Lury 2014). It was developed at a time when the accuracy of the Argentinian state-sponsored statistical measure, the Consumer Price Index, was under attack: there were public protests, and many rival measures came into existence as academic economists and representatives of various private institutions started their own inde-pendent, non-governmental projects to monitor prices. One of these, the Billion Prices Project (www.thebillionpricesproject.com), provided a challenge not simply to the figures that were being produced by the Argentinian national statistics office but also to the production of the CPI as a national statistic at all, since it involved the use of online or big data from many countries rather than the primary data of national statisticians. Created by two Argentinian economists, the BPP methodology can be used to monitor the daily price fluctuations of approximately 5 million items sold by approximately 300 online retailers in more than 70 countries. It is currently used to inform the production of, among other things, a daily price index and inflation rate for many countries, including Argentina.

The BPP originated in 2007 as part of Cavallo's PhD thesis at Harvard University which compared the online price variations for Argentina, Chile, Brazil and Colombia. What started as an academic exercise, however, was later subsumed in a company that describes itself as 'the leading source of daily inflation statistics around the world', trading under the registered trademark of PriceStats (www.pricestats.com/). The BPP website refers those seeking 'more high-frequency inflation data across countries and sectors' to PriceStats, 'the company that collects the online data we use in our research initiatives and experimental indexes'. The company itself makes its data available in commercial, academic and 'public' forms, defined by varying terms of access. For example, the PriceStats website says that it distributes (that is, presumably, sells) its daily inflation statistics through an exclusive partnership with State Street Global Markets, whose target clients manage hedge funds, pension funds and sovereign wealth funds. It also describes itself as in partnership with the BPP, introduced by PriceStats as 'an academic initiative that uses high frequency price information to conduct breakthrough economic research'. Finally, PriceStats says it collaborates with public institutions to improve decisions in public policy: for example, they create special indices that measure the price of specific goods across countries to 'anticipate the impact of commodity shocks on low income, vulnerable populations'. In short, the scraped data is used by PriceStats for multiple kinds of calculative operations, in relation to diverse clients. In this regard it appears as if, in contrast to the methodology developed by national statisticians to produce the CPI, the scraping methodology has the potential to offer new and different opportunities for many kinds of economic actors.

The second example concerns what Alberto Corsín Jiménez (2014b) calls beta or open source urbanism, a phenomenon he understands in relation to an ongoing transformation in the infrastructural landscapes of cities across the globe. While not ignoring the embedded economic and political interests that continue to dominate such landscapes, he focuses on the ways in which citizens are

> wiring the landscape of their communities with the devices, networks, or architectures that they deem worthy of local attention or concern. From community urban gardens to alternative-energy microstations or Wi-Fi networks, open source hardware projects wireframe the city with new sociotechnical relations. Such interventions in the urban fabric are transforming, and even directly challenging, the public qualities of urban space. Public spaces become technomaterial artefacts that citizens take upon themselves to service and maintain.
>
> *2014b: 342*

Corsín Jiménez locates these developments in relation to the 'new economy of open knowledge' emerging from novel organizational forms, such as peer-to-peer networks of collaboration. In relation to these histories he sees the possibility of what he describes as a right to infrastructure.

He identifies three dimensions to this right. (1) Conceptual: projects in open source urbanism populate urban ecologies with novel digital and material entities whose emergence destabilizes classical regulatory distinctions between public, private or commercial property forms, technologies and spaces. (2) Technical: open source urban projects are built on networks of expertise and skills that traverse localized boundaries: 'Decentralised communities working in open source projects have to reach prior consensus over the methods, protocols, and standards to be applied. These decisions often generate new designs, techniques, and rules for certification' (2014b: 343). (3) Political: open source projects transform the stakes in and modes of urban governance. In an open source project, a community may assume political and expert management of its infrastructure (see, for example, www.intheair.es/). By bringing these three dimensions together, he suggests, it becomes possible to read the right to *infrastructure* as a verb, not a noun: 'The process of *infrastructuring* makes visible and legible the languages, media, inscriptions, artefacts, devices, and relations – the betagrams – through which political and social agencies are endowed with any expressive capacity whatsoever' (2014b: 357).

The two examples given here provide contrasting uses of the opportunities offered by new knowledge infrastructures. However, both attest to the way in which such infrastructures afford the potential to redistribute methodological expertise, and bring a variety of actors into competition with each other for epistemological legitimacy. In relation to this transformed research landscape, interdisciplinarity is unavoidably an integral part of what Mike Savage describes as a 'messy, competitive context' in which 'the roles of different kinds of intellectuals, technical experts, and social groups are at stake' (2010: 237).

## Futures thinking

Many recent discussions of interdisciplinary research are preoccupied with the relation of research to the future; that is, they are linked to futures thinking in powerful ways. Of course, this relation has a long methodological history, including debates on the possibility and value of establishing causality and its relation to prediction, practices of foresight, innovation, projection (Connor, this volume) and expectation (Brown and Michael 2003). Mike Savage (2010) has, for example, described the ways in which the use of the survey in social science in the second half of the twentieth century involved statistical analysis designed to produce predictions about the future through the sequencing of (static or discrete) cross-sectional data collected at repeated intervals of time. In contrast, the anthropologist Marilyn Strathern describes ethnography as having a non-linear methodology. Ethnography, she says, deliberately attempts to 'generate more data than the investigator is aware of at the time of collection': that is, rather than devising research protocols

> that will purify the data in advance of analysis, the anthropologist embarks on a participatory exercise which yields materials for which analytical protocols are often devised after the fact. In the field the ethnographer may work by indirection, creating tangents from which the principal subject can be observed (through 'the wider social context'). But what is tangential at one stage may become central at the next.
>
> *Strathern 2004: 5*

The outcome, she claims, is 'anticipation by default'.

Recent changes to contemporary research infrastructures – including, notably, the increase in computational capacity, the growing availability of 'real-time' data and transformations in what Carolin Gerlitz (this volume) calls the ecosystem of data retrieval – have stimulated discussion as to whether the future can somehow be brought into the present. That is, for example, whether the future can be not just predicted but in some sense anticipated (see also Speculating by Alex Wilkie, this volume). This is part of the contemporary resurgence of interest in design as a methodology (see Michael, DiSalvo, and Akama and Pink, this volume). Traditionally, design and design processes have been seen to be purely practical, that is, as merely applied knowledge, and therefore were not considered as basic research. However, as anticipated by Herbert Simon (1996), design is increasingly being deployed across many disciplines as a method(ology) to make relations to the future a primary concern (Schaffner 2010). This is linked to a concern with the performativity of method, and has contributed to a turn toward praxis, critical making, and synthetic thinking and led to debates on the value of pre-emption, anticipation and speculation (Dunne and Raby 2014; Yelavich and Adam 2014). The appropriation of design methodology is also associated with the making of artefacts of all kinds, including epistemic artefacts, synthetic materials or even smart cities. In research that aims to activate the present in terms of 'smartness', for example, such as research linked to smart materials, smart cars or smart cities, design methodology contributes to the creation of cognitive and epistemic artefacts that have the capacity to modulate the present in what is described as real time, bypassing more human forms of governance.

In other methodological practices, multiple futures are anticipated by algorithmic optimization techniques, while in still others, possible futures are pre-empted through methods that extend and revise traditional statistical techniques of probability (Amoore 2013). In some research, the methodological potential of recognizing alternative futures by exploring the possibilities of 'alternation in phase over time' is formulated in explicitly political terms. The anthropologist Bill Maurer describes his own practice as 'building the alternative in the now'. In interview, he says:

> I've always been trying to get a sense of not 'where is the alternative?' but '*when* is the alternative?' How do we see and hear and feel alternative moments that spring up right in the practice of the everyday, even if they fall out of phase again and back into the conventional? For me, the task is both to elucidate those moments and then from a political standpoint to see if there are things that are worth dilating a little bit, expanding or having last a little bit longer. It's part of building the alternative in the now, and also part of this commitment to a rich empiricism that is attentive to the alternatives that are going on all around us, at least some of the time. . . . if . . . there are moments or pockets that are coming into phase, where alternatives happen and they work, then you start to get people having a sense of 'oh, you know, another world is possible, another world is actually already here if we care to pay attention to it'.
>
> *Tooker 2014*

As a making active of the present, interdisciplinary methods cannot avoid futures thinking.

Having outlined some of the dimensions of the contemporary configuration of interdisciplinarity, I want now to consider their implications for interdisciplinary methods through a discussion of four -ings.

## Essaying

In the introduction to a collection titled *Essaying Essays* (Kostelanetz, Di Ponio and Bebout 1974), Richard Kostelanetz gives a definition of the essay as 'a short composition on any particular subject'. Instances of this genre include, he says, not only short pieces of continuous prose, but also lists, aphorisms, interviews, and expository pastiche. *Essaying* essays, he says

> are those printed pieces that essay (i.e. attempt, try) to redefine the genre. . . . By realizing other ways of doing what 'essays' have traditionally done, they are essays twice over, confronting not only their particular subjects but, by implication, alternative possibilities for the form itself.
>
> *1974: vii–viii*

In this instance, the doubling of (essaying) essays is the (literary) activity of problematizing:

> Innovative essays are those that confront not just dimensions of extrinsic reality but also the intrinsic, literary problem of how else essays might be written. . . . As essays depend upon organization rather than fabrication, formal changes instill a re-essaying of a chosen subject; precisely because a different form reveals connections and perspectives that were not previously available, structural invention can change the essayist's thoughtful perception.
>
> *1974: xvi*

Of course, essaying might not even seem to describe a method from the perspective of some disciplines: matters of form, if considered at all, are considered secondary. But if we think about methods as -ings we can see that methods always involve the composition of a form: the activation of the present as the *present*-ation and re-*present*-ation of what is given.

## Mapping

Mapping is a group of methods that is evidently and reflexively undergoing rapid change (Batty *et al.* 2013; Dodge and Kitchin 2011; Thrift 2008; Hind and Lammes 2016). These changes, notably the move from the desktop to the web, have meant that 2-dimensional mapping has become key to many kinds of spatial (re-)presentation. In these changes, other -ings, notably visualizing have come to play an increasingly important role.

The recent move from desktop to web has had several stages, including the increase in availability of computable devices able to display data; the augmentation of the longstanding use of social data with space-time series generated in real time; the rise of graphic interfaces and the multiplication of software that allows a user to interact with data; and the possibility of remote storage of data in the Cloud. These developments have allowed for the emergence of computationally intensive methods of mapping. They have also contributed to a situation in which relations between mapping and maps are opened up as a site of methodological experimentation. Of course, these relations have always been available for experiment. However, the nature of the current changes has been so profound, and their implications so consequential that previously consolidated methodological relations have been destabilized, and new possibilities given a chance to emerge. Perhaps most significantly, these changes mean that mapping can be put into time (and space) in new ways: that is, different techniques for sequencing time have been developed to supplement static, cross-sectional methods. In addition, however, these changes have also meant both the possibility of more and different actors being involved in the activity of mapping – including commercial actors as well as 'crowds' of various kinds – and the possibility for mapping to be linked to more and other methods in new ways (Albuquerque, Herfort and Eckle 2016; see also the discussion of iBordering by Holger Pötzsch in this volume). This is especially clear in the use of a variety of techniques of visualizing as part of mapping, which now routinely also involves data-driven modelling and simulating. As Batty *et al.* note, 'the fact that the data volumes are large and the models often compute[r] intensive means that visualisation is of the essence' (2013: 20).

Perhaps most significantly, as well as being a way to display findings, visualization now plays a generative role in mapping, that is, visualizing is not simply representation, but, recursively, presentation and re-presentation (Calvillo 2012). In his discussion of Visualizing in this volume, Greg McInerny uses the term 'Design Space' to envisage 'a hyper-volume of *all* possible visualization designs with as many dimensions as there are ways to visualize data using coordinate and mapping systems, visual encodings and formatting, scales and sizing, sampling and aggregation methods'. To consider specific visualizations in relation to such a space makes it possible, he suggests, to see them as both 'arbitrary and specific': they could always be otherwise as the necessarily contingent outcome of visualizing, but they are as they are for specific methodological reasons. In such a space, the doubling of visualizing as part of the process of mapping offers the opportunity of a kind of virtuous methodological circle of presentation and re-presentation: the workings of a map or a model can now be exposed in ways that are highly amenable to visualization with the consequence that observations and predictions can be exhaustively compared across many combinations of model calibrations while, 'linking model processes to outcomes generates novel ways of visualising the relationships between processes and spatial outcomes' (Batty *et al.* 2013: 20; McInerny this volume).

## Folding

The collection *Essaying Essays* (Kostalanetz *et al.* 1974) previously discussed was published in the 1970s, at a time when the status of paper as the principal medium or material of print and its associated forms such as the book was just beginning to be challenged. On the one hand, the status of the essay as a form of print media is taken for granted by Kostelanetz in the unremarked oscillation between 'essay' and 'printed essay'. On the other, he notes that some forms of essay cannot be easily reproduced in the material form of a book:

> One alternative form that could not be reprinted here, alas, is the modular essay whose physically separate parts can be rearranged, the relationships among them continually changing; for the variable kind of exposition that is possible, say, in a packet of printed cards is simply impossible in a spine-bound book.
>
> *1974: ix*

Now, however, the materials of essays are no longer routinely assumed to be paper, bound as books. Other materials are possible, even perhaps more likely: 'Uncreative or conceptual writing is a type of literature that is born of and made possible by the digital' (Goldsmith 2011; see also Salazar-Sutil (2015) and the chapter in this volume on Notating by Moritz Wedell).

To ensure we do not take for granted the media or materials of methods, we can consider the activity of *folding*, at least as described by Michael Friedman and Wolfgang Schaffner. The fold, they say, 'is a material operation or/and an operative material starting at the molecular and ending at the macro level' (2016: 8). It occurs in nature as when frogs fold leaves to secure their eggs, when chimpanzees fold leaves to swallow them or when trees fold their leaves while it rains. In human culture, folding is constitutive of weaving, knotting, braiding and calculating as well as writing. Friedman and Schaffner suggest that folding understood in this way has the potential to open new relations across disciplines. Discussing examples from materials science, biology, architecture and mathematics to literature and philosophy, they suggest that folding overcomes old dichotomies, such as the organic and the inorganic and nature and culture, and blurs the boundaries between experimental, conceptual and historical approaches.

Each of weaving, knotting, braiding, calculating and writing, they say, 'might be thought as dealing with a sequential series of signs' (Friedman and Schaffner 2016: 8). But, so they argue, to understand them as examples of folding allows us to consider the consequential effects of the *materiality* of the sequencing of signs. Their case is made through a discussion of the relations between folding, folded material and code. Digital code, they say, is normally understood as 'what codifies operations and processes into an alphanumeric series of signifiers, enabling us to view operations – such as code – as a linear series of transmissible, discrete operations, which can be repeated over and over' (2016: 9). This abstracted understanding of code derives from a particular history in which, they note:

> Starting from the end of the 19th century, code was no longer perceived as what stems from a codex, but rather as what codifies and externalizes thought or meaning through the codification of difference itself. Transmitted alphanumeric code points towards deciphering, reading and writing as what belongs to digital one-dimensional code.
>
> *2016: 9*

In relation to this history, 'Digital code is ultimately denuded of any sign of materiality; it represents pure form overcoming not only materiality but also the surrounding material environment and space' (2016: 9).

To consider code as folding, they claim, overturns this view, since it forces an understanding of the sequencing of signs as always and inevitably materially spatialized. They quote the mathematician and philosopher Rene Thom who says: 'If, as Paul Valery said, "Il n'y a pas de géométrie sans langage," it is no less true that there is no intelligible language without geometry, an underlying dynamic whose structurally stable states are formalized by the language' (Friedman and Schaffner 2016: 14).

Explicating this rather abstract claim, they say that to codify folding by imposing on it a single language or one-dimensional linear codification is reductive. Instead, they say, Thom proposes to fold code, a material process in which code enters into more than one dimension. Understanding code as an instance of folding, they conclude, has three benefits:

> First, it suggests that codification should be communicated through physical and material effects. Second, it transcends the dichotomy of the codifying symbols, namely, that these symbols can *either* represent objects or execute actions. A folded code can do both. . . . Third, the materiality of [for example] bending, stretching or twisting does not refer to ideal operations, but rather to mechanics, to the manner in which material itself changes.
>
> *2016: 15*

To explain further, they say we need to recall the following 'simple fact': 'folding is a spatial operation'. The line (as a kind of series or sequence) that appears as a consequence of folding a piece of paper is a spatial effect. It might appear in two dimensions but, so they suggest, understanding a line in terms of folding allows us to see it as an effect of 'three-dimensional folding processes that go beyond the mere sequential chain of equally-connected symbols, to include three-dimensional operations, such as bending, stretching, twisting and translating'. In this way – by unfolding folding – they show it to be a constitutive, multi-dimensional activity.

## Rendering

These three examples draw attention to the twists in time and space that enable compound methods to activate the present. Importantly, however, this activation is not normally something that is a one-off, a discrete procedure, but rather, something that itself is conducted as part of a sequence or series in which a problem is (collectively if not collaboratively) individuated. As Hans-Jorg Rheinberger says, 'research is a cultural configuration endowed with its own temporality' (Rheinberger, 2010: 21). In other words, problems are the methodologically induced property of what we might call practical fields (one example of which Brighenti (2017) describes as measure-value circuits or measure-value environments), themselves comprising researchers, methods, materials and media, connected to each other in time and space in diverse ways as part of a constantly changing research infrastructure (see also Law 2004).

This has always been the case. Studies of science suggest that single experiments have rarely been the site of significant discovery: 'What are more typical are series of exploratory experiments that communicate among themselves with varying intensity, constituting an experimental texture in which equally unexpected condensations and eradications can occur at any time' (Rheinberger 2010: 20).

In other fields too, the sequencing of repetition has been a site of methodological innovation. For example, Zeilinger (2014) proposes that we recognize sampling as a methodological intervention, not simply as borrowing or stealing, but as a purposeful replacement of a recognizable original. He gives two cases. The first concerns a record of the Bobby Darin 1959 hit song *Dream Lover*, which as a result of the repeated playing of selected passages, is scratched so that the needle gets stuck, repeating certain grooves: 'the jumping needle transforms the line "Dream lover, where are you— with a love oh so true?" into a loop that sounds like: "Dreamlo-lo-lo-lover where are yo-u-u-u-u . . .?"' (2014: 163–164). The effect of this stuttering, he says, subtly changes the line's connotation: 'The scratched record takes on the quality of a new utterance . . ., and, from its inscribed involuntary repetitions and stammering, the listener may discern the longing for love, the insecurities, and the unfulfilled desires of a whole generation of listeners' (2014: 164). The second case concerns the film *Alone*. Zeilinger says:

> By my estimation, Arnold's *Alone* appropriates a total of around two minutes from several source films and, by inserting countless repetitions of sampled snippets, stretches the source material to roughly eight times its original duration. This intervention allows the filmmaker to focus on a number of archetypical character constellations (such as Father–Son, Father–Mother–Son, and Mother–Son–Love Interest) and to foreground in these constellations psychological issues that the cultural mainstream tends to gloss over.
>
> *2014: 164*

Through these two cases, Zeilinger argues that sampling can uncover, foreground and repurpose the meanings of original materials. He shows how sampling – or repetition more generally – can *return* the past to the present and the future. In making this argument he is very much concerned with the ways in which repetition is sequenced, that is, in the terms being developed here, how the organization of the iterative and distributed engagement with the given can individuate a problem in very particular ways:

> When we sample, we do not necessarily produce anti-authoritative ruptures (that would be the legal action of the sampling artist as pirate); rather, sampling allows us to become part of circuits of meaningful repetition that can create new intimacies, new rapports between us, the original work, and the sampling piece itself. Sampling simultaneously dismantles and reinstates a work, an idea, or a unit.
>
> *2014: 169*

However, while the methodological value of repetition has always been central to knowledge production, I want to conclude this section by suggesting that the changes happening now in the research infrastructures are reconfiguring this value in new ways. This is linked to the ways in which iteration or repetition can now more easily be configured as *recursion* (Totaro and Ninno 2014).

Let me proceed by example once again. Fluidity is an open source computational code that is the key methodological resource of a large group of scientists at the Applied Modelling and Computation Group at Imperial College in London. In his ethnography of this group, Matthew Spencer describes the complex temporalities that are involved in the transformation in use of this code by scientists in different disciplines, working both independently and collaboratively, in syncopated rhythms with each other. He writes:

> Research projects carry with them the whole weight of their past. While the trajectory of construction may move from a mathematical model of an analytical solution to a model of a well-studied experiment, the results of these previous stages become concretised in the apparatus as part of a testing system. When a scientist moves on to model something new, it is important to be assured that changes made in doing this have not undone earlier successes that built the foundation for the project. So as a test incorporated into the automated build and test suite, the earlier result will be run every time modifications are made to the code, ensuring that confidence from past success can still hold. . . . When a model is under active development, it is never enough to cite validations and verifications that have been made in the past, because these have been made with respect to a different code. All past verification and validation accreted in the present system of research is thus carried forwards with current research projects, applied over and over again to every new iteration of the code.
>
> *Spencer 2013: 107*

As Spencer acknowledges, scientific practice has always been recursively distributed in space and time but his study shows the importance of the changing organization of that repetition in a shared computational infrastructure *that is itself changing* as does the (changing) object of study – fluidity – and the nature of interdisciplinary collaboration. This characteristic – the operationalization of repetition in ways that actively engage and exploit a context that is itself changing – is fundamental to the innovations that are highlighted here.

A second example relates to the use of computer-generated images (CGIs) in urban planning. Rose, Degen and Melhuish (2014) argue that rather than seeing them as still images, as static representations of urban space, they should be understood as interfaces circulating through a dynamic software-supported network space:

> the action done on and with CGIs as they are created takes place at a series of interfaces. These interfaces – between and among humans, software, and hardware – are where work is done both to create the CGI and to create the conditions for their circulation.
>
> *2014: 386*

Crucially, understood as interfaces, the circulation of a CGI is not secondary to its creation, but both a condition and a consequence of its methodological value. Indeed, Rose *et al.* propose that the study of interfaces demands, 'a new methodology attentive to the work done as they circulate through software-supported spatialities' (2014: 402).

A third example concerns the increasing importance attached to search in the conduct of re-search. David Stark observes a shift in the ways in which (so-called dynamic) networks are transforming the processes of classification that are fundamental to many kinds of research (2011: 169). Things changed, he says, when the founders of Google reorganized search from a classificatory to a network logic:

[N]ew social technologies exploit, radically in recombination, the three basic activities of life on the Web: *search, link, interact*. . . . [S]earch based on the structure of the links . . . Interact based on the structure of searches . . . [L]ink based on the structure of the interactions. (2011: 171)

Stark introduces Luis Rocha's TalkMine, an adaptive recommendation system, as an example of the methodological possibilities opened up by the recombinatorics of search, link and interact. Rocha's aim is to 'achieve an interesting coupling [of a recommendation system] with users'. Stark emphasizes the capacity of TalkMine to correct the key deficiency of programs that conceive data bases as 'passive' and model search as information retrieval, that is, that assume that the existing, often static, structure of an information resource contains all the relevant knowledge to be discovered.

In contrast, Stark says, 'Once the vast databases are seen as an associative knowledge structure, the goal is to make them accessible as evolving knowledge repositories' (2011: 171). New categories emerge by treating users themselves as information resources with their own specific contexts. While the concept of category is not abandoned it is reconceived in relation to contexts produced in relation to circulation or movement:

> A category is temporarily constructed by integrating knowledge from several information resources and the interests of users expressed in the interactive process. As a temporary container of knowledge, it resembles transient, context-dependent knowledge arrangements characterized by Andy Clark as 'on the hoof' category constructions. Such short-term categories bring together a number of possibly highly unrelated contexts, which in turn create new associations in the individual information resources that would never occur with their own limited context.
>
> *Stark 2011: 173*

In each of these instances, it is the organization of iteration or repetition as recursion that provides new kinds of methodological affordance, including, perhaps most significantly, supporting practices of contexting (including users or participants) to provide new kinds of resource for knowledge production (Seaver 2015). As such, the ability to configure repetition as recursion challenges those forms of knowledge that are accomplished in forms of repetition (such as replication) that secure their value in terms of context-independence. Certainly, the rise of recursion provides many opportunities for researchers to participate in what Martin Savransky (2016) calls 'the adventure of relevance'. In short, the examples indicate the profound implication of contemporary changes in the research infrastructure for inter-disciplining,[4] that is, for the securing of both the autonomy and accountability of interdisciplinary methods in relation to what we might call the infra-empirical (to adopt and adapt Patricia Clough's (2008) use of the term), an empirical that is always caught up in infrastructuring.

It is because of the significance of this inter-disciplining that I want to point to the importance of what I will call *rendering* (Day and Lury 2017). Rendering or rendition is a term with many everyday as well as technical definitions, including: a performance, a translation, an artistic depiction, a representation of a building or an interior executed in perspective, as well as meaning to return, to make a payment in money, kind or service as by a tenant to a superior, to pay in due (a tax or tribute) and, in legal terms, to transfer persons from one jurisdiction to another. The origin for all these uses of the term is the Latin *reddere*: 'to give back'. Its use here thus speaks directly to the understanding of methods as -ings: it directs attention to the notion of the present as *the given*, not simply as that which is fixed or obdurate, but, in anthropological terms, as a gift, and thus makes it possible to see all -ings as emergent in circuits of giving or giving back. However, the contemporary salience of extraordinary, irregular or forced rendition also suggests we need to be especially sensitive to the political dimensions of what it means to 'give back' in contemporary research infrastructures.

Consider, in this respect, the artist Hito Steyerl's description of how the practice of film editing is being transformed in relation to changes in the knowledge infrastructure. It is, she says,

> being expanded by techniques of encryption – techniques of selection – and ways to keep material safe and to distribute information. Not only making it public, divulging or disclosing, but really finding new formats and circuits for it. I think this is an art that has not yet been defined as such, but it is, well, aesthetic. It's a form. . . . Now it's not only about narration but also about navigation, translation, braving serious personal risk, and evading a whole bunch of military spooks. It's about handling transparency as well as opacity, in a new way, in a new, vastly extended kind of filmmaking that requires vastly extended skills.
>
> *Steyerl and Poitras 2015: 311*

Steyerl proposes that the question of how information is 'stored, secured, circulated, redacted, checked, and so on . . . [the] entire art of withholding and disseminating information and carefully determining the circumstances' is a 'formal decision'. She emphasizes that this decision has an unstable temporality:

> When I'm working with *After Effects*,[5] there is hardly any real-time play back. So much information is being processed, it might take two hours or longer before you see the result. So editing is replaced by rendering. Rendering, rendering, staring at the render bar. It feels like I'm being rendered all the time.
>
> What do you do if you don't really see what you edit while you're doing it? You speculate. It's speculative editing. You try to guess what it's going to look like if you put key frames here and here and here. Then there are the many algorithms that do this kind of speculation for you.
>
> *Steyerl and Poitras 2015: 312*

In dialogue with Steyerl, the filmmaker Laura Poitras discusses the program TREASUREMAP used by the US National Security Agency (NSA) to provide analysts with 'a near-real-time map of the Internet and every device connected to it'. She suggests that at the core of the NSA's approach to data collection is another kind of activation of the present, a 'retrospective querying – how to see narrative after the fact' (Steyerl and Poitras 2015: 312).[6]

Given the significance of the issues that Steyerl raises for how (interdisciplinary) methods make time and space, it is apparent that one of the most important questions facing inter-disciplinary research today is how the doing of methods can activate the present in ways that do not render each instance 'merely' historical, but instead develop what Tahani Nadim calls the 'talent to return'. Suggestions put forward in this handbook include Arranging or *enchaînement*, what Harmony Bench describes as the 'crafting of relationships of contingent interdependency'. *Enchaînement*, she says, is

> a type of classroom exercise in ballet technique in which a series of travelling steps are linked (literally chained) together as a phrase or combination that moves across the dance floor. *Enchaînements* are built, or arranged, from a discrete vocabulary and syntax that determines

which movements or steps may logically precede or follow others. Connections among individual step-units are not wholly predetermined, nor are they wholly open to any connection whatsoever.

Sensitised to the importance of such connections, participating in interdisciplinary enquiry in a time of rendition requires researchers to be attentive to how they participate in a relay, as Jellis describes it, and to think about how they can pass on the baton as Olutoyosi Tokun puts it in the chapter on Dirty methods in this volume. As Abdoumaliq Simone points out, 'A practice is more than a particular way of doing something, more than simply technique, for it entails obligations to others who have also "practiced"' (2015: 18). As such interdisciplinary research can only benefit from – and perhaps especially now requires – the cultivation of what Yoko Akama and Sarah Pink describe as companionship, what Corsín Jiménez describes as trajectories of apprenticeship, and what Michael calls the staging of care.

## Conclusion: lateral methods and the autonomy and accountability of interdisciplinarity

This essay has proposed that considering methods as -ings enables us to consider them as ways to intervene in and make the present active. It further suggests that to use methods in this way contributes to both the autonomy and accountability of interdisciplinarity. Let me conclude by pulling together the different strands of this approach in a discussion of interdisciplinary methods as methods of the lateral.

As Gad and Bruun Jensen (2016) observe in their discussion of lateral concepts, in a broad sense the lateral observes a many-to-many relation between domains of knowledge and practice (see also Maurer 2005; Dalsgaard 2016). To describe interdisciplinary methods as methods of the lateral is thus to draw attention to the many-to-many relations that are made across and between disciplines. In one sense, the term speaks to the general observation that domains of knowledge and practice influence each other in unpredictable ways. However, in the presentation of interdisciplinary methods made here, this influence is shown to be immanent to the individuation of problems. In other words, as -ings, interdisciplinary methods are not mere links or associations between disciplines that somehow stand above or outside their objects of study, but dynamic conduits for relations of interference in which differences and asymmetries between disciplines are explored and exploited in relation to specific problems, in specific places, with specific materials.

What Gad and Bruun Jensen say of lateral concepts is also true of lateral methods:

> their development begins with the recognition of specific kinds of movement between forms of knowledge within a particular field of concern. If movements and modifications of forms of knowledge happen continuously, the lateral question becomes one of *activation*: how might researchers draw energy from something that is happening in front of our eyes anyway.
> *2016: 5*

In particular, the approach to interdisciplinary methods presented here proposes that they be understood in terms of their potential to activate the present. Highlighting how methods make as well as take time and space, the emphasis has been on the form of methodological relations in which the practices of method take place. This was understood as a process of composition: the folding of relations to person, place, material and process into the doing of a method, or the folding of a compositional methodology into the practice of method (Lury forthcoming).

Importantly, while activating the present was understood as a way to develop the autonomy of interdisciplinarity, the approach proposed here does not understand autonomy in terms of a complete independence of methods from their context of implementation, or indeed a desire for a lack of accountability. Instead the autonomy of interdisciplinary research is understood as an achievement of the real time of practice, and as such is always situated, always also a matter of accountability. But – and this is perhaps especially important in what one might call a time of rendition, by real time is not (only) meant the computationally driven temporalities of digital computation, but all the times made real in situated methodological practices of presentation and re-presentation. As Gad and Bruun Jensen observe, 'The risks and possibilities of the lateral . . . are about nurturing an attention to what it takes to establish relative forms of compatibility between divergent forms of knowing and acting in . . . a decentered world' (2016: 12). In other words, developing the autonomy and accountability of interdisciplinary methods involves an obligation to pursue a particular path, precisely insofar as it is 'a means to induce thinking, to build up a perspective over time, to generate a sense of efficacy, and a sense of belonging to something capable of absorbing individual action and effort' (Simone 2015: 18). As an activation of the present, interdisciplinary methods may contribute to the autonomy of interdisciplinary research but this self-direction need not be at odds with at least some forms of accountability. Autonomy and accountability can both be accomplished if questions of epistemology are reflexively and recursively implicated with ontological awareness. To understand interdisciplinary methods as -ings is to understand them as an activation of the present that is 'a practice of rootedness in processual awareness that can give shape amidst the unpeaceful, uninhabitable and unknowable state of crisis in which living is also taking place'. And while 'giving shape is not the same as solving the problem of crisis [interdisciplinary methods have the potential] to find form without distracting from the gravity of the real' (Berlant 2008: 7).

## Structure of the handbook

The field of interdisciplinary research methods is not new. However, it has few shared texts that give researchers specific guidance and an empowering sense of what is possible in interdisciplinary enquiry. This is the ambition of this handbook. Yet, given the dispersed and dynamic nature of this field, comprehensiveness is not best provided by either a premature standardization of terms, or the separation out of methodological issues from those of either theory or method. For this reason, following this introduction, the contributory -ings are grouped in the following sections, each of which is contextualized by one or more of the co-editors: 1. Making and Assembling – Rachel Fensham and Alexandra Heller-Nicholas; 2. Capturing and Composing – Emma Uprichard; 3. Engaging and Distributing – Sybille Lammes; and 5. Valuing and Validating – Mike Michael. The entries in each section address theory and methodology as well as methods, in order that the issues they illustrate are not disconnected from wider debates. They are typically presented as -ings, but some (also) speak to the relation between verbs and states or conditions. Importantly, however, they all focus on a specific research practice, ensuring that what can be quite abstract matters of epistemology and ontology are addressed concretely. The fourth section – 4. Of Interdisciplinarity is introduced by Angela Last, and consists in a series of interrogations of recent collaborative projects, relating to some of the most significant recent sites of interdisciplinarity. All the co-editors introduce their section with an essay, outlining their own distinctive perspective on interdisciplinary enquiry.

I want to end this introduction by thanking them, the editorial assistant, Tuur Driesser, Steven Connor for permission to use his translation of the poem by Serres that is the Preface to this volume, and all the contributors for giving their time(s) – their patience, their drive, their persistence, imagination and care – to the project of interdisciplinarity.

## Notes

1 As Rachel Fensham observed in our discussion of this term, not all languages have a similar syntax: a reminder that serves to draw attention to the complex geo-politics of methodology and their relation to the historical present.

2 For Serres an engagement with an expanded vocabulary of prepositions is necessary at a time when 'the milieu [the middle or the in-between] arises in every place' (Serres 1994: 128). For Serres – like many other contemporary thinkers, milieu does not just refer to media, but speaks instead to the ways in which the occupation of in-between has become a defining characteristic of contemporary life. The discussion of rendering below builds on this view to suggest that rendition is one of the most powerful ways in which milieux are occupied today.

3 In this respect, interdisciplinary methods as described here do not contribute to what has been called technological solutionism (Morozov 2013), an approach in which solutions are identified for problems that do not yet exist.

4 Andrew Pickering stresses 'the disciplining of human agency' that is part of scientific practice, referring to the ways in which practices are 'interactively stabilised together with other cultural elements in practice' (1995: 102).

5 Adobe After Effects is a software tool for video compositing, motion graphics design and animation.

6 Following this line of thought, phenomena such as 'fake news' can be understood as epistemic artefacts of an era of rendition.

## References

Albuquerque, J., Herfort, B. and Eckle, M. (2016). The tasks of the crowd: a typology of tasks in geographic information crowdsourcing and a case study in humanitarian mapping. *Remote Sensing*, 8: 859.

Amoore, L. (2013). *The Politics of Possibility: Risk and Security Beyond Probability*. Durham, NC: Duke University Press.

Apostel, L., Berger, G., Briggs, A. and Michaud, G. (eds) (1972). *Interdisciplinarity: Problems of Teaching and Research in Universities*. Paris: OECD.

Back, L. and Puwar, N. (2012). *Live Methods*. Oxford: Wiley-Blackwell.

Barry, A., Born, G. and Weszkalnys, G. (2008). Logics of interdisciplinarity. *Economy and Society*, 37(1): 20–49.

Barry, A. and Born, G. (Eds.) (2013). *Interdisciplinarity: Reconfigurations of the Social and Natural Sciences*. Abingdon: Routledge.

Batty, M., Gray, S., Hudson-Smith, A., Milton, R., O'Brien, O. and Roumpani, F. (2013). Visualising Spatial and Social Media. *UCL Working Paper Series*, Paper 190.

Berlant, L. (2008). Thinking about feeling historical. *Emotion, Space and Society*, 4–9.

Boix Mansilla, V., Lamont, M. and Sato, K. (2016). Shared cognitive-emotional-interactional platforms: markers and conditions for successful interdisciplinary collaborations. *Science, Technology & Human Values*, 41(4): 571–612.

Brighenti, A. B. (2017). The social life of measures: conceptualizing measure-value environments. *Theory, Culture and Society*, 35(1): 23–44.

Brown, N. and Michael, M. (2003). A sociology of expectations: retrospecting prospects and prospecting retrospects. *Technology Management and Strategic Analysis*, 15(1): 3–18.

Buchloh, B. (2000). Process sculpture and film in the work of Richard Serra. In H. Foster and G. Hughes (Eds.) *Richard Serra* (pp. 1–19). Cambridge, MA: The MIT Press.

Büscher, M., Urry, J. and Witchger, K. (Eds.) (2010). *Mobile Methods*. London: Routledge.

Calvillo, N. (2012). The affective mesh: air components 3D visualizations as a research and communication tool. *Parsons Journal for Information Mapping*, 4(2): 1–8.

Chen, H.-K. (2010). *Asia as Method: Toward Deimperialization*. Durham, NC: Duke University Press.

Chow, R. (2006). *The Age of the World Target: Self-referentiality in War, Theory and Comparative Work*. Durham, NC: Duke University Press.

Clough, P. (2008). The affective turn: political economy, biomedia and bodies. *Theory, Culture and Society*, 25(1): 1–22.

Corsín Jiménez, A. (2014a). Introduction. *Journal of Cultural Economy*, 7(4): 381–398.

Corsín Jiménez, A. (2014b). The right to infrastructure: a prototype for urban source urbanism. *Environment and Planning D: Society and Space*, 32: 342–362.

Crenshaw, Kimberlé W. (1995). Mapping the margins: intersectionality, identity politics, and violence against women of color. In Kimberlé Crenshaw, Neil Gotanda, Gary Peller and Kendal Thomas (Eds.) *Critical Race Theory: The Key Writings that Formed the Movement* (pp. 357–384). New York: The New Press.

Dalsgaard, S. (2016). Carbon valuations: alternatives, alternations and lateral measures. *Valuation Studies*, 4(1): 67–91.

Day, S. and Lury, C. (2017). New technologies of the observer: #BringBack, visualization and disappearance. *Theory, Culture and Society*, 34(7–8): 51–74.

Dodge, M. and Kitchin, R. (2011). *Code/Space: Software and Everyday Life*. Cambridge, MA: The MIT Press.

Dunne, A. and Raby, F. (2014). *Speculative Everything: Design, Fiction and Social Dreaming*. Cambridge, MA: The MIT Press.

Fraser, M. (2009). Experiencing sociology. *European Journal of Social Theory*, 12(1): 63–81.

Friedman, M. and Schaffner, W. (2016). On folding: introduction of a new field of interdisciplinary research. In M. Friedman and W. Schaffner (Eds.) *On Folding: Towards a New Field of Interdisciplinary Research* (pp. 7–30). Bielefeld: Verlag.

Gad, C. and Bruun Jensen, C. (2016). Lateral concepts. *Engaging Science, Technology, and Society*, 2: 3–12.

Garrels, G. (2011). An interview with Richard Serra (2010). In *Richard Serra Drawing: A Retrospective* (pp. 58–83). Houston, TX: The Menil Collection.

Goldsmith, K. (2011). *Uncreative Writing*. New York: Columbia University Press.

Gross, A. and Lury, C. (2014). 'The downs and ups of the consumer price index in Argentina: from National Statistics to Big Data', in Statactivism: State Restructuring, Financial Capitalism and Statistical Mobilizations, Special Issue, *Partecipazione e Conflitto – Rivista Scientifica di Studi Sociali e Politici*.

Hayles, N. K. (2005). *My Mother was a Computer: Digital Subjects and Literary Texts*. Chicago, IL: University of Chicago Press.

Hind, S. and Lammes, S. (2016). Digital mapping as double-tap: cartographic modes, calculations and failures. *Global Discourse*, 6(1–2): 79–97.

Kimbell, L. (2015a). Rethinking design thinking, Part I. *Design and Culture*, 3(3): 285–306.

Kimbell, L. (2015b). Rethinking design thinking, Part II. *Design and Culture*, 4(2): 129–148.

Kostelanetz, R., Di Ponio, J. F. and Bebout, N. M. (Eds.) (1974). *Essaying Essays: Alternative Forms of Exposition*. New York: AC Institute.

Krauss, R. (1985). *The Originality of the Avant-Garde and Other Modernist Myths*. Cambridge, MA: The MIT Press.

Law, J. (2004). *After Method: Mess in Social Science Research*. Abingdon: Routledge.

Lezaun, J., Marres, N. and Tironi, M. (2017). Experiments in participation. In C. Miller, E. Smitt-Doer, U. Felt and R. Fouche (Eds.) *Handbook of Science and Technology Studies* (Volume 4, pp. 195–222). Cambridge: MIT Press.

Lury, C. and Wakeford, N. (Eds.) (2012). *Inventive Methods: The Happening of the Social*. London: Routledge.

Lykke, N. (2010). *Feminist Studies: A Guide to Intersectional Theories, Methodologies and Writing*. London, New York: Routledge.

Marres, N. (2012). *Material Participation: Technology, the Environment and Everyday Publics*. London: Palgrave Macmillan.

Maurer, B. (2005). *Mutual Life, Limited: Islamic Banking, Alternative Currencies, Lateral Reason*. Princeton, NJ: Princeton University Press.

Morozov, G. (2013). *To Save Everything, Click Here: The Folly of Technological Solutionism*. PublicAffairs.

Niranjana, T. (2000). Alternative frames: questions for comparative research in the Third World. *Inter-Asia Cultural Studies*, 1(1): 97–108.

Niranjana, T. (2013). Introduction to Genealogies of the Asian Present: Situating Inter-Asia Cultural Studies. Retrieved March 2017 from: www.academia.edu/14934670/Introduction_to_Genealogies_of_the_Asian_Present_Situating_Inter-Asia_Cultural_Studies.

Nowotny, H., Scott, P. and Gibbons, M. (2001). *Re-thinking Science: Knowledge and the Public in an Age of Uncertainty*. Cambridge: Polity Press.

Osborne, T. (2013). Inter that discipline. In A. Barry and G. Born (Eds.) *Interdisciplinarity: Reconfigurations of the Social and Natural Sciences* (pp. 82–98). London and New York: Routledge.

Pickering, A. (1993). Antidisciplines or narratives of illusion. In E. Messer-Davidow, D. R. Shumway and D. J. Sylvan (Eds.) *Knowledges: Historical and Critical Studies in Disciplinarity* (pp. 103–123). Charlottesville: University Press of Virginia.

Pickering, A. (1995). *The Mangle of Practice: Time, Agency and Science*. Chicago, IL: University of Chicago Press.

Rabinow, P. (2003). *Anthropos Today: Reflections on Modern Equipment*. Princeton, NJ: Princeton University Press.

Rheinberger, H.-J. (2010). *An Epistemology of the Concrete: Twentieth Century Histories of Life*. Durham, NC and London: Duke University Press.

Riles, A. (Ed.) (2006). *Documents: Artefacts of Modern Knowledge*. Michigan: University of Michigan Press.

Rittel, H. W. J. and Webber, M. M. (1973). Dilemmas in a general theory of planning. *Policy Sciences*, 4: 155–169.

Rose, G., Degen, M. and Melhuish, C. (2014). Networks, interfaces, and computer-generated images: learning from digital visualisations of urban redevelopment projects. *Environment and Planning D: Society and Space*, 32: 386–403.

Salazar-Sutil, N. (2015). *Motion and Representation: The Language of Human Movement*. Cambridge, MA: The MIT Press.

Savage, M. (2010). *Identities and Social Change in Britain since 1940: The Politics of Method*. Oxford: Oxford University Press.

Savransky, M. (2016). *The Adventure of Relevance: An Ethics of Social Inquiry*. London: Palgrave Macmillan.

Schaffner, W. (2010). The design turn: Eine wissenschaftliche Revolution im Geiste der Gestaltung. In C. Mareis *et al.* (Eds.) *Entwerfen – Wissen – Produzieren. Designforschung im Anwendungskontext*, Bielefeld, 33–46.

Seaver, N. (2015). The nice thing about context is that everyone has it. *Media, Culture and Society*, 37(7): 1101–1109.

Serra, R. (1967–68). *Verb List*.

Serres, M. (1994). *Atlas*. Paris: Éditions Julliard.

Shah, N., Sneya, P. P. and Chattapadhyay, S. (Eds.) (2016). *Digital Activism in Asia Reader*. Leuphana: Meson Press.

Simon, H. (1996). *The Sciences of the Artificial*. Cambridge, MA: The MIT Press.

Simone, A. (2015). Relational infrastructures in postcolonial urban worlds. In S. Graham and C. McFarlane (Eds.) *Infrastructural Lives: Urban Infrastructure in Context* (pp. 17–39). Abingdon and New York: Routledge.

Spencer, M. (2013). Reason and Representation in Scientific Simulation, PhD thesis, Goldsmiths College.

Stark, D. (2011). *The Sense of Dissonance: Accounts of Worth in Economic Life*. Princeton, NJ and London: Princeton University Press.

Stenner, P. (2014). Transdisciplinarity. In T. Teo (Ed.) *Encyclopedia of Critical Psychology* (pp. 1987–1993). New York: Springer.

Steyerl, H. and Poitras, L. (2015). Techniques of the observer: Hito Steyerl and Laura Poitras in conversation. *Artforum*, May, pp. 306–317.

Strathern, M. (2004). *Commons and Borderlands: Working Papers on Interdisciplinarity, Accountability, and the Flow of Knowledge*. Oxon: Sean Kingston Publishing.

Thrift, N. (2008). Movement-space: the changing domain of thinking resulting from the development of new kinds of spatial awareness. *Economy and Society*, 33(4): 582–604.

Tooker, L. (2014). Conversation with Bill Maurer. *Exchanges: Warwick Research Journal*, 2(1) http://exchanges.warwick.ac.uk/article/view/99

Totaro, P. and Ninno, D. (2014). The concept of algorithm as an interpretive key of modern rationality. *Theory, Culture and Society*, 31(4): 29–49.

Whatmore, S. J. and Landström, C. (2011). Flood apprentices: an exercise in making things public. *Economy and Society*, 40(4): 582–610.

Yelavich, S. and Adam, B. (Eds.) (2014). *Design as Future-Making*. London and New York: Bloomsbury.

Yoshimi, T. (2005[1960]). Asia as method. In Richard F. Calichman (ed. and trans.) *What is Modernity? Writings of Takeuchi Yoshimi* (pp. 149–165). New York: Columbia University Press.

Zeilinger, M. J. (2014). Sampling as analysis, sampling as symptom: found footage and repetition in Martin Arnold's *Alone. Life Wastes Andy Hardy*. In D. Laderman and L. Westrup (Eds.) *Sampling Media* (pp. 159–171). Oxford Scholarship Online.

# Section 1
# Making and assembling

# 1

# Making and assembling

## Towards a conjectural paradigm for interdisciplinary research

*Rachel Fensham and Alexandra Heller-Nicholas*

---

Making and assembling produce an odd pairing of terms. Making derives from the short vocalization 'mek' from an Anglo-Saxon word, and hearkens to the Germanic verb *'machen'* meaning to do or to make. It has both a universal application, in the sense that everyone from children to adults makes sound, while on the other hand, it aligns with specialists who form unique or distinctive works, such as a fine machine or a beautiful painting. Making, or doing, also leads us directly to processes whereby materials become transformed by an action, such as a person making a cocktail or the weather making us feel hot. On the other hand, assembling is Latinate, as with the French verb *'assembler'*, in English also to assemble, as in the putting together or gathering of people, objects or things. Etymologically, this latter term relates then to the important notion of 'assembly' as a site of public cultural enactment, as well as to the assembly line of Fordist manufacturing. We might also have a more prosaic view of assembling in contemporary culture when we consume the plasticity of a robotics toy, a piece of IKEA furniture, or a Facebook page. Thus, we might conceive of a sharp contrast between making and assembling as methods, in the sense that making suggests creating, and something more primal, fashioned even from mud, whereas assembling tends towards order, and something more civilized, or institutional. We do not assert this dichotomy in any cultural hierarchy because we prefer to examine these terms operating in relation to one another, and as moving generatively between the social and linguistic, or human and non-human, in contemporary research.

It is possible to assemble a range of theorists – experimental thinkers – who become touchstones for different ways of making and assembling in interdisciplinary research, and scholars have become increasingly attuned to the ways in which theories of practice might differ from one discipline to another (Schatzki 2001: 11). Thomas Kuhn, for instance, acknowledges this when he writes about both the specificity of science as a knowledge practice and the generalizing power of a paradigm shaped by practice. In the process of narrating interdisciplinary research, and its potential admixture of norms equating to disciplines, alongside the entanglement of the practical, social and conceptual for this essay, we have arrived at Carlo Ginzburg's notion of the conjectural paradigm – what he has called 'the lightning recapitulation of rational processes' (1989: 117). This is a concept that allows us to find a space in-between disciplines, as well as to do research work that makes and assembles the ethical, political, creative, socially engaged and fun.

Such research may also be productive of a distinctive ontology, one that embraces both the history and power of representations as well as embodied social relations. To return to our key terms by way of exemplifying this ontology, the tensions between the primal and the civil recall in part Elaine Scarry's foundational book *The Body in Pain: The Making and Unmaking of the World* (1985). Scarry's privileging of the word 'making' and its antonym in her subtitle offer a fruitful starting place to rethink these opposing ideas and their impact on systems of thought more broadly. In her analysis of war and torture, for instance, Scarry notes that 'physical pain . . . is language-destroying' (1985: 19): thus invoking the political dimension inherent when pain is deliberately employed as an ideological tool designed to dehumanize. For Scarry there is ample evidence of how regimes seek to 'unmake' their victims and their experience of the world. The act of making, then – specifically through the act of creative expression (song, literature, film) – also has a concrete ideological function of *re*-making, or *re*-assembling, of putting together and creating anew that which has previously been unmade. Making, in Scarry's terms, involves both imaginative work, as well as an 'activity extended into the external world, and has as its outcome a verbal or material artefact' (1985: 177). Whether the making of a political structure, encompassing legal texts and border police, or the making of structural or sensory objects, this complex proposition includes 'obligations':

> For made things do incur large responsibilities to their human makers (and their continued existence depends on their abilities to fulfil those responsibilities: a useless artefact whether a failed god or a failed table, will be discarded); just as, of course, human makers also incur very large obligations to the objects they have made.
>
> *1985: 182*

When we make – whatever we make – the 'responsibility' we take for the act itself is, from Scarry's perspective therefore, intrinsically ideological.

Going beyond this humanistic concept of creation, the notion of assemblage is regarded as central to the ontology of Gilles Deleuze and Félix Guattari's *A Thousand Plateaus* (2004). The use of the term assemblage, as Ian Buchanan points out, has however been derived from the (mis)translation by Brian Massumi of their term, *agencement* from French and he proposes a more suitable term might be arrangement (Buchanan 2015: 383). In this sense, an assemblage might involve reorganizing diverse elements from across disciplines to create something unpredictable or with new valorizations. Emphasizing the dynamic process rather than the final product itself, for Deleuze and Guattari, all assemblages are historicized through combination, and as such they also have agency: for instance, any given political formation authorizes the circulation of bodies, the expression or repression of affects, and the production or reproduction of collectives and institutions. An *assemblage*, such as a research problem or task, can therefore be driven both by processes of territorialization and deterritorialization, and in its analysis, by coding and decoding such arrangements. An *assemblage* functions then through the deliberative fusion of multiple aspects of a situation. For Deleuze and Guattari: 'An *assemblage*, in its multiplicity, necessarily acts on semiotic flows, material flows, and social flows simultaneously (independently of any recapitulation that may be made of it in a scientific or theoretical corpus)' (2004: 25).

Guattari's experiments in the psychiatric clinic, for instance, 'used the grid as a tool to transversalise' and to lead all staff, residents and visitors 'towards the apprehension of the singular scenes composed and modified by each participant in relation to the relevant collective constraints and institutional matters that emerged in the process' (Genosko 2009: 61). For the ethico-political development of how we conceive methodologies to be 'made' in this introduction, Deleuze and Guattari vitally consider an assemblage to involve re-conceptualizing the components of a research project by allowing for a critical heterogeneity, both of logic and aesthetics, to emerge.

## The movement of interdisciplinarity

In comparison with these radical deconstructive methods of making and assembling, the conventional heuristics of academic knowledge claim the actualization of fields of knowledge, constituted by historically defined sets of relationalities, positions, or lines of argument. Scholars of disciplinary modes of researching mostly occupy a discursive space, an institutional or subject position in relation to other thinkers via established patterns of citational linking. Within a discipline, these arguments are supported by evidence, which in turn develop lines of argument. In interdisciplinary discourse, however, these elements that make spatial sense of theory-making are sometimes characterized by an in-between-ness, and a not-quite-belonging. So, rather than making a space of knowledge for ourselves from a centralized location of discursive action, or in terms of unidirectional lines and stable shapes that serve as basic elements of a rather geometric way of modelling an argument, the interdisciplinary turn leads, we would suggest, towards a more dynamic, spatialized understanding of what a field of knowledge is and, by extension, who the specialists in the field might be, such as the authors assembled here, some well known and others less so.

If in response to new conditions, the research processes of interdisciplinarity foster the particular space of ideation as limen, as threshold, or as interstice, then they also function like nodes in a network, as spaces inter or in-between. Movement, as Nicolas Salazar-Sutil argues, might be about the physical locatedness of human movement, but it might also be conceived 'in terms of electronic location within global networks' (2015: 211). In this shifting of positions beyond the linear accumulation of ideas, our argument goes further: interdisciplinarity implies more than space, it implies movement, what we might define as the multi-dimensional properties of images, objects or thought changing in time and space. As such, this alternative paradigm involves a spatial and temporal realization of any movement that precedes or follows it, as well as a recognition of mobility as a process of sensitizing research to a particular situation within discourse, within art, within the social.[1] We are not alone in mounting this argument for movement, and experience, as critical to the new formations of interdisciplinary research in cultural studies, science studies, social and political theory. For Bruno Latour:

> In my view, ecology is only very rarely a politicized form. Usually the questions I am interested in — about sensitizing, about an Anthropocenic recognition of mobility, of process — these questions are sealed off by politics, and, surrounded by well-meaning self-righteousness … [as a] metaphor for complete control, the puppet actually makes its puppeteer carry it somewhere else. It gets modified, mobilized, or moved — and you are then moved by the thing you move, which is the most interesting relation we have with the world.
>
> *2016: 321*

This realization of academic thinking as movement in terms of interdisciplinary research is dependent, then, upon acts of making or assembling (or, likewise, unmaking and disassembling).

In this essay, the notion of movement as a pre-eminent paradigm and method also serves as an invitation to adopt a kinetic way of 'doing' ideas in the academy that will allow scholars to build new and specific models for interdisciplinary research that are appropriate to the challenges of our time. In the movement from one field to another, such research will test and corroborate theory, enable comparison of different forms of evidence, and require the construction of new kinds of research artefact, instance or residue. The contributions assembled in this section argue and exemplify a range of approaches to interdisciplinarity that have been encouraged by a fundamental change in the way they make theory and assemble alliances through research: they are conjecturing projects and processes within a spatiality of communication, and a horizon of social transformation, that sees knowledge production happen in electronically and increasingly mediated ways with, however, an ongoing sense of the unevenness of distributions of power, wealth and access. They therefore practise research through making and unmaking, assembling and disassembling.

All the authors here stake a claim on interdisciplinarity, flagged by backgrounds that link and diverge. In spite of the interdisciplinary framework, these people are also located within disciplines – even if and when the focus is to work beyond 'just' finding something new or different. Their methods include the attempt to interrogate/interrupt/interpolate the restrictions of disciplines to work towards remaining open to new practices, ideas and methodologies, by embracing the conjectural paradigm over the restrictions implied by any traditional privileging of the general. We know this not only based on the entries themselves, but because we asked some of these authors by email to articulate their relationships to interdisciplinarity and how it helps to model their identity as researchers and we found their answers most illuminating.

For Catherine Ayres:

> Being an interdisciplinary researcher to me requires an ethos of generous critique. I try as hard as I can to appreciate the contributions various disciplines are trying to make to our intellectual world, even when this means directly challenging the core tenets of my 'home' discipline.

Thomas Jellis offers a different approach in terms of how he moves beyond disciplinarity in his professional practice. 'I'm inclined to think in terms of disciplinary matters of concern – or refrains – and how these might also speak to other fields; the task is to work out how to enable temporarily shared trajectories between them.' Harmony Bench reflects upon how this process changes her academic identity:

> My understanding and framing of myself as within a discipline or as adhering to a methodology has shifted over the past several years. Since all of my training has been interdisciplinary (with the exception of a degree in ballet), disciplinarity was not a concern of mine until going on the job market, at which point I described myself as a generalist in the field of dance.

In contrast, Margaret Wertheim has spent a career outside of academia, peripatetically challenging the disciplines of science and mathematics to rethink its models, calculations and designs as hand-made, collective formations. And yet she notes:

> [W]e all benefit in our daily lives from the knowledges produced and acquired by these specialists, and we should all applaud the dedication and commitment it takes to achieve this kind of work. Every academic discipline has been subject to such diversification and subdivision, which seems to be one of the characteristics of our intellectual age.

Matthew Reason, on the other hand, acknowledges an important subjectivity in the very term discipline, noting that it 'looks very different according to where you are standing'. He continues, 'I have found the real challenge of cross-disciplinary work is when the methodologies don't align, when there are not only different discourses or points of reference but different understandings of what knowledge is.' So, whether a researcher works in the physical or social sciences, the arts or the humanities, there are paradigms that constrain and constitute a disciplinary subjectivity and methodologies of practice. Ramon Lobato positions his own research as being based in one discipline, while reaching out productively to others: 'I work between media and cultural studies, and also draw a lot on economic and geographic modes of analysis and thinking.' Moving from cultural texts to hard data, Lobato continues,

> I feel my core disciplines provide a useful home-space that can be moved through and pushed back at when needed, so I have a fairly comfortable and pragmatic relationship with those – and tend to view other disciplines as providing useful ideas to be ransacked and raided as needed.

We concur with Lobato, because it is how we ourselves 'do' ideas. While many of our research interests and practices overlap, notably around bodies, feminism and mediated genres, we have found that we make and assemble in strikingly different ways. The field of performance studies, for Rachel, straddles forms of analysis that read historicized alignments between embodiment and culture, text and agency, nature and representation, while on the other hand, she is concerned with the messiness of experience, akin to what Jane Bennett has called 'vibrant matter': the 'earthy, not-quite-human capaciousness' of things (2010: 3). As such, contemporary performance research maps relations that are simultaneously semiotic (between concepts) and material (between things), or both. Derived from theatre studies (initially literary studies of drama), anthropology, film and cultural studies, the interdisciplinarity of performance research has never asserted fixed relations between subject and object, nor one singular perspective on reality. On one level, its method evolves like the Freudian analysis of signs, and on another it examines a Foucauldian archaeology of power within disciplinary structures, however, performance studies also gives agency to artists and scholars to improvise and play with the dynamics of theatrical forms of expression and communication. An event, and its multiple unfoldings, might thus involve intransigent actors, accumulated objects and hybrid structures, whether it is a choreography, a festival or the arduous ascent of a mountain range. In research terms, in this contested field, both participant and observer become challenged by the presence of diverse subjects and the mediation of experience.

Rachel's own research addresses modernist and contemporary theatre aesthetics, particularly in relation to dance histories, with an acknowledgement that both archival and repertory sources provide valuable insights into the transnational significance of cultural production.[2] Since Diana Taylor's (2003) interleaving of documents and embodied performance that transmit 'cultural memory' in Latin America, many performance scholars have stressed the remaking of political, cultural and social traditions as well as the complexity of performances that transmit new and embodied meanings, whether in the performance of protest or, as dancer Deborah Hay (2000) would contend, in the choreography of cellular movement in the body.

This background came to the fore in a recent digital arts project Rachel conducted, which involved archivists, a dancer, cultural theorists, computing engineers, a games designer and dance scholars in the making of a digital avatar from a silk 1920s' dance costume. The making of this 'daffodil dress' provoked consideration of why we seek to collaborate in interdisciplinary or multi-disciplinary projects (Fensham and Collomosse 2015: 148–161). Of course, such scientist-artist collaborations are increasingly common but such projects require the assembling of a team who can find some common ground over different questions. For instance, why do historians think about the costume as a singular representation in a dance repertoire? Or, why do archivists inscribe its details in a catalogue entry? Or, how do computer scientists abstract it as algorithms in a software program? In each of these sites, what happens to the affective labour or embodied history of the garment? Given that there is always something at stake in a method, these questions of how the garment might deliver an experience of movement, rather than be relegated to tissue paper in an archive box, could only be answered by an interdisciplinary methodology.

Formatively the dress itself became the agent of a shared enquiry that moved between historical contexts, folding and interacting with women's bodies, escaping from corsets and constraining social values in the early twentieth century, into the refrain of liquid, slower, on-screen mobilities. Compared favourably with more masculinist avatars and surveillance technologies, the dress also offered its own reconfiguring of computer vision research. As an inquiry into the visual regimes of cultural history (early twentieth century and the present), the methodology confronted materialities that arise from wearing a particular dance costume at a given historical moment as steps towards the virtuality of an idea about movement with wider consequences. 'The ruffle on a dress', as Walter Benjamin suggests somewhat cryptically, might produce an image, or conceptualization, of the secular desire for an 'eternal' (1989: 69), and so the research presses against the temporal and sensory properties of costume in relation to performance history.

While Rachel's primary area of focus is on Dance and Theatre Studies, Alexandra is primarily a Cinema Studies scholar, whose research incorporates aspects of art history, anthropology, performance studies and gender studies in work that primarily focuses on horror, cult and exploitation cinema traditions. As, however, a practising radio and film critic, her ideas about and around cinema manifest in a less formal way than academia traditionally dictates, and as part of this more public facing engagement with screen cultures she maintains an active Twitter account that she uses as a forum to post film stills: a kind of informal digital scrapbooking. The image sets that receive the most attention are those constructed loosely around the idea of Aby Warburg's *Mnemosyne Atlas* project (1924–1929): four film stills are arranged in a grid, connected by motifs pertaining to composition, colour or *mise en scène*.

As Cornelia Zumbusch (2010) notes, 'between 1926 and 1939 Warburg's work on the Atlas consisted of arranging and mounting photographs of artworks, as well as commercial art, playing cards, or stamps, in various ways on canvas covered wooden panels'. Zumbusch continues that, like Benjamin's *Passagen-Werk* (1927–40), 'both view visualization – not only on the level of the object, but also in terms of a representational principle – as an irreducible aspect of historical research' (2010: 119). From this perspective, although initially intended to be a fun, light-hearted way to present eye-catching film-related visual material, these sets have opened seemingly endless new ways into thinking about film, not only in terms of its formal qualities, but also in regard to its myriad histories and the ideological mechanics of representation. In the context of social media, it has also opened up new collaborative possibilities with fields she would not previously have thought to have located in a shared critical space: most recently, philosophy, fashion theory and political science.

From a formal perspective, the discipline of Cinema Studies has had a long-held bad habit of reducing the representation of sexual violence in cinema to simplistic generic notions of codes and conventions relating to this trope. A broader historical overview of how screen images of sexual violence and retribution fit into art historical traditions, particularly the so-called 'heroic' rape imagery of the Italian Renaissance has been missing and Alexandra has sought to redress this gap in her earlier monograph on the broadly dismissed category of rape-revenge cinema (Heller-Nicholas 2011). For most critics, images of rape on film were historically understood in relation to the so-called 'media effects' model about the potentially harmful influence of screen violence on its audiences,[3] or – more foundationally – to rely heavily on the psychoanalytic model instigated by Laura Mulvey's essay 'Visual Pleasure and Narrative Cinema' (1975).

Although art history is not a radically new interdisciplinary combination with Cinema Studies, it is particularly uncommon in terms of representations of gender on film, and through this perspective Alexandra re-assembled an alternative history: through art iconography, she could investigate what part these films played in wider confusion about sexual violence across a range of different cultures more generally. Adopting Diane Wolfthal's observation that 'diverse notions' of sexual violence in medieval and early modern art 'coexisted contemporaneously' (1999: 182) beyond the privileged domain of Italian painting, Alexandra found a similar phenomenon in the contemporary rape-revenge film. Wolfthal's critical model of sampling the contemporaneous thus offered a framework that permitted a remaking of how the intersection of rape and revenge could be conceived, right back to cinema's earliest days. The act of exploring differences between rape-revenge films from Japan, Argentina, Turkey, Canada, Australia, Germany, France and Britain as well as the dominant Hollywood film industry constructs an act of critical assembling and making. Aligned with insights from the playful and popular assemblages of her Twitter account, these complementary discursive frames have encouraged a more geo-political reading, well beyond traditional psychoanalytic and media effects approaches, of how violent films become manifest in a range of cultural and historical contexts.

In these accounts of interdisciplinarity from our contributors, we have identified a shifting of researcher as subject, a kind of conceptual movement, whereby disciplines and their methods undertake subtle realignments. Nowhere is this more demanding, and potentially exhilarating, when making and assembling the materials, problems and persons for a research project, than when the demands of practice shape the development of concept-formation.

## Making and assembling as practice

To a certain extent, the methodology of assembling and making is predicated upon the 'practice-turn' in social and applied research. Theodore Schatzki has defined this as 'the belief that such phenomena as knowledge, meaning, human activity, science, power, language, social institutions and transformations' occur within and are aspects or components of the *field of practices*' (2001: 11). Beyond the social sciences, however, this shift towards recognition of embodied human activity has dominated the discourse that has arisen in art schools around 'practice-as-research', a now well-theorized approach to undertaking research in and through the materials and historically attuned methods of distinctive art practices (Roms 2010). Without rehearsing at length these arguments, which have ranged from the phenomenological to the critical, to the interweaving of ideas with processes of production, they lead above all to an emphasis on iterative experimentation. The concept that is most productive in this context arises not from the recognition of research as a distinct form of creative thinking, or even from the institutional imperatives that have required the establishment of a separate qualifier for artistic research, but rather for its revaluing of practice as a form of knowledge production that can be creative and critical, affective and cognitive (Borgdorff 2012).

The Marxist notion of praxis complements this question of practice as it relates to making or performance as research because it shifts the focus from the individual artist's productive performance of technique and back to the role of human practices as a form of reproduction or assembly in the workings of a social system or order of knowledge. Additional approaches to practice draw from the emphasis in Bourdieu's educational sociology on fields of practice, an acknowledgement that cultural production is itself always embedded in a social formation that is not fixed but forms an interlinked network of practices, ways of moving, being and shaping the world, or habitus (Bourdieu and Passeron 1977: 31). Practice, for Bourdieu, connotes a 'durable training' and 'internalisation' of cultural values that produce both the inculcation of ideas, as well as the agency, that is important not only to artists but social theory.

Another move that might be made from practice-as-research is to the mastery of those practices acquired through the repetition and learning requirements of a new or higher level of technical competency or performance capacity. Appropriate scholarly skills need also to be 'supplemented by some combination of perception, propositional knowledge, reasons and goals' that formalize order or express attitudes and positions, in short as an assemblage, whether making a painting or establishing an ethnographic study (Schatzki, 2001: 17). Practice is therefore not accidental, but rather a form of doing which requires attention to the activity of acquisition, the necessity of rendering and re-rendering in order to shape the outlines of knowledge over time. From a critical perspective, a practice may become stultifying and lead to the passivity of a normative horizon of understanding, or the acquiescence to a social and political ideology and regime, but alternatively it can generate new alignments and distancing from habit, pattern and variation.

In his book on *Material Thinking* (2004), the cultural theorist, Paul Carter writes of method-ologies of creative practice aligned both with 'craft', the technologies required and acquired through working with materials, whether they be celluloid film, dance gestures, clay or paint; and that of the thinking in and with artists whose own methods rub up against the sociality of knowledge production in such a way that materials and their signs become discursive, as they enter into historical, social and political formations. As he writes, there is 'a propensity of materials to form into significant spatio-temporal groupings. Some of these are instantaneous and registered eidetically; others depend on calibrating the relations between things that happen, and holding in mind (and place) the Brownian notion of multiplicity' (2004: 180).

For Carter, this relationality of materials demands of the researcher a range of responses, that include remembering, waiting, tracing, assembling, mimicry, evaluating, decoding, etc. And the researcher may or may not find their labours are successful, since there will be events and 'non-events', as well as institutional structures that 'bracket(ed) off the environ-ment of making' (2004: 52). Carter seeks to reinstate a value within the practice of making that is respectful of these attributes of sensory and critical juxtaposition; for the artist is always situated within formal constraints, they have a body, as well as methods for making and remaking, placing and replacing or recording and re-arranging.

For the researcher who is not an artist, the need to emphasize practice as a method sounds like a tautology since all processes of research arise not only from practising thought, but they also involve the technologies and techniques of a disciplinary or interdisciplinary set of knowledges. Examples of this in science might be the biochemist who studies the contents of their Petri dish but also feeds the insects that go under the microscope, organizes the coloured dyes into bottles and jars, and takes the photographs that record changes on the surface of the glass. Organizationally, these methods of assembling and disassembling precede the critical tools required to read and interpret variations within a system. How equally intricate the method-ologies of the humanities or social science researcher, with their assembling of file cards, bibliographic notes, interview schedules, film footage, and the like. These practices involve finding a method or series of methods to work on or with materials (paper, sand, crochet, people, postcards, data, maps, etc.) within a critical framework that has now displaced the dichotomy between objects of study and the subject as researcher.

## Towards a conjectural paradigm

To formalize this notion of method as movement and practice more fully, we contend that what is at play in the doing of interdisciplinary research is what Italian historian Carlo Ginzburg has identified as the 'conjectural paradigm'. Ginzburg's 'conjectural paradigm' might appear to fly against dominant scientific trends for cultural and social analysis, but he has identified its precedents: from primitive hunters to modern scientists. This conjectural paradigm allows the utilization of a broad range of critical tools, in which the selection and application of these various approaches are governed by a defining, dynamic instinct. While Ginzburg identifies intuition as being essential to this process of deduction, he uses the term cautiously: 'I have scrupulously refrained up to now from bandying about this dangerous term, *intuition*. But if we really insist on using it, [it is] . . . synonymous with the lightning recapitulation of rational processes' (1989: 117). Ginzburg states that this type of intuition lies at the heart of a vast range of eclectic but equally rigorous and important intellectual enquiries – from Sigmund Freud's psychoanalysis to Francis Galton's fingerprint technology, from Sherlock Holmes to art historian Giovanni Morelli, from modern medicine to traditional folkloric practices. The conjectural paradigm is therefore an 'epistemological model' (1989: 96) positioned in opposition to the dominant 'scientific paradigm' that he claims has governed the

> quantitative and anti-anthropocentric orientation of natural sciences from Galileo on (that) forced an unpleasant dilemma on the humane sciences: either assume a lax scientific system in order to attain noteworthy results, or assume a meticulous, scientific one to achieve results of scant significance.
>
> *1989: 124*

As demonstrated by his eclectic range of examples, this conjectural paradigm is not new, and 'it is very much operative in spite of never having become explicit theory' (Ginzburg 1989: 96). Its value, he claims, is that 'such a study may help us break out of the fruitless opposition between the "rationalism" and "irrationalism"' (1989: 96) dichotomy that marks the scientific paradigm.

The importance of Ginzburg's conjectural paradigm to making and assembling in research lies as much in its 'rational' aspects as its 'lightning' component. The notion of *speed* is vital to the conjectural paradigm but it is not that there is a total absence of rational thought processes involved; rather they are identifiable *only* after those initial connections have been made in the 'lightning' flash of cognition. Once the skills are attained (be they the deductive skills of Sherlock Holmes, Giovanni Morelli or the Neolithic hunter), the conjectural model does not reject reason and rationality as such, but insists that real intellectual insight can be most effectively produced in this instinctive, instinctual flash. Spawned from rational knowledge, and the labour of a rigorous chiselling at an intractable problem, the effort of making has a 'gut' moment of clarity that defines the conjectural paradigm. Ginzburg's approach asks the researcher not to merely trust the process, but specifically to *trust our own* process as it emerges from what has been assembled.

This claim does not imply simply an individual justification of an intuitive idea, dramatized in an instant of revelation, but rather a conceptual willingness to engage with new understandings of how we do research, and of how the social manifestations of making a research culture require the assembling of new configurations of people, skills and technologies. In the methodological acts of making and assembling of our contributors to this collection, we see the conjectural paradigm at work, manifesting as a 'general impulse', to invoke Ramon Lobato's term, that drives each researcher towards their chosen methodologies, not naively but with an orientation towards what might mould or stratify their research paradigm.

Each of the contributors to this section offer, through their use of a singular present participle, a concept (and sometimes more than one) that builds a vocabulary and introduces skills that might shape the kind of research we are proposing. Catherine Ayres and David Bissell, for example, champion the notion of 'suspending' as a valuable methodological intervention in interview methodologies – of putting on hold, of pausing, of acknowledging acts of repression typical of traditional research practices. 'Introducing "suspending" as part of a researcher's toolkit may enable radically different practices, politics, and ethics of research', they state. 'But doing so also in some ways demands *more* of researchers'. This 'more' requires an opening up to the critical potential of 'discomfort', and a turn towards insight over learned professional practice. Their willingness to conjecture fights against the idea of an in-process research topic being somehow 'unfinished'.

By employing the ballet classroom exercise *enchaînement*, Harmony Bench considers the methodological benefits of arranging in data visualizations, which itself requires an intuitive combination of a range of other practices: gathering, collecting, generating, evaluating, filtering, sorting, cleaning, charting, scoping, curating, assembling, visualizing, correcting, testing, juxtaposing, modelling and crafting. For Bench, mapping a touring repertoire demands a combination of scholarly composition with 'arrangements . . . [that] offer internally coherent, yet potentially inexhaustible combinations'. As Bench argues, this interdisciplinary methodology enables the global circulation of social, political and cultural mobilities.

For Thomas Jellis, his model of experimentation shifts away from both the fashionable deployment of the term and its scientific origins. His approach is twofold: first through 'the invocation of attentive participation' and the researcher's experimental incorporation of a range of activities in their practice (for him, these include 'talking, reading, designing, cooking, walking, foraging, choreographing'). Second, the researcher as impresario – one who makes unusual connections – potentially shifts 'the energies of a field in productive ways'. Jellis issues an invitation to become an experimental subject in interdisciplinary research so that we 'amplify the ways in which experimental hubs exceed particular locales'.

Margaret Wertheim's entry on figuring usefully produces a gendered analysis of symbolic language and its use in scientific research. Extending feminist critiques of science, Wertheim notes that 'figuring calls our attention to the wisdom of embodied objects, whose qualities are not merely reducible to, or predictable from, their descriptive codes'. For Wertheim, this critique informs a methodology of tactile geometric construction akin to handiwork in an eco-critical method. While natural forms such as corals, kelp and sea-slugs have long existed, mathematicians spent centuries formalizing structural logics that could not describe such phenomena. Even now, with our greatly expanded understanding of geometry, many natural structures still cannot be wholly articulated by this apparatus.

Rebecca Coleman seeks to rethink imaging as an encounter with her research subjects, and as a site of establishing the researcher explicitly as a maker. Coleman explores 'some of the ways that the social sciences might work with practices developed in, and/or inspired by, art and design, and as a consequence might draw attention to making images as a research practice'. Developing a series of practical and conceptual questions, she focuses on two of her own research projects – one around making collages and the other on making and sending postcards. By doing so, Coleman raises urgent questions around participation and the status of the participant, in terms of both researchers *and* their subjects. Coleman frankly confronts the complexities of participatory research.

For Matthew Reason, drawing as an act of making is at its most practical level itself a methodological process. 'Marks made on paper – with pencil, crayon, ink, pen – appear instantly, they are real and absolute', he says. They demand we 'spend time with our thoughts, memories or experiences as we begin, develop and complete a drawing'. The phenomenology of drawing – of making an impression – for Reason is not a philosophical abstraction but linked to an immediate awareness of presence and possibility. Rather than a pure theory of affect, he values the discomfort and pleasures of drawing both in and for his research. Experiences that appear ephemeral, intangible and ineffable through this methodology are brought into being by the process of reflection: crucially, 'experience here isn't only had, but also made'.

Accompanied by found objects, gestures and vocalizations, for Jennifer Green sand drawings constitute a multimodality that communicates 'important information, coding movement, habitation and histories' of Australian Indigenous communities. In interdisciplinary research, linguists require a greater responsiveness to the complexity of 'verbal art-forms' and their narrative locality. Green recognizes that while traditional methods might disassemble sand stories into a series of semantic units, taken as a whole, they become a 'small repertoire of linear, curvilinear, circular and spiral forms represent[ing] people, plants, artefacts, domestic items and other aspects of local environments'. For Green, 'delineating the similarities and differences between sand story songs and other song repertoires from Central Australia leads to a more sophisticated understanding of the ethnopoetics of the verbal arts'.

Ramon Lobato employs the concept of 'rescaling' to address methodological tensions between micro and macro in relation to his research on media industries. For Lobato, rescaling 'involves manipulating notions of scale in research design', an approach that he has found useful in addressing a tension in his area of research where traditionally 'methods used to study industry do not always work well in the world of media' (for example, textual analysis or reception studies), and vice versa. Lobato steps back to address how the logic of contexts (corporate, legal, commercial, cultural, institutional, etc.) affect methods and the research we make. Through contrast and inversion, rescaling reveals the problematic and reductive logic of binaries such as big/small and micro/macro.

We are struck in reading through the essays in this section, how much the conceptions of assemblage and making are resistant towards the certainty of historical conceptions of singular truth, let alone the notion that an objective research paradigm consists only of a rationality. In harnessing intuitive responses in the research encounter, the interdisciplinary dynamic of the methods explored by our authors implies something deliberative, in allowing the unexpected to emerge from the activity of the collaborators – again privileging the notion of process over end-product. Indeed, as Luciana Parisi has suggested there is a 'topological notion of physical uncertainties defined by the directly lived, the gestured, the felt, the danced, or generally by experienced contingencies' (Dawes 2013). We would like to call this the movement of doubt, a valuation of doing that installs an aspect of modesty and humility into research, because the process is emphatically and necessarily framed around what each researcher does not know, rather than a ritualistic posturing and repeating the motions of what we can already assert.

While the fields of practice in which many of us research might increasingly overlap, in relation to new problematics and expanded social networks, our theorists – and we ourselves, in this essay – seek to offer possible approaches for rejecting the reductionist demands of scientific methods that can be imposed on humanities and social sciences research by research funding bodies, citation factors and other structures that dominate academic institutions today. We are perhaps united in a rethinking of disciplinary research as an interdiscipline that involves an assembling and making which remains dynamic, linked to a shifting of perspectives, and a picking up of tools and a downing of preconceptions as they relate to new and unanticipated situations. We see the challenge of an interdisciplinarity of methods to be its mobility, or its emphasis on movement between one state and another. Although the research subjects in these contributions might, at first, look disconnected, what unifies them is the idea of movement. Making and assembling as an interdisciplinary methodology is an act of hovering, of moving towards, shifting away, being drawn to and pulled back. This almost magnetic sense of attraction and repulsion sparks the creation of research: the making of ideas, and the assembling of knowledge.

## Notes

1 See, for instance: Ingold (2011).
2 For further consideration of archival theory, see Fensham (2013).
3 This was typified in a famous Roger Ebert review of the notorious rape-revenge film *I Spit on Your Grave* (Meir Zarchi, 1978) from 16 July 1980, which can be found online at www.rogerebert.com/reviews/i-spit-on-your-grave-1980 (accessed 21 August 2016).

## References

Benjamin, W. (1999). *The Arcades Project*. Translated by Howard Eiland and Kevil McLauglin. Cambridge: The Belknap Press of Harvard University Press.

Bennett, J. (2010). *Vibrant Matter: A Political Ecology of Things*. Durham, NC: Duke University Press.

Borgdorff, H. (2012). *The Conflict of the Faculties: Perspectives on Artistic Research and Academia*. Amsterdam: Leiden University Press.

Bourdieu, P. and Passeron, J. C. (1977). *Reproduction in Education, Society and Culture*. London: Sage.

Buchanan, I. (2015). Assemblage theory and its discontents. *Deleuze Studies*, 9(3): 382–392.

Carter, P. (2004). *Material Thinking: The Theory and Practice of Creative Research*. Melbourne: Melbourne University Press.

Dawes, S. (2013). Interview with Celia Lury, Luciana Parisi and Tiziana Terranova on topologies. *Theory, Culture & Society*. January 15. Retrieved 21 August 2016 from: www.theoryculturesociety.org/interview-with-celia-lury-luciana-parisi-and-tiziana-terranova-on-topologies/

Deleuze, G. and Guattari, F. (2004). *A Thousand Plateaus: Capitalism and Schizophrenia*. Translated by Brian Massumi. New York: Continuum.

Fensham, Rachel (2013). Choreographic archives: the ontology of moving images. In Rune Gade and Gunhild Borggren (Eds.) *Performing the Archive/Archiving Performance* (pp. 146–162). Chicago, IL: University of Chicago Press/Museum Terculanum Press.

Fensham, R. and Collomosse, J. (2015). Digitizing dance costumes: a case study of movement and materiality in iWeave. In N. Salazar-Sutil and S. Popat (Eds.) *Digital Movement: Essays in Motion Technology and Performance* (pp. 148–161). Houndmills, Basingstoke: Palgrave.

Genosko, G. (2009). *Felix Guattari: A Critical introduction*. London: Pluto Press.

Ginzburg, C. (1989). *Clues, Myths, and the Historical Method*. Baltimore, MD: Johns Hopkins University Press.

Hay, D. (2000). *My Body the Buddhist*. Middletown, CT: Wesleyan University Press.

Heller-Nicholas, A. (2011). *Rape-Revenge Films: A Critical Study*. Jefferson, NC: McFarland and Co.

Ingold, Tim (2011). *Being Alive: Essays on Movement, Knowledge and Description*. Abingdon: Routledge.

Latour, B. (2016). Sensitizing. In C. Jones, R. Uchill and D. Mather (Eds.) *Experience: Culture, Cognition and Common Sense* (pp. 315–324). Cambridge, MA: MIT Press.

Mulvey, Laura (1975). Visual pleasure and narrative cinema. *Screen*, (16)3: 6–18.

Roms, H. (2010). The practice turn: performance and the British Academy. In J. McKenzie, H. Roms and C. J. W. L. Lee (Eds.) *Contesting Performance: Global Sites of Research* (pp. 51–70). New York, NY: Palgrave.

Salazar-Sutil, N. (2015). *Movement and Representation: The Language of Human Movement*. Cambridge, MA: MIT Press.

Scarry, E. (1985). *The Body in Pain: The Making and Unmaking of the World*. Oxford: Oxford University Press.

Schatzki, T. (2001). Introduction: practice theory. In T. Schatzki, K. Cetina and E. von Savigny (Eds.) *The Practice Turn in Contemporary Theory* (pp. 9–23). London: Routledge.

Taylor, D. (2003). *The Archive and the Repertoire: Performing Cultural Memory in the Americas*. Durham, NC: Duke University Press.

Wolfthal, D. (1999). *Images of Rape: The 'Heroic' Tradition and its Alternatives*. Cambridge: Cambridge University Press.

Zumbusch, C. (2010). Images of history: Walter Benjamin and Aby Warburg. In C. Emden and G. Rippl (Eds.) *ImageScapes: Studies in Intermediality* (pp. 117–144). Bern, Switzerland: Peter Lang.

# 2

# Arranging (*enchaînement*)

## Harmony Bench

Within the classical ballet tradition, the concept of choreography was preceded by that of arranging: dance masters availed themselves of a codified vocabulary of movements and steps, which they 'arranged' into dance compositions. This essay focuses on arranging as a method or practice suited to contemporary interdisciplinary research with an emphasis on the digital humanities. Broadly construed, arranging refers to any process of collecting, ordering, fitting-together, and displaying objects – from flowers to furniture to puzzle pieces. My own thinking is informed by the sequencing of steps known in ballet as *enchaînement*, a type of classroom exercise in ballet technique in which a series of travelling steps are linked (literally chained) together as a phrase or combination that moves across the dance floor. *Enchaînements* are built, or arranged, from a discrete vocabulary and syntax that determines which movements or steps may logically precede or follow others. Connections among individual step-units are not wholly predetermined, nor are they wholly open to any connection whatsoever.

While pre-existing the computational, ballet as an aesthetic vocabulary and a mode of training participates in a digital episteme. It cultivates habits of mind and body that resonate with contemporary database logics, valuing, for example, modularity, reversibility, and symmetry within a system of units that can be broken down into ever-smaller components, or conversely, built into an almost infinite number of compositions. Though my own understanding is informed by *enchaînement*, I find in the larger concept of arranging a productive model for my recent forays into data-driven scholarly projects such as the *Mapping Touring*[1] project I present here. Arranging as a method is concerned with crafting relationships of contingent interdependency; it draws attention to internal coherence and sense-making through the juxtaposition and co-articulation of units of information and their relations.

## The given to be arranged

*Mapping Touring* is a digital humanities research project that tracks and maps the appearances of early twentieth-century dancers, choreographers and dance troupes with an emphasis on touring. My concern is to represent the dates of performance, cities and venues, and repertory performed, giving shape to economies of movement through which gestural vocabularies and 'kinesthetic legacies' (Srinivasan 2007: 2) flowed prior to our current media-saturated era.

With this project, I am interested in the movement of movement – how it moves, what it moves, where it moves, why it moves, and who or what moves it. To craft the mapping dimension of the project, however, I have also been required to generate the datasets that can then be plotted and visualized.

Media theorist Lev Manovich cautioned in *The Language of New Media*, 'data does not just exist – it has to be generated. Data creators have to collect data and organize it, or create it from scratch' (2001: 224). The term data comes from the Latin word *dare*, which means 'that which is given' (Galloway 2015). As a mass noun, 'data', which are given unordered or unorganized, attain their value as a collection or a set that can be parsed and displayed. Data are given to be arranged, and it is the process of arrangement, first as a set and later as a visualization, that makes data usable and therefore meaningful. Arranging is the process of forging relationships among the given data. Though data have the aura of neutrality, they already carry the weight of interpretation within a system of cultural values that guide identification and assembly within scholarly research. In arranging data, one shapes and re-shapes the relations within a collection such that analysis and further interpretation may follow.

In my *Mapping Touring* project, we are gathering data about performance engagements from concert dance programmes.[2] Although archives contain scrapbooks, route books, correspondence, newspaper reviews, and other artefacts with which we corroborate the performance data, it has been necessary to focus primarily on programmes to delimit the parameters of the project. However, in selecting concert programmes as our source material, we also limit ourselves to performance events that subscribe to a cultural logic in which the creators, performers and audiences of dances are human; dance pieces are identified as unique entities (notwithstanding variability over time); and the dates, times and locations of each event have been documented. Because this project renders the historical record anew, it cannot change that record or offer a substantive challenge to the ideologies and biases that produced it. But in moving from the case study to the dataset, the gaps and absences that historians and cultural scholars have already identified in the record should appear in even greater relief. For this reason, as theatre scholar Debra Caplan (2015) has compellingly argued, macro-histories are most productively arranged in tandem with micro-histories: as methodological complement rather than competitor. *Mapping Touring* certainly pursues a macro-history of twentieth-century concert dance, but the universalizing tendencies of such a digital humanities project can be interrupted when researchers arrange canonical figures and standardized data alongside micro-histories, local performance cultures and marginalized performance practices.

## Arranging as composing and discerning: datasets and visualization

Initial stages of arranging as a method might include gathering, collecting, generating and evaluating, followed by activities such as filtering, sorting, cleaning, charting and scoping. Later steps might be curating, assembling, visualizing, correcting, testing, juxtaposing, modelling and crafting. Though I have presented these as stages, in point of fact I have found it necessary to travel back and forth along this spectrum of activities. First attempts at data visualization offer new insights as well as rehearsal for the fuller project, and also make errors or a need for additional information apparent. Arranging data through visualization is thus a mode of discovery as well as display. Inevitably, visualizations reveal flaws and deviations in the underlying data, such as misspelled names or titles. Visualizations thus occasion a return to previous steps to correct, refine and clean.

However, cleaning data risks erasing important information in favour of standardization. In this way arranging data is also like arranging steps in ballet, where cleaning refers to a process of bringing dancers into alignment and producing a desired level of uniformity prior to performance. Both forms of cleaning are subtractive rather than additive: they take away 'impure' features. It is only when seeing an arrangement together that one can see what is out of place – what is 'noise' or what does not make sense and requires further investigation or scholarly analysis. Not all anomalies, however, are impurities to eradicate; many lead to important discoveries, add nuance, or open new lines of scholarly or aesthetic inquiry. Though cleaning is necessary to make data meaningful by making comparison possible, premature or overly zealous cleaning can erase the very differences one hopes to find.

For example, in a small project helping me test the larger *Mapping Touring* idea, I assembled performance data from Anna Pavlova's tours to Central and South America during the First World War (see Figure 1.2.1).[3] In plotting the performance locations on a map I discovered an anomaly. It would appear that Anna Pavlova and her company are in Puerto Rico and San Francisco at the same time. Human error is the most logical explanation for the uncertainty as to where Pavlova actually was. However, the dataset is incomplete, and company members did not always travel together due to injuries or visa problems. Furthermore, the turn of the twentieth century is full of copycat performers who stole each other's routines and sometimes performance identities. What is clear is that this particular arrangement of location information revealed an anomaly that requires further attention, whether that ultimately means correcting the data or following its lead toward a new analysis of Pavlova's global movements. While I could clean this data by simply erasing the unlikely San Francisco appearances, doing so without first determining how they came to be listed there would be a missed opportunity. Arranging can reveal outliers, but the difference between a mistake and a discovery can only be discerned with further exploration.

## Arranging as interpreting, or data hermeneutics

Creating new arrangements of old information can help scholars re-examine assumptions and inherited narratives. In concert dance history of the twentieth century, for example, New York City quickly emerges as a global cultural capital. Yet, creating an arrangement of performance data that weights all locations equally – regardless of the number of engagements – paints a very different picture of what it meant to be a performing artist or performance audience in the first half of the twentieth century. Suddenly unexplored mid-sized cities and small towns emerge as important destinations for the arts. How might evaluations of audience sophistication shift if, for example, we discover that artists perform the same repertory in small towns as they do in large cities? Or if it appears instead that artists select different repertory for different cities, what might that indicate about how they relate to audiences, or their understanding of audience perception? Arranging touring information alongside railway routes and other modes of transportation (Wilke 2014; Elswit 2015) promises to shed additional light on performance engagements in smaller towns, and might further assist in re-evaluating the geography of aesthetic cosmopolitanism in light of long-distance networks (Latour 1987) that connect movement vocabularies across continents. Arranging the domestic touring pathways of concert dance with popular vaudeville and burlesque circuits will also give historians and cultural theorists a better understanding of how movement vocabularies circulate in relation and in direct response to each other.

For the sake of this essay, I have only focused on a small portion of Pavlova's touring data, but the possibilities of arranging as method lie not only in rendering a single strand of data in different visual arrangements, but also arranging touring information from different artists alongside each other. For example, understanding Pavlova's movements during the First World War would be further amplified by creating an arrangement that included the performance

**Anna Pavlova's company tours to South America 1917-1918**
**(Gran Compañia de Ballets Clásicos Anna Pavlowa)**

Program by Country and Date

| Year.. | Month .. | Day.. | Country | City | Program |
|---|---|---|---|---|---|
| 1917 | July | 12 | Chile | Valparaiso | Giselle (Adam); Hungarian Rhapsodie (Liszt); Pizzicato (Drigo); Gavo |
| | | | | | The Magic Flute (Drigo); Walpurgis Night (Gounod); Waltz of the Ros |
| | | 13 | Chile | Valparaiso | Giselle (Adam); Hungarian Rhapsodie (Liszt); Pizzicato (Drigo); Gavo |
| | | 14 | Chile | Valparaiso | Flora's Awakening; Invitation to the Dance (Weber); Obertass (Lewan |
| | | 15 | Chile | Valparaiso | Coppelia (Delibes); Waltz of the Roses (Delibes); Pierrot (Dvorak); Ni |
| | | 22 | Chile | Santiago | Flora's Awakening; Invitation to the Dance (Weber); Mazurka (Glinka) |
| | | | | | The Fairy Doll (Beyer); Walpurgis Night (Gounod); Primavera (Helmu |
| | | 24 | Chile | Santiago | Amarilla (Drigo and Tchaikovsky); Chopiniana (Chopin); Gopak (Tcha |
| | | 25 | Chile | Santiago | The Fairy Doll (Beyer); Invitation to the Dance (Weber); Mazurka (Gli |
| | | 26 | Chile | Santiago | Coppelia (Delibes); Greek Dance (Brahms); Rose Mourante (Tchaiko |
| | | 29 | Chile | Santiago | Coppelia (Delibes); Greek Dance (Brahms); Rose Mourante (Tchaiko |
| | | | | | The Magic Flute (Drigo); Invitation to the Dance (Weber); Waltz of the |
| | August | 2 | Chile | Santiago | Raymonda; Noche Buena (Tchaikovsky); Obertass (Lewandowski); P |
| | | 3 | Chile | Santiago | Snowflakes (Tchaikovsky); Egyptian Dance; Waltz of the Roses (Deli |
| | | 4 | Chile | Santiago | Oriental Impressions; Orpheus and Eurydice (Gluck); Hungarian Rha |
| | | 5 | Chile | Santiago | Snowflakes (Tchaikovsky); Egyptian Dance; Waltz of the Roses (Deli |
| | | | | | The Magic Flute (Drigo); Orpheus and Eurydice (Gluck); Pizzicato (D |
| | | 15 | Argentina | Buenos Aires | The Fairy Doll (Beyer); Walpurgis Night (Gounod); Primavera (Helmu |
| | | 19 | Argentina | Buenos Aires | Coppelia (Delibes); Greek Dance (Brahms); Gavott Pavlowa (Lincke) |
| | Septe.. | 5 | Argentina | Buenos Aires | The Fairy Doll (Beyer); Walpurgis Night (Gounod); Mazurka (Glinka) |
| | | 23 | Argentina | Rosario | The Fairy Doll (Beyer); Egyptian Dance; Obertass (Lewandowski); Pi |
| | | | | | The Magic Flute (Drigo); Snowflakes (Tchaikovsky); Obertass (Lewar |
| | | 30 | Argentina | Rosario | Coppelia (Delibes); Hungarian Rhapsodie (Liszt); Night (Rubenstein); |
| | | | | | The Magic Flute (Drigo); Walpurgis Night (Gounod); Mazurka (Glinka |
| | October | 4 | Argentina | Rosario | Snowflakes (Tchaikovsky); Primavera (Helmund); Pierrot (Dvorak); P |
| | | 10 | Argentina | Buenos Aires | Flora's Awakening; Invitation to the Dance (Weber); Mazurka (Glinka) |
| | | 11 | Argentina | Buenos Aires | Oriental Impressions; Chopiniana (Chopin); Primavera (Helmund); Pi |
| | | 13 | Argentina | Buenos Aires | Snowflakes (Tchaikovsky); Orpheus and Eurydice (Gluck); Waltz of th |
| | | 14 | Argentina | Buenos Aires | The Fairy Doll (Beyer); Walpurgis Night (Gounod); Obertass (Lewand |
| | | 16 | Argentina | Buenos Aires | Amarilla (Drigo and Tchaikovsky); Les Preludes Symphonic Poem; M |
| | | 18 | Argentina | Buenos Aires | Orpheus and Eurydice (Gluck); Hungarian Rhapsodie (Liszt); En Sou |

(a)

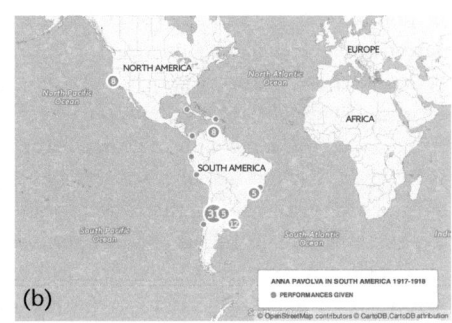

(b)

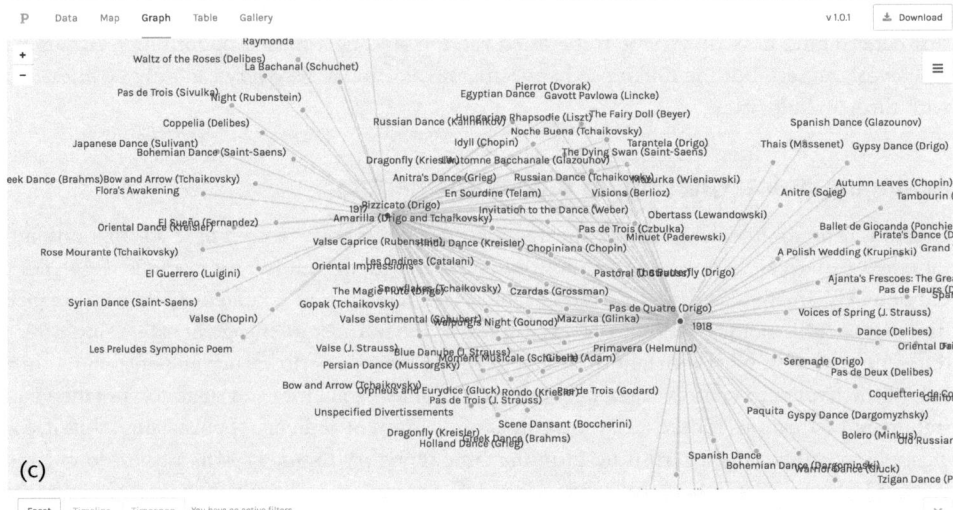

(c)

*Figure 1.2.1*    These screenshots represent three arrangements of the same data as (a) a list, (b) a map, and (c) a network, which I have opted to present in three different visualization platforms. Each arrangement highlights a different component of the underlying data from Pavlova's touring in Central and South America during the First World War: the list in Tableau shows a chronological progression in tandem with repertory performed; the map in CartoDB represents an a-temporal geo-spatial accumulation of events; the network in Palladio sorts repertory into works performed in the 1917 and 1918 calendar years, and those shared between the two as part of the same performance season, which generally runs July–June.

engagements of other artists or companies during the same time period, such as Diaghilev's Ballets Russes, Isadora Duncan, or Ruth St. Denis and Ted Shawn. How do their touring patterns relate to each other? Are the movements of one contingent on the movements of another? Does one seem to exhibit more freedom of movement than another?

Too often, data visualizations are presented as rhetorical black boxes – assertions presented without substantiating data, which nevertheless claim status as truth through quantification. A hermeneutic approach to data, however, considers the arrangement at hand, supporting the critical evaluation of what Orit Halpern describes in *Beautiful Data* as the 'aesthetic crafting to this knowledge, [the] performance necessary to produce value' (2014: 5). What does a specific arrangement suggest or argue, and how do the aesthetics of the arrangement help or hinder that argument? What other arrangements might counter its claims? What is the ideological bent of the data or its process of generation? What role does software play in foregrounding or obscuring certain interpretive possibilities?

Arranging, which enables combinations and re-combinations of a given set of data, establishes contingent interdependencies among (in)dependent variables. In the case of Anna Pavlova, one is reminded (or learns for the first time) of the diversity and extensiveness of Pavlova's repertoire. Not only did Pavlova and her dancers perform familiar classics from the Russian ballet canon, they performed social dance pieces, orientalist works, indigenous-themed dances, operas, and dances composed in response to local dance traditions. *The Dying Swan* might have captured the public's imagination, but arrangements of Pavlova's repertory tell a more nuanced story. Furthermore, Pavlova is remembered for the global reach of her touring, but the business and mechanics of travel, along with competition among dance artists for venues and audiences, has not been fully researched. Nor have scholars delved into comparative analyses of thematic or stylistic similarities in repertory across companies. Yet when looking at a map of touring, a list of performance locations, or the spread of repertory performed from season to season, such considerations come to the fore. Arranging performance data into sets and visualizations opens new avenues for analysing networks and geographies of influence, political economies of touring, and the global circulation of movement practices and aesthetics.

As a method for contemporary interdisciplinary research, arranging combines experimentalism with composition in scholarship. Researchers arrange data, working back and forth through processes of gathering, cleaning, displaying, and analysing, crafting relationships among pieces of information to determine which arrangement might offer a new perspective or prompt a new question. Arrangements, like the *enchaînements* found in ballet studios, offer internally coherent, yet potentially inexhaustible combinations. They can expand or contract in scope but remain rule-bound; they are flexible but logical. A shift of relation can produce new interpretations and understanding.

## Notes

1 *Mapping Touring* is supported by the Office of Research's Grants for Research and Creative Activity in the Arts and Humanities at The Ohio State University. See Bench 2015.
2 Archives frequently contain both performance programmes and souvenir programmes. Souvenir programmes are generally produced for an entire tour and contain extensive contextual information such as biographies, librettos and photographs, but do not reflect the locales in which an engagement was held. Performance programmes, in contrast, document the specifics of each performance event within a local context, including local advertisements and announcements.
3 This project was undertaken in conversation with dance scholar Kate Elswit for a co-authored presentation. See Bench and Elswit.

## References

Bench, H. (2015). *Mapping Touring: Dance History on the Move*. www.harmonybench.wordpress.com.

Bench, H. and Elswit, K. (2015). Mapping Dance Touring: Onstage and Backstage. Presentation at the American Society for Theatre Research, Portland, OR, Nov. 5–8.

Caplan, D. (2015). Big Data and the 'Obscurity' of Yiddish Theatre. Presentation at the American Society for Theatre Research, Portland, OR, Nov. 5–8.

Elswit, K. (2015). Merging Tables and Layering Maps. *Moving Bodies, Moving Culture*. Last modified 28 July 2015. https://movingbodiesmovingculture.wordpress.com/2015/07/28/merging-tables-and-layering-maps/.

Galloway, A. R. (2015). From Data to Information. Last modified 22 September 2015. www.culture andcommunication.org/galloway/from-data-to-information.

Halpern, O. (2014). *Beautiful Data: A History of Vision and Reason since 1945*. Durham, NC: Duke University Press.

Latour, B. (1987). *Science in Action: How to Follow Scientists and Engineers through Society*. Cambridge, MA: Harvard University Press.

Manovich, L. (2001). *The Language of New Media*. Cambridge, MA: MIT Press.

Srinivasan, P. (2007). The bodies beneath the smoke or what's behind the cigarette poster: unearthing kinesthetic connections in American dance history. *Discourses in Dance*, 4(1): 7–47.

Wilke, F. (2014). *Performance, Transport and Mobility: Making Passage*. London: Palgrave MacMillan.

# 3
# Drawing

*Matthew Reason*

What am I asking, if I ask a research participant to draw me a picture?

In different guises this request has been asked of participants across psychology, health, education, marketing and my own context of arts audience research. In all these areas, and others, the invitation to draw has become one of a range of visual methodologies recognized as having great potential when conducting research with people.

The use of drawing is often described as a 'projective technique', motivated by the perception that when responding to direct questioning research participants may either be reluctant or not consciously able to reveal their true attitudes or deepest feelings. To address this difficulty, researchers use methods designed to enable participants to 'project' their feelings through mediating activities. Examples include the use of word association tests, sentence and story completion, photo sorts and indeed drawing. Drawing is utilized in this manner in contexts as diverse as market research – to circumvent participants' self-consciousness and 'delve below surface responses to obtain true feelings, meanings or motivations' (McDaniel and Gates 1999: 152) – and art therapy – where its use is 'based on the accepted belief that drawings represent the inner psychological realities and the subjective experiences of the person who creates the images' (Malchiodi 1998: 5).

There are, however, a variety of contentious propositions here, both ethical and methodological, particularly in the assertion that a drawing methodology reveals truths below the surface of the participant's consciousness. Art therapy, in particular, has moved away from an overly 'diagnostic' use of drawing, where the expert therapist asserts their ability to know the experience and feeling of the participant through the content of the drawing alone. Instead, drawing as a research methodology has become fundamentally connected to participants' talk about drawing. That is, participants are perceived as 'expert' in their own experiences and positioned as the first and most important interpretation of their own drawings. This is the case both in therapeutic (Malchiodi 1998) and research (Gauntlett 2007) contexts.

As a 'draw and talk' methodology, drawing is no longer an almost magical process that is revelatory of inner or authentic truths. If we ask a research participant to draw we cannot presume that this necessarily makes the responses more insightful, or more complete, or more anything else than those communicated by talk alone. It is, however, a request that constructs a specific dynamic between researcher, participant and the subject/object of enquiry that has the potential to produce different kinds of insights and understandings (see for example Elliot Eisner (2008) on visual knowledge).

When I ask a participant to draw me a picture I am inviting a different dynamic than if I had simply asked them to talk. I do not expect them to respond instantly. Instead drawing imposes a slowing down, a pause for reflection in the returning to memories. Yet, and this is vital, it is not simply a pause for thought (and I might very well ask a participant to have a think about something for a few minutes before answering). More specifically, it is a pause to draw.

Drawing is an activity in which the marks made on paper – with pencil, crayon, ink, pen – appear instantly, they are real and absolute, but which also requires us to spend time with our thoughts, memories or experiences as we begin, develop and complete a drawing. This combination of duration *and* immediacy is the unique quality of drawing as a research methodology. It is at once an active doing and also a quieter, reflective thinking. Through this duration it

*Figure 1.3.1*   Lost in drawing. A research workshop in progress. Photograph Brian Hartley. Copyright Matthew Reason

is possible for thoughts, realizations or insights to come into knowing in a manner that is less about discovering something pre-existing and more about constructing knowledge through the process itself.

While the nature of drawing is much debated within art theory, 'drawing' as an arts-based or creative research methodology is typically framed fairly broadly and rarely narrowly conceived in terms of media or materials. In my own practice, I have tended to provide participants with a diverse range of arts materials and allowed them to select those that most suit their temperament or the nature of what they want to communicate. The results often go beyond the qualities of line to include elements such as colour and texture, producing artefacts that might at times be closer to painting or even collage (when participants elect not to draw on paper but rather rip it up and 'draw' *with* paper). Visual or expressive 'mark making' might, therefore, be a more accurate if less evocative description for the diversity of creative processes that participants employ. Considered either as drawing or mark making, participants are required not only to spend time with memories, thoughts and feeling but also to start to externalize these visually. The mark is both of the participant, and yet also separate from them; it is, as Joseph Beuys puts it, the changing point at which ideas become 'the visible thing' (Rose 1993). Mark making therefore operates for participants as both a form of research (what is it I know about this memory? What did I see? What do I remember?) and of expression (how did it make me feel? What did I think about it?). The result is a process of reflective contemplation, as participants' responses are mediated through the act of making a mark – a line, a shape, a trace, a shade – with this active doing enabling a dialogic process between participant and an art work that is exterior to them.

My own research focuses on audiences' memories, experiences and perceptions of theatre and dance performances. A typical scenario might involve accompanying a group of child or adult audience members to a performance and then facilitating an arts workshop with them. Audiences often like to talk about the performance they have seen, and implicit within the drawing workshops is the desire to slow down and extend this desire to share and externalize the experience. In this context, the particular request is for the participants to 'draw something you remember from the performance'.

In my own practice this request is framed and structured with care. The workshops begin with warm-up drawing games – getting the participants to 'take a line for a walk' or draw portraits of themselves with their 'wrong' hand. As researcher, I join in these exercises, using all of our drawings as an opportunity to explain that I am not concerned with the relative quality of the pictures. Here it is worth noting that drawing as an activity is experienced very differently by adult and child participants. As Pia Christiansen writes, children themselves consider 'all' children to be competent at drawing, which to them is an ordinary rather than specialized activity (Christensen and James 2000: 167). When working with children, therefore, drawing feels a natural activity and following the introductory exercises the workshops consist of an extended period of largely free drawing. As the children draw I will talk to them, individually or in pairs, asking them to tell me about their drawing and through that about the performance they witnessed.

For most adults, in contrast, drawing is not an everyday activity and many grown-ups will automatically assert that they 'can't draw'. For this reason, running art workshops with adults requires additional structured elements, continuing the drawing exercises through the whole process with further suggestions designed to distance participants from their self-consciousness about drawing. So, for example, participants might be invited to draw with their eyes closed, without taking their pencil off the paper, or on a large sheet of paper while using a pen taped to the end of a two-metre-long bamboo cane.

Whether with children or adults, the process of drawing invites participants to spend time with their memories and experiences. In making the decision of *what* to draw and *how* to draw it, participants are required to invest effort and project themselves into their art work and into their memories. Drawing represents a prompt to consider what things looked like, what things meant and how things felt. Or, more accurately, to *discover* this knowledge *through* the process of drawing. Here a couple of examples are useful.

A seven-year-old girl draws a puppet goose from the theatre performance she saw in her school hall. But she does so not as a puppet and not from the perspective given to her in the performance. Instead this goose is a fully fledged bird, depicted flying high above the stage, which is shown as small and far below. In the picture the goose is crying. The drawing depicts something not literally seen in the performance and from a perspective different from that of the audience. While prompted by the performance, the detail and specificity of these acts of imagination and empathy are only fully realized through the requirement to draw.

In another example, a woman in her 40s constructs a collage by slowly and laboriously cutting and sticking long red strips of card to a black sheet of paper. As she works she comments of the dance performance she had seen: 'I mostly got a feeling from it, I got a kind of sense of it being very, very angular and full of energy. But with a real kind of tension there.' Here the content and materiality of the picture echoes the memory, constructing a visceral, sensory response to the performance. Both the picture and the performance are an expression of angles and staccato tension.

In very different ways – one abstract and the other figurative – these examples display a sympathy between the visual depiction and the emotional meaning invested into the memory. In other words, the process of responding visually requires the participants to look again, look closer and through investing themselves into the memory depict more than they had

*Figure 1.3.2*  Crying goose. Child's drawing of *Martha* by Catherine Wheels Theatre Company. Copyright Matthew Reason

*Figure 1.3.3*  Angular and full of energy. Adult's collage of *Ride the Beast* by Scottish Ballet/ Stephen Petronio. Copyright Matthew Reason

seen – adding, interpreting, imagining, playing. The verbal responses, therefore, are the product not just of the original experience but also of the artistic intervention into that experience.

It is important to note that the transformative impact of the creative process is not generic but specific to the particularities of the methodology of visual expression, as manifested, for example, in the challenge of representing a theatre or dance performance on paper and the manner by which the participants engaged with the materiality of art making. The making of a picture takes time: thoughts and intentions at the beginning change and evolve as the image develops. Participants did not simply decide what to depict, but rather discovered both the what and how of the (re)presentation in the act of doing. As one of the adult participants in a research workshop commented:

> I thought the interesting thing for me when doing the art work was that I started with something but I started a dialogue with my image. So I started to put in other things that weren't there. But they were thoughts and feelings, so it opened up a conversation with myself about my memory of what I'd seen.

While the possibility of invention might concern a researcher seeking some externalized truth, for me this notion of a dialogue between the research participant and their experience is both striking and useful. In my research, I am interested in exploring audiences' aesthetic, embodied and emotional experiences of theatre and dance. These are by definition ephemeral and often seem both intangible and ineffable. Indeed, John Carey suggests that not only are the aesthetic experiences of *other* people impossible to access, but in truth our own aesthetic experiences are largely unapparent even to ourselves (2006: 23). Aesthetic experiences, therefore, can be considered not as something that participants know and simply have to tell, but instead as something that is brought into being by the process of reflection. Experience here is not only had, but also made.

The transformative process of drawing as a creative expression should, therefore, be considered a process through which new and different kinds of knowledge are generated and communicated. This is central to much artistic research and can be seen in terms of what Henk Borgdorff (2010: 61) describes as 'the pre-reflective, non-conceptual content of art' that 'creates room for what is unthought, that which is unexpected'. It is a process that is not linear or fully planned, but equally not fully unintentional. Along with other forms of visual expression, drawing is not simply a projective technique, but a form of structured exploration and generation. It is an approach particularly suited to researching emotional or affective memories, where the interest is in the lived phenomenological experience of the participant. Drawing, potentially at least, is a process that exposes the unthought and the unexpected, often not just to the researcher, but also to the participants themselves.

## References

Borgdorff, H. (2010). The production of knowledge in artistic research. In M. Biggs and H. Karlsson (Eds.) *The Routledge Companion to Research in the Arts* (pp. 44–63). London: Routledge.

Carey, J. (2006). *What Good Are the Arts?* Oxford: Oxford University Press.

Christensen, P. and James, A. (Eds.) (2000). *Research with Children: Perspective and Practices*. London: Falmer Press.

Eisner, E. (2008). Art and knowledge. In J. G. Knowles and A. L. Cole (Eds.) *Handbook of the Arts in Qualitative Research* (pp. 3–12). Thousand Oaks, CA: Sage.

Gauntlett, D. (2007). *Creative Explorations*. Abingdon: Routledge.

Malchiodi, C. (1998). *Understanding Children's Drawings*. London: Jessica Kingsley.

McDaniel, C. and Gates, R. (1999). *Contemporary Marketing Research*. Cincinnati, OH: South Western College.

Rose, B. (1993). Joseph Beuys and the language of drawing. In A. Temkin and B. Rose (Eds.) *Thinking is Form: The Language of Joseph Beuys* (pp. 73–118). London: Thames and Hudson.

# 4

# Experimenting

*Thomas Jellis*

The notion of experiment is increasingly attracting a certain cachet within and beyond academia. Perhaps what is most striking about this interest in, and proliferation of, experiments is that they are no longer understood to be the domain of the sciences, nor of a narrowly exclusive range of avant-garde aesthetic practices. Although experiments are still sometimes looked upon as ethically ambiguous (if not dangerous), various spaces, practices and events are increasingly being described as experimental. This is as pervasive in pop culture as it is in academia. For instance, turn on the TV and witness the fawning over the latest experimental chef. Or look at the huge grant for the recently established social science 'lab' at your university. It might even be claimed that, like the injunction to be critical in the 1990s, the injunction to be experimental has become one of the defining refrains of the early twenty-first century and certainly contemporary research in the social sciences.

Within this context, it becomes important to investigate the ways in which experiment is mobilized and to what ends. There are many things at stake here, not least of which is the value of experiment, especially when its currency is in danger of being devalued through the proliferation of the term. That is to say, if the notion of experiment is expanded to include ever more things, what happens to its specificity? Moreover, what kind of analytical purchase does it provide? My own research has sought to bear witness to this ever-increasing plurality of experimenting, a 'something/happening' that is gathering force and gaining traction through diverse energies. Crucially, I have sought to position myself as both investigator of how this experimental inflection is taking place and, in a very real sense through my own experimenting, to become a vector of the inflection itself. As much as my research seeks to foreground the new spaces and logics of experiments that are emerging, I am also interested in putting these ideas to work – to see how I too can experiment.

I am drawn to the notion of reclamation to think through this burgeoning experimentalism. By this I do not mean 'taking back what was confiscated, but rather learning what it takes to inhabit' (Stengers 2008a: 58); learning what can be done anew, with an experimental approach that is no longer tethered to the sciences. Crucial to what follows is an acknowledgement that these alternative experiments, however considered – 'extra-scientific' (Vasudevan 2007), 'ethico-aesthetic' (Guattari 1995), 'wild' (Lorimer and Driessen 2014) or simply uncategorizable – do not conform to the model of experiment as concerned with hypothesis testing. Such experiments disrupt the very notion of what it means to experiment, making no 'clear distinction between the terms "experience" and "experiment"' (Stengers 2008b: 109).

Although it carries with it a good deal of epistemic baggage, not least its association with positivism, reclaiming experiment is an opportunity to reflect on the ends of experiment and to think about how certain forms of experimentation serve to redefine problems for researchers. In this essay I want to outline how experiment may facilitate a flexing, or disruption, of ways of thinking. Following calls to document and reflect on 'innovative forms of methodological experimentations' (Dwyer and Davies 2010: 95), what follows is an attempt to consolidate outputs from an ever-increasing methodological repertoire to suggest how an experimental approach might be cultivated. I do so by reflecting on two examples of my own modest experiments, which might be characterized as participating and relaying. This comes out of a research project that examined the relations between geography and experiment (Jellis 2013, 2015). In large part, the impetus for this was my contention that there are new spaces of experimenting that are worthy of examination as a part of a renewal of experimentation within geographical thinking. The empirical remit of such a project consisted of ethnographic investigations of a loose constellation of laboratories, across Berlin (Insititut für Raumexperimente), Brussels (FoAM), London (Office of Experiments) and Montreal (SenseLab and Topological Media Lab), through which to examine experimenting.

The first way I want to explore my own attempts at experimenting is by way of participation. Much has been written about participatory research and related ideas of co-production, not to mention engagement and inclusivity. Yet this kind of work has rarely theorized how any research is always already an ongoing participation with the world, rather than something that can be chosen or selected. This is a participation that unfolds by 'becoming affected and inflected by encounters' (McCormack 2008: 2). Participating at sites, for me, involved a range of activities – talking, reading, designing, cooking, walking, foraging, choreographing – and I came to embrace the awkward role of not knowing quite what I was attending to, which others who have undertaken ethnographic research may relate to. My affirmative stance was quite literal; I always said 'yes' to suggestions and became involved in all kinds of projects. As such, I was enrolled into these experimental spaces in a number of ways. Some of my work appeared at an art exhibition in Berlin; an essay I had written was published in a collaborative book; I became part of an editorial board for one of the lab's journals, where I was also involved in the translation of texts; I represented another collective at a book launch and produced internal reports of events; I was also involved in copy-editing on a manuscript from another experimental group. Much of this work, then, was textual, but nevertheless it indicates the ways in which I became part of – if only temporarily – these experimental groups and practices. More than this, though, it is about not imposing arbitrary limits on participation; we do not need to stick to what we are comfortable with (and for me, this included, in particular, foraging and dancing). It is also important to remind ourselves that participation precedes recognition, it preceded me saying 'yes', as 'our awareness is always of an already ongoing participation in an unfolding relation' (Massumi 2002: 231). Whether we like it or not we are always already participating, and so the very question of what it means to participate is something with which one can but experiment.

How then to respond to such a suggestion? My way of working this through has been with the invocation of attentive participation, which builds on work to recalibrate participant observation (see Thrift 2000). Although we might look to extend what counts as participation, it is crucial that it is not participation just for the sake of it. As such, I tried to experiment with participant observation by questioning what I could participate in and how this might be refashioned to foreground attention in particular ways. This follows through any research process, such that any attempt to outline the unfolding relations and emergent events involves constructing lures for attention – where attention is a means of 'becoming able to add, not subtract' (Stengers 2008b: 99) – to make more of 'the feelings, the codes, the awkward intensities, the architected space, the architecture of time' of fieldwork (Dewsbury 2009: 326). Building on ethnographic research and its descriptive qualities, such an approach looks to amplify the manifold experiences of any kind of fieldwork in a way that does not seek recourse to fidelity but to re-animation. This is part of an 'ethos of stretching the means by which research is done and striving to continue as experiments fail or always come short in the attempt' (Dewsbury 2009: 323). Research outcomes might, then, be less about what we have found or extracted, and more about what we have done – and struggled with – and the affective swash of these encounters and their after-effects.

The second, and related, way I have been thinking about experimenting is about the researcher as a relay, or an impresario. In my own research on experimental spaces, one thing that I could contribute to the sites was my awareness of other, similar organizations that I was already working with or in the process of negotiating access of some sort. Something that emerged as an important way of experimenting was the modest undertaking of *connecting* disparate experimental groups. This kind of work does not fall within the criteria for measured outputs that many of us worry about – or at least those of us based in the UK – and yet it can serve to radically reconfigure the working practices of hitherto only loosely associated sites. Of course, this might well not be the case; there is every chance that an encounter will result in silence; not necessarily failure but certainly no follow-up or future connections. The researcher as impresario might, at first blush, sound grand – portentous even – and yet it is a risky, uncelebrated, decidedly unsexy endeavour that offers a profoundly different way of thinking about comparative research.

In this sense, this process of connecting made explicit that these various labs are not so much distinct sites as different ways in which the things I am interested in can relate to one another. And it was through this process of connecting that my research started to trouble the notion of a coherent 'experimental space' across these seemingly distinct 'field-sites'. By putting in touch one experimental lab with another, they were able to seek out synergies, highlight differences, and pool techniques for experimenting. But it also helped me to think differently about these groups – through what I termed experimental ecologies. To think experiment ecologically is to attend to the ways in which experiment always escapes particular sites. And so by fostering these connections between 'my' different field-sites, or put differently, relaying the various matters of concern from one to the next, I was able to amplify the ways in which experimental hubs exceed particular locales.

While it might be a big claim for an interdisciplinary method, experimenting can serve as an instance of – and an ethos for – reclaiming or shifting the energies of a field in productive ways. To be sure, experimenting has the potential to be disruptive of repertoires of practices and of modes of thinking. While it has been invoked in sometimes uncritical ways, which can suggest a certain heroism, a focus on experiment and the experimental cannot be dismissed as a passing intellectual fad as it raises crucial questions for how the social sciences proceed methodologically. More specifically, my concern has been with how experiment is a 'searching for a new way of going on' (Thrift 2008: 223). The question of experiment might continue to remain elusive, but this is no bad thing. Instead, in this age of 'experimentality', to experiment in thought and method is to reconfigure what constitutes the world.

## References

Dewsbury, J. D. (2009). Performative, non-representational, and affect-based research: seven injunctions. In D. Delyser, S. Atkin, M. Crang, S. Herbert and L. McDowell (Eds.) *Handbook of Qualitative Research in Human Geography* (pp. 321–334). London: Sage.

Dwyer, C. and Davies, G. (2010). Qualitative methods III: animating archives, artful interventions and online environments. *Progress in Human Geography*, 34(1): 88–97.

Guattari, F. (1995). *Chaomosis: An Ethico-Aesthetic Paradigm* (P. Bains and J. Pefanis, Trans.). Sydney: Power Publications.

Jellis, T. (2013). Reclaiming Experiment: Geographies of Experiment and Experimental Geographies, DPhil thesis, University of Oxford.

Jellis, T. (2015). Spatial experiments: art, geography, pedagogy. *cultural geographies*, 22(2): 369–374.

Lorimer, J. and Driessen, C. (2014). Wild experiments at the Oostvaardersplassen: rethinking environmentalism in the Anthropocene. *Transactions of the Institute of British Geographers*, 39(2): 169–181.

Massumi, B. (2002). *Parables for the Virtual*. Durham, NC: Duke University Press.

McCormack, D. P. (2008). Thinking-spaces for research-creation. *Inflexions*, 1.

Stengers, I. (2008a). Experimenting with refrains: subjectivity and the challenge of escaping modern dualism. *Subjectivity*, 22(1): 38–59.

Stengers, I. (2008b). A constructivist reading of process and reality. *Theory, Culture and Society*, 25(4): 91–110.

Thrift, N. J. (2000). Non-representational theory. In R. J. Johnston, D. Gregory, G. Pratt and M. Watts (Eds.) *The Dictionary of Human Geography* (p. 556). 4th ed. Oxford: Blackwell Publishing.

Thrift, N. (2008). *Non-Representational Theory*. London: Routledge.

Vasudevan, A. (2007). Symptomatic acts, experimental embodiments: theatres of scientific protest in interwar Germany. *Environment and Planning A*, 39(8): 1812–1837.

# 5

# Figuring

*Margaret Wertheim*

Figuring is a word with deep interdisciplinary resonances in mathematics, literature and science; and as an activity it encompasses diverse histories and contexts: from textile makers weaving patterned figures with jacquard looms to courtly dancers spinning dynamical figures across a ballroom floor. The human figure, long a staple of artistic representation, is now a locus of constant measurement as we count steps and calories to figure our physiology. As cognitive beings, of course, we are continually figuring things out. In this essay I want to focus on the act of making figures that instantiate scientific and mathematical principles, a practice-based methodology that lies at the heart of the Institute for Figuring (IFF), a Los Angeles based organization I co-founded and direct with my twin sister Christine Wertheim. The IFF – its acronym being the logical symbol for 'if and only if' – is an enterprise dedicated to the aesthetic and poetic dimensions of science and mathematics (www.theiff.org). With my background in physics, and Christine's in literature and philosophy, the institute was born from our entwined desire to explore processes of figuring as a way of thinking beyond symbolic form.

Although Western thought has long privileged symbolic modes of representing information – the mathematical equations of physics, the DNA code of molecular biology, the binary codes of computation – we suggest that the insights illuminated by material figuring extend traditional disciplinary approaches and often reveal surprising facets of knowledge. In our work, we explore how sign systems can manifest in concrete figurative forms. For instance, fractals can be constructed out of business cards, platonic solids can be woven from bamboo sticks, hyperbolic surfaces can be crocheted, tessellations can be cross-stitched, the projective plane can be knitted. Even the logic underlying digital computing has a geometric analogue that can be represented by a three-dimensional network. Not infrequently, abstract relations have correspondences in material objects that lend themselves to concrete play, and to a consequent playing with ideas. As a practice of making concrete figurative forms, figuring calls attention to the wisdom of embodied objects, whose qualities are not merely reducible to, or predictable from, descriptive codes.

Partly stimulated by Friedrich Froebel's revolutionary nineteenth-century 'kindergarten' system of pedagogy (Brosterman 1997), with its focus on tactile geometric construction, another inspiration for our practice has been the field of chemistry. As the study of atomic assemblages, chemistry is inherently combinatoric; there are only a hundred or so atoms but these can be arranged in an infinite variety of molecules. Not everything, however, is possible, for various

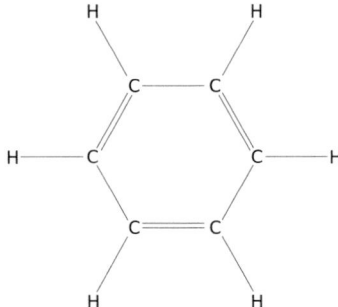

*Figure 1.5.1*  Hexagonal structure of benzene. Image courtesy of the Institute for Figuring.

laws, patterns and regularities assert themselves. To articulate such relationships, chemists have developed a variety of symbol systems, including a lexical notation for specifying any particular compound as well as a graphical notation for representing the arrangement of its atoms in space. Take the case of benzene, a vital organic molecule made up of six carbon atoms and six hydrogen atoms. Benzene's chemical formula is $C_6H_6$, and its graphical representation is shown in Figure 1.5.1.

In figuring out how molecules work and how they can be assembled, chemists also make three-dimensional models to represent how chemicals occupy physical space. In the past, these were hand-crafted out of balls and sticks, for example Watson and Crick's famous model of DNA, yet contemporary chemical modelling is mostly now done on computers, including state-of-the-art virtual reality set-ups, such as the CAVE where researchers can move around and through a simulation to explore the physical figure of their molecule (Cruz-Neira, Sandin and DeFanti 1993). The design of new drugs, for example, is largely premised on understanding the *shapes* molecules make and the specific shapes of the body's receptors into which they must fit. A good deal of pharmacy is applied geometry and here function literally follows form.

At the IFF we aim to create circumstances where participants can experiment with similarly embodied activities, generating objects that delight the eyes and stimulate our haptic sensibilities while also illustrating formal sets of relationships. Such acts of figuring allow for creative exploration within a context of rules and constraints having their own internal logics. Interested in the dance between codes and forms, like chemists, we also seek insight by figuring out problems through the structures of our models. In one key project, our *Crochet Coral Reef* (Wertheim and Wertheim 2015), we examine hyperbolic geometry through crochet, an unlikely conjunction between mathematics and feminine handicraft inspired by a discovery by Daina Taimina, a mathematician at Cornell (Taimina 2009). In hyperbolic space (an alternative to the Euclidean space we learn about in school), geometric forms behave in novel ways: parallel lines can diverge while the angles of a triangle may sum to $0°$. In Taimina's models such theorems may be visually illustrated by sewing diagrams onto the woollen surface, thereby concretizing abstruse mathematical concepts (Wertheim 2005).

Geometrically precise, Taimina's models are generated from a simple algorithm – 'crochet n stitches then increase one', where 'n' may be any fixed number. Additional stitches increase the surface area, generating the geometric opposite of a sphere. But what happens if we deviate from this rule? Let us say we increase one in every three stitches, then one in every ten? Here we no longer get a hyperbolically exact surface, for it is no longer geometrically regular. Just as there is only one *sphere* (an object with constant *positive* curvature), so there is only one pure hyperbolic surface (an object with constant *negative* curvature). By morphing the code and

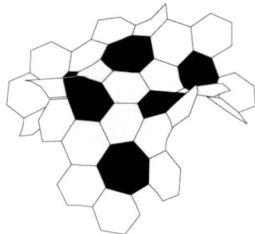

*Figure 1.5.2*   Model of the hyperbolic plane constructed from hexagons and heptagons. Image courtesy of the Institute for Figuring and *Cabinet* magazine.

manually introducing deviations, we crochet Reefers move away from mathematical 'perfection' into a domain of organic possibility. In the *Crochet Coral Reef* project, we are indeed exploring negative curvature analogues of wonky spheroid forms such as the oblate shapes of sea urchins and the asymmetries of eggs. Nature itself has been playing with such forms in the frilly surfaces of corals, kelps and sea-slugs for hundreds of millions of years. Yet mathematicians, with their formalized rules, spent hundreds of years trying to prove that such structures were impossible.

Across the academic spectrum, we see a growing interest in and sensibility towards embodiment and the qualities of material being. The philosopher of science, Evelyn Fox Keller, for instance, has noted the limitations inherent in the 'master molecule' theory of DNA and drawn attention to the dynamic role of the cell cytoplasm in embryonic development (Fox Keller 1995). Feminist science studies scholar Donna Haraway celebrates entangled webs of 'sympoesis' (a word meaning 'to make with'), in which embodied critters together create and nurture environmental health (Kenney 2015). In mathematician Brian Rotman's radical proposal for a non-Platonist account of mathematics, articulated in his book *Ad Infinitum* (Rotman 1993), he suggests that doing maths is itself an active embodied process. Much like the act of crochet, Rotman declares that even mathematics results from cognitive acts carried out by physical agents subject to physical limits, and thus, he says, there is no such thing as a perfect sphere or perfect straight line. For him, *all* mathematical objects – including numbers – are finite entities that arise when concrete actors construct them. Rather than a static and transcendent domain waiting to be discovered, Rotman considers mathematics as an evolving landscape of forms continuously brought into existence by communities of practising mathematicians.

By morphing a crochet code, and exploring the potentialities of a woolly DNA, we at the IFF have created communities of localized knowledge and expertise who branch out from geometric perfection to generate vast simulations of coral reefs. To date, more than 10,000 women in a dozen countries on five continents have collaboratively stitched a crochet 'tree of life'. 'Iterate, deviate, elaborate' has been the motto for this handiwork, which now constitutes a globally extensive experiment in applied geometry and emergent algorithmic complexity. As life on earth begins with simple cells and evolves into ever-more complex forms, so our crochet forms have evolved. Figuring with our fingers – a literal *digit-al* technology – has opened a diversity of patterns that provide a yarn-based analogue for thinking through Darwinian ideas.

These crafted objects are underpinned by a DNA-like code (the pattern of stitches that can be written down in symbols, much like the symbols articulating molecules). Yet the code is not wholly determinant. When figuring with materials, the properties of substances impress themselves on structures, causing chains of consequence that often cannot be predicted in advance. A form figured in stiff acrylic thread might stand pert like a stony coral, but constructed in silk it might flop like a piece of kelp. Real figures – as opposed to idealized mathematical ones – result not just from the codes and equations scientists use to describe them, but also from the qualities of their components. As in chemistry, *matter* matters, and the structural properties of molecules also result not merely from their chemical formulae but, critically, from physical interactions between their parts. This is why proteins are so hard to model on computers; here scientists must engage with the real-world physics of complex atomic interactions. As a methodology, our insight is that figures themselves come into being through acts of figure-ing that involve rules and deviations, material substances and dispersed communities of practice.

## References

Brosterman, N. (1997). *Inventing Kindergarten*. New York, NY: Harry N. Abrams Inc.

Cruz-Neira, C., Sandin, D. J. and DeFanti, T. A. (1993). Surround-Screen Projection-based Virtual Reality: The Design and Implementation of the CAVE. *Siggraph'93: Proceedings of the 20th Annual Conference on Computer Graphics and Interactive Techniques*, 1993, pp. 135–142.

Fox Keller, E. (1995). *Refiguring Life: Metaphors of Twentieth-Century Biology*. The Wellek Library Lecture Series at the University of California, Irvine. New York, NY: Columbia University Press.

Kenney, M. (2015). Anthropocene, capitalocene, chthulhucene: Donna Haraway in conversation with Martha Kenney. In H. Davis and E. Turpin (Eds.) *Art in the Anthropocene: Encounters among Aesthetics, Politics, Environments and Epistemologies* (pp. 255–270). London: Open Humanities Press.

Rotman, B. (1993). *Ad Infinitum . . . The Ghost in Turing's Machine: Taking God Out of Mathematics and Putting the Body Back In*. Palo Alto, CA: Stanford University Press.

Taimina, D. (2009). *Crocheting Adventures with Hyperbolic Planes*. Wellesley, MA: A. K. Peters.

Wertheim, M. (2005). *A Field Guide to Hyperbolic Space*. Los Angeles, CA: Institute for Figuring Press.

Wertheim, M. and Wertheim, C. (2015). *Crochet Coral Reef*. Los Angeles, CA: Institute for Figuring Press.

# 6

# Imaging

*Rebecca Coleman*

A focus on imag*ing* enables a consideration of the ways in which images might be the subject or outcome of a research project, and also an integral part of *doing* it. In this contribution, I consider some specific images that have been made in my research, both by research participants and myself. However, the main focus of the chapter is on understanding imaging as a research methodology that involves processes of making, assembling and circulating. In this sense, I am interested here not so much in how images may be considered as data, but more in how they may participate in the creation and dissemination of research; in what images might do in and for the research process. In particular, I place emphasis on the 'ing', in order to indicate the processual character of making, assembling and circulating – these are dynamic and transformational practices – and of images – which are themselves understood as potentially unfinished, sensory and affective experiences (rather than static objects or texts).

Such an understanding of the role of images in research might appear rather obvious to those in the performing and visual arts involved in practice research, where arts and media practices are recognized as generators of knowledge, and perhaps as performative interventions. However, with some notable exceptions, in the social sciences images have largely been considered as representations to be analysed in order to make sense of patterns and inequalities in visual culture, and/or as means of documenting encounters and experiences with the social world. While these approaches remain important, this contribution aims to explore some of the ways that the social sciences might take up practices developed in, and/or inspired by, art and design, and as a consequence might work with imaging as a research practice. Hence, I make interdisciplinary methodologies and methods the focus of my discussion, paying particular attention to the conceptual and practical issues that such approaches provoke.

I discuss two examples of how my research has worked through different imaging practices: the first involves research participants making collages, and the second involves me, as researcher, making and sending postcards. The images at stake here are thus broadly understood. They include collages and postcards both as images themselves, and as images that are made through a range of materials including other images, as I discuss below. My understanding of imaging is similarly broad. I consider how imaging can potentially involve multiple and diverse aims and practices, how different participants within a research project might (or might not) produce and circulate images, and how imaging as interdisciplinary methodology raises a number of questions that require further attention. While this chapter explores two examples from my own research, it is important to note that this focus is not prescriptive; other still and moving images and imaging practices (e.g. those involving videos, painting and drawing) might also be relevant to the suggestions I make here. That is, my understanding of images as open-ended experiences and of imaging as processual and transformational might lend itself to a range of approaches that are concerned with the non-representational, sensory, and inventiveness of the social world.

## Imaging as process and practice

The understanding of imaging that I outline above can be unpacked through a first example of a project with young women on how they experience their bodies through images (Coleman 2009). In this research, I wanted to empirically study how these girls' experiences of their bodies emerged and were arranged through relations with different kinds of images. Much feminist research on girls' bodies and images concentrates on media images. While these images were raised as significant, what also became apparent in my research was that other images – including photographs, mirror images, and comments about their bodies from other people – were also important. Methodologically, I included image-making sessions alongside the more traditional sociological methods of individual and group interviews.

The aim of these sessions was to try to integrate images into the research, so that images were not just the subject of the research (what it was about), but part of how images were themselves studied (a methodology). I was also interested in encouraging participants to visualize and examine their experiences of their bodies.

The sessions involved the girls collaging images of their bodies through materials from different sources, including magazines, a Polaroid camera, craft materials, make-up and sweet wrappers (see Figure 1.6.1, and also Coleman 2009). To begin to think through these collages, the notion of assembling is particularly helpful. An assemblage refers to a temporary and changing arrangement of multiple parts (Deleuze and Guattari 1987). The collages made by the girls can thus be understood as assemblages; they are constituted by materials taken from various sources, which are arranged in ways that demonstrate how they have been transformed in the move from one source and setting to another (e.g. from a magazine to a collage, from mass media to a classroom to various academic publications), and in the relations the parts have with each other (e.g. through how they may be juxtaposed, and/or organized so as to create a particular impression).

Furthermore, in these collages, issues concerning change are highlighted. For example, Anna's collage (Figure 1.6.1) highlights how understandings of a person might change depending on whether they are based on looks and appearance or 'what's inside'. Other participants juxtaposed photographs of themselves with images from mainstream women's magazines. Fay, for example, assembled a photograph of herself with magazine images and wrote 'I wish', indicating what she experiences her body to be, and what she would like it to become. Here, then, imaging as research methodology is a process through which specific images are created, which can then be analysed as social science data.

However, as well as the images themselves being critically analysed, the research methodology of imaging is itself worth considering. The sessions were productive in that the girls clearly enjoyed participating, with some asking for the sessions to be extended so they could continue working on their collages. While making their images, some of the participants also began working together in informal ways. For example, discussing what materials and techniques others had used for inspiration for their own collages, and having seen how others had included them,

Figure 1.6.1   Anna's image

a number of the girls incorporated pipe cleaners into their images towards the end of the session. As well as belonging to an individual participant, the images produced were therefore clearly shaped by the group experience, enabling me to reflect upon how different methods produce different kinds of knowledges and data (a topic that commonly features in discussions about the strengths and weaknesses of individual as opposed to group interviews, for instance).

The imaging methodology also raised a number of challenges. After they had made their collages, I asked the girls to explain them to the group. This was not so much because I think it is necessary for visual, sensory or imaging research methodologies to be translated into words, but rather to encourage the girls to reflect on their images, and the experiences of their bodies with which they had engaged (see also the chapter on Drawing for a similar point) and to share them with the group. However, the girls found it difficult to verbalize their experience of making the images and the images they made. Imaging as methodology, therefore, raises questions regarding whether and how articulations about and assessments of such practices are necessary, possible and/or desirable. Relatedly, although I had wanted to make images a central part of the research, as a sociologist more familiar with writing about textual data, I struggled to know what to do with the images once they had been made. Indeed, in the book on the research, I discuss the images in one section of one chapter, and treat them similarly to extracts from the interviews I conducted. This technique demonstrates how visual and sensory as well as textual data are valid, and how imaging is a productive methodology; however, incorporating the images into a relatively traditional written publication was perhaps not the most appropriate means of attending to either the specificity of the images that were produced, or the imaging methodology deployed.

A second example is a project that attempts to put speculative visual methods to work to explore a recent patent by Amazon for 'speculative shipping', where goods will be shipped in advance of order to geographically distributed hubs to minimize the time between online order and delivery (see Coleman 2016). To solve the problem of returning speculatively shipped products to the warehouse if the item is not subsequently ordered, the patent proposes to deliver the package 'to a potentially interested customer as a gift' to 'build goodwill' (Spiegel et al. 2013). This notion of creating goodwill through the delivery of unexpected packages in the post is noteworthy, given how the patent for speculative shipping was described in the press as 'delightful and exciting [. . .] We like getting things in the mail, even if we didn't ask for them' (Kopalle 2014).

In one iteration of the project, I attempted to develop my own system of speculative shipping, drawing on mail art; an artistic movement aimed at creating international networks of artists based on gift rather than commercial exchange, and challenging distinctions between high and low culture in terms of what materials might be used in the practice. Using materials bought from Amazon as well as the packaging they were delivered in, I made postcards that I sent in the UK Royal Mail postal system to unsuspecting recipients, which I asked them to write and/or draw on and return to me (see Figures 1.6.2 and 1.6.3). In some ways then, the research created images, in that the postcards can be understood as collages which could be analysed both before they were mailed and after they were returned. They can therefore be treated as research data.

However, the broader aim of the project was to explore the implications of this imaging methodology for how images may be circulated. Would it be possible for this methodology to create a 'delightful and exciting' means of exchange between people who did not know each other? In many ways, the project failed: of the 26 postcards I sent, only one was returned to me – and it was left blank. However, in other ways, this failure enabled me to understand the imaging methodology I had deployed as 'less a case of answering a pre-known research question

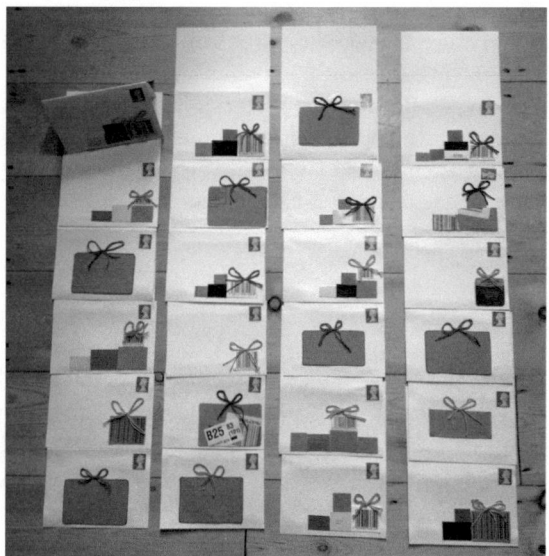

*Figure 1.6.2*   Selection of postcards made

[. . .] than a process of asking inventive [. . .] questions' (Wilkie, Michael and Plummer-Fernandez 2014: 4). For example, I have asked, might sending unexpected packages in the post be unwelcome, rather than delightful and exciting? Are the postcards I made recognizable as gifts, in the ways that unexpected packages received from Amazon are? What happens when data are not produced in a research project in the way that they were designed to be?

Furthermore, the process of making the postcards led me to learn about mail art, and to consider how it might transform into a sociological method, which necessarily includes making

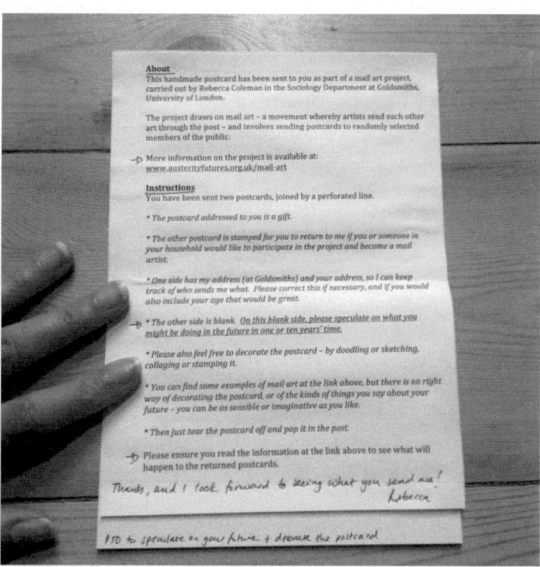

*Figure 1.6.3*   Outline of research and instructions included on postcards

decisions about research ethics. How might hitherto unknown participants of research be involved in circulating and disseminating research in ways that are ethical? What kinds of ethical questions regarding recruitment, inclusion, anonymity and 'impact' might this kind of research raise? Still further, presenting on this research project in different contexts has illuminated how boundaries between different disciplines remain monitored. Sociologists have asked me how the project is sociological: What are my data? How will I analyse them? What is my understanding of the social and the role of sociology in relation to the data? Artists have critiqued both my artistic skills and understanding of mail art.

## Conclusion

This chapter has introduced some indicative cases of how imaging might be of value to a wider project of developing interdisciplinary methodologies. In terms of my second example, while the tone of some of the questioning around disciplinary boundaries has been dispiriting, the questions draw attention to both the difficulty of doing interdisciplinary work – what happens when methodology itself becomes that which is in focus in a project? What audience is the research engaging? – *and* the necessity of developing such interdisciplinary projects. For example, if Amazon's speculative shipping might elicit feelings of delight and excitement, it seems reasonable to ask how social science research might provoke and engage such feelings. Would imaging be able to produce positive affects? Are text-based methods and modes of analysis most suitable to grasp what Les Back calls the 'the fleeting, distributed, multiple, sensory, emotional and kinaesthetic aspects of sociality' (2012: 28)? Questions emerging from the first example concern how images produced in imaging research might be disseminated, and relatedly, whether the production of such data requires the social sciences to move 'beyond text'. Perhaps, as Puwar and Sharma (2012) argue, the social sciences might revive notions of curation, taking up practices more widespread in the arts, and requiring new forms of auditing to account for non-textual outputs.

Such questions indicate how interdisciplinary methodological practices might cultivate what, in a discussion of speculative design and sociology, Mike Michael calls the 'common byways' along which seemingly different and distinct methods, practices and approaches travel; 'How can the engagements between these be rendered open, multiple, uncertain, playful?' (Michael 2012: 177). Such engagements are necessary, I would suggest, in light of broader shifts that see methods as entangled with the becoming of the social world, and as a means of engaging both research participants and potential audiences in creative, meaningful and affectively enriching ways.

## References

Back, L. (2012). Live sociology: social research and its futures. *The Sociological Review*, 60(S1): 18–39.

Coleman, R. (2009). *The Becoming of Bodies: Girls, Images, Experience*. Manchester: Manchester University Press.

Coleman, R. (2016). Developing speculative methods to explore speculative shipping: mail art, futurity and empiricism. In A. Wilkie, M. Savransky and M. Rosengarten (Eds.) *Speculative Research: The Lure of Possible Futures* (pp. 130–144). London: Routledge.

Deleuze, G. and Guattari, F. (1987). *A Thousand Plateaus: Capitalism and Schizophrenia*. London and New York: Continuum.

Kopalle, P. (2014) 'Why Amazon's Anticipatory Shipping Is Pure Genius'. *Forbes*, 28 January 2014. Retrieved 16 February 2015 from www.forbes.com/sites/onmarketing/2014/01/28/why-amazons-anticipatory-shipping-is-pure-genius/

Michael, M. (2012). De-signing the object of sociology: toward an 'idiotic' methodology. *The Sociological Review*, 60(S1), 166–183.

Puwar, N. and Sharma, S. (2012). Curating sociology. *The Sociological Review*, 60(S1): 40–63.

Spiegel, J. et al. (2013) Method and system for anticipatory package shipping. United States Patent, 24 December 2013. Retrieved 16 February 2015 from http://patft.uspto.gov/netacgi/nph-Parser?Sect1=PTO1&Sect2=HITOFF&d=PALL&p=1&u=/netahtml/PTO/srchnum.htm&r=1&f=G&l=50&s1=8615473.PN.&OS=PN/8615473&RS=PN/8615473

Wilkie, A., Michael, M. and Plummer-Fernandez, M. (2014). Speculative method and Twitter: bots, energy and three conceptual characters. *The Sociological Review*, online first, 14.08.2014: 1–23.

# 7

# Rescaling

*Ramon Lobato*

This essay reflects on a tactic I call *rescaling*, which involves manipulating notions of scale in research design. It is not a formal methodology; rather, it is a sort of general impulse, or perhaps a flexible methodological device, that can be useful for illuminating relations between apparently disconnected phenomena. In what follows I discuss how rescaling can potentially be used within my own field of media and communications research, and perhaps in some other fields as well.

The context in which I have come to think about these issues is through undertaking research projects on digital media industries, in particular, on those productive elements of the media that exist outside the boundaries of 'industry' per se (Lobato and Thomas 2015) – examples include free software, piracy and user-generated content. Such research presents specific methodological challenges. For example, there are obvious difficulties with data collection, reliability and sampling when studying systems that are, by nature, informal and ephemeral. Second, there is the problem of finding the right analytical tools for the job. Methods used to study industry do not always work well in the world of media, and methods used to study media do not always work well for industries. I found it necessary to cobble together a framework from a variety of disciplines, drawing on ideas from economic anthropology, political economy, and cultural studies as needed, and trying to work across social science and humanities approaches. Undertaking this research has been difficult, but it has also been generative from a methodological perspective because of the need to improvise.

One consequence of this research is that I have begun to think a lot about scale, and how it can be tweaked and inverted for the purposes of research design. Scale in research is about, among other things, the interdependence between the big and the small, and how we conceptualize this relationship in our data collection and interpretation. When studying industries, for example, we make certain kinds of choices about what counts most in the analysis and at what level of 'the economy' these things are located. In other words, we enact an imaginary vertical ordering of the elements that make up an economy. But as geographers including Agnew (1993) and Allen (2011) remind us, notions of scale are always socially constructed and have real-world effects that can be illuminated in and through moments of scalar tension. As Marston (2000: 220) writes, 'scale is not necessarily a preordained hierarchical framework for ordering the world – local, regional, national and global' but rather 'a contingent outcome of the tensions that exist between structural forces and the practices of human agents'.

Scale as a critical concept can be traced back through various intellectual traditions. Feminist and postcolonial theory, for example, have given us powerful tools with which to understand the relationship between big and small; so have anthropology, social history and development studies, in their emphasis on the interdependence of structural forces and everyday lifeworlds. And then there are the many literary and artistic provocations that ask us to see the big and the small in new ways (including surrealism and the writing of Georges Perec). So we have an array of conceptual resources available to us when thinking about scale.

Within social science research on industries, the problem of scale often finds expression in the tension between macro and micro levels of analysis. At the top end, social science is of course well served (or poorly served, depending on your view) by structural theories of economic change. These approaches often mirror a particular kind of relationship of scale: experts use 'big' frameworks – including political economy and regulatory theory – to study 'big' things, like multinational corporations and government policy development. In media industry research, this often translates into 'a political–economic perspective that emphasizes macrolevel structural issues of regulatory regimes, concentration of media ownership, historical change, and their larger connection to capital interests' (Havens, Lotz and Tinic 2009: 234).

Like all ways of knowing, such methods have limitations. Tools such as the Herfindahl–Hirschman index, for example, which measures levels of industry concentration using a scale from 0 to 1, rely on the kind of financial data that is normally only available for publicly listed companies. From this perspective, a lot of what passes for industry research has an in-built bias towards large, listed firms, which are often taken to be constitutive of industry more generally. In other words, standard tools of economic analysis tend to reflect and produce certain ideas about what industries are – ideas that are culturally and historically specific. This particular model does not always translate well to cultural and media industry research, for example, because some parts of these industries have quite different organizational logics.

A parallel tradition of research takes a micro-level view of industries using micro-level methods. This approach is associated more with economic anthropology, certain forms of cultural sociology, and, in the humanities, cultural studies and cultural history, which tend to think of industries as structured formations of people, power and discourse. From this perspective, industries can be investigated by looking at the people and practices that constitute them, so the first step in research is to generate detailed and contextualized accounts of everyday practices at ground level. In this 'bottom-up' tradition we therefore often find situated studies of cultural workers and professionals, grounded in ethnographic and qualitative interview traditions (Caldwell 2008; Hesmondhalgh and Baker 2011), as well as close analysis of industry discourses and 'trade talk'. This is the arena of modest claims, rigorous theorization, reflexivity and grounded speculation, where knowledge is understood as provisional and situated.

These patterns will be familiar to many readers as reflecting the epistemological cleavage between empirical social science and interpretive humanities traditions – something regularly attacked within the flattened ontologies of actor-network theory, non-representational theory and ecological and topological epistemologies. This cleavage is rarely so stark in practice because the correspondence between macro and micro methods and objects of study can be – and often is – productively inverted by juxtaposing different scales. For example, across existing media industry research one can find many examples of research projects that use *bottom-up methods to study large things* (as in the ethnography of media institutions), or *top-down methods to study small things* (as when researchers use methods like network analysis to study small online communities).

Let me provide some examples to illustrate the point. In recent research projects, I have studied a range of actors in media industries (including executives, fans, pirates and geeks); observed activities in retail sites and online spaces; scrutinized web forums; read archives of leaked material; analysed pricing patterns; compiled biographies of key individuals and case studies of typical companies; and so on. Over the course of this research I learned through trial and error that interesting findings tend to emerge when I applied the 'wrong' scalar approach, sometimes by accident. For example, when investigating the economic dynamics of pirate DVD vending, it was quite useful to spend part of the interview with 'pirates' asking the kind of questions one would ask a corporate executive. Who is your competition? How much do you charge for your products, and why? Where do you get your stock from? Who works for you, and how do you manage them? This helped to cut quite quickly to the commercial realities structuring what is, after all, a commercial operation. In other words, it was necessary to *rescale* my assumptions about the methods appropriate to these actors.

A second example comes out of a project I have recently been doing on the anonymization and personal privacy software industry. This is a most unusual industry to study because it is highly fragmented, with very low barriers to entry, and many companies operate as bedroom enterprises. It has no industry associations and no central sources of data. Yet it has all the dynamics of any other commercial sector, including fierce competition, differentiation, professionalization and labour mobility. To analyse this industry we did some of the things that one would do when studying a 'big' industry: compiled an international database of known companies; collected data on server locations, price points and marketing strategies; and interviewed company representatives willing to speak on the record. But we also had to use the micro-level methods associated with qualitative research: hanging out in online spaces, signing up for various services and trying them out, discourse analysis of promotional materials, and so forth. And during interviews, it was helpful to mix up the micro and macro scales by interspersing data-oriented questions ('how many staff do you have, and where are they located?') with questions inviting more textured, qualitative responses ('take me through your average working day').

What I want to suggest is that these inversions of scale can be understood as a kind of methodological tactic. In other words, juxtaposing the big and the small – and the methodological orientations associated with each scale – often produces useful insights. This is not a radical proposition; nor is it new. But I find that playing around with notions of scale can be useful as an exploratory process when devising a research programme, even when we are investigating the 'hard' stuff of firms, industries, revenues and so on. Translating these ideas into research practice provides a framework for experimenting with macro and micro scales and the methodologies associated with each – a framework that I have found well suited to studying the unfamiliar institutional forms of digital media, but which might also be useful as an exploratory device in other fields as well.

## References

Agnew, J. (1993). Representing space: space, scale and culture in social science. In J. Duncan and D. Ley (Eds.) *Place/Culture/Representation* (pp. 251–271). Routledge: London.

Allen, J. (2011). Topological twists: power's shifting geographies. *Dialogues in Human Geography*, (1)3: 283–298.

Caldwell, J. (2008). *Production Culture: Industrial Reflexivity and Critical Practice in Film and Television*. Durham, NC: Duke University Press.

Havens, T., Lotz, A. and Tinic, S. (2009). Critical media industry studies: a research approach. *Communication, Culture & Critique*, 2(2): 234–253.

Hesmondhalgh, D. and Baker, S. (2011). *Creative Labour: Media Work in Three Cultural Industries*. Abingdon: Routledge.

Lobato, R. and Thomas, J. (2015). *The Informal Media Economy*. Cambridge: Polity.

Marston, S. A. (2000). The social construction of scale. *Progress in Human Geography*, 24(2): 219–242.

# 8

# Sand drawing

*Jennifer Green*

Human interactions consist of a creative bricolage of the resources a culture brings to its communicative tasks. As well as speaking, people might point to real and imagined entities and locations, manipulate fictive objects in the air with their hands, create traces, maps and diagrams, use writing systems, and make permanent or semi-permanent marks on a range of surfaces. Sign languages are the primary mode of communication for some, and for others sign is used either alongside speech, or instead of speech in particular cultural contexts. Tangible items such as tools, technologies and other aspects of the social and environmental context might also be employed (Goodwin and LeBaron 2011; Nevile, Haddington, Heinemann and Rauniomaa 2014). Understanding why particular ways of communicating meaning take precedence over others, and modelling exactly how this complexity is orchestrated remains one of the challenges for understanding human language.

Recent work in linguistics and in a variety of related disciplines has led to a growing recognition of the multimodal nature of human communication (Kendon 2004; Enfield 2009; Jewitt, Bezemer and O'Halloran 2016). Moving away from speech or text-only perspectives it is now the norm to see language as 'embedded within an interactional exchange of multi-modal signals' (Levinson and Holler 2014: 1; Vigliocco, Perniss and Vinson 2014). Although usage of the term 'multimodal' varies greatly, the ways that people produce and perceive communicative signals can be envisaged as falling within two major modality divisions. Speech utilizes the vocal/auditory modality, and sign languages, gesture and systems of graphic representation utilize the kinesic/visual one. Within each modality are various systems or 'potentials' that convey meaning. The fundamental units of communication can thus be envisaged as 'composite utterances' in which elements of different semiotic systems work together (Enfield 2009).

This essay considers several of the semiotic parameters that are employed in the telling of Indigenous sand stories from Central Australia. Sand stories are a traditional narrative form in which skilled storytellers, primarily women and girls, incorporate speech, song, sign, gesture and drawing (Wilkins 1997; Munn 1973; Green 2014). In sand stories a small set of conventionalized graphic symbols are embedded in a complex semiotic field that includes various other types of actions, for example lexical manual signs, as well as speech. Such narrative practices emerge in a particular cultural and ecological niche where soft sand is readily at hand. The significance of the ground is seen on many levels. It is a locus of important information, coding movement, habitation and histories, like a vast notice board. It is a place of day-to-day habitation and relaxation, and for the seated person invites inscription. The surface of the ground is valued for its rich palette and observed in the minutiae of its seasonal variation.

Bringing an analytic perspective to understanding how the complexity of sand stories works to convey meaning raises methodological and conceptual challenges. An interdisciplinary approach to these questions can draw upon insights and methodologies from sign language and gesture studies, ethnomusicology, semiotics, psychology and anthropology, to name a few. Such an approach also presents an opportunity to 'move beyond the empirical boundaries of existing disciplines' (Jewitt et al. 2016: 2) and develop new approaches for data analysis that can account for phenomena that in some instances fall below the radar of academic inquiry if a narrow perspective is taken. At the methodological level, and as appreciation for the role of 'visible bodily action' (Kendon 2004; Seyfeddinipur and Gullberg 2014) in communication grows, many linguists are using video technologies for language documentation. This enables consideration of multiple aspects of complex utterances, and the ways that various types of action – for example, gesture, sign, drawing and eye-gaze – work together.

The thematic content of sand stories ranges from accounts of day-to-day events to performances of traditional narratives that are closely associated with the topography of the land and its ancestral progenitors or 'Dreamings'. In remote Indigenous Australia, particularly in the pre-television era, sand stories were a form of popular entertainment. For children, they provide a context for developing a range of spatial and graphic skills, as well as training and practice in the verbal arts. Although some knowledge of the ways these stories were told in the past is highly endangered, the form is still used, even as the practice is adapting to meet changing social contexts.

Sand stories begin with the clearing of a space on the ground in front of the narrator. The resultant drawings and mini-installations of objects are both product and process, and involve a complex interplay between dynamic and static elements. The semi-permanence of the graphic marks made on the ground is subservient to broader rhetorical aims as the story unfolds. Between 'scenes' or 'episodes' the seated narrator wipes the space on the ground in front of them clean before beginning to draw again. Drawing on the ground is done with the hand, or with sticks or wires (Figure 1.8.1). A story-wire is multi-functional, used as a drawing instrument and as a tool to point with. Story-wires are also used to provide rhythmic accompaniment, and repeated tapping of the wire on the ground embellishes the soundscape, sustains the attention of interlocutors and propels the narrative forward. Small leaves are sometimes used to represent story characters and the narrator choreographs these embodied objects on the sand space in front of them, which some have likened to a miniature stage set (Figure 1.8.2).

In sand stories a small repertoire of linear, curvilinear, circular and spiral graphic forms represent people, plants, artefacts, domestic items and other aspects of local environments. One of the most common of these is the 'U' shape which represents 'person', and this graphic form bears an iconic resemblance to the imprint on the sand made by a seated person. The ways that these simple elements are combined to generate new meanings suggests that the graphic forms of sand drawing have a rudimentary syntax. As can be seen in Figure 1.8.2, a semi-conventionalized oval shape drawn on the ground and representing a wooden dish can co-occur in the context of an arrangement of leaves that represent people. Their age and gender are conveyed by the relative size of the leaves and by the faint traces inscribed on them, derived from the painted designs that women wear when they perform ceremonies. Other lines drawn in the sand are traces of movement, re-enacting ancestral pathways or representing everyday journeys (Ingold 2007; Munn 1973; Green 2014). Varying the speed and rhythm with which a line is drawn or a leaf is moved in these narratives evokes particular types of action. For example, the graphic

*Figure 1.8.1*   Using the story-wire (Photo: J. Green)

*Figure 1.8.2*   Small objects such as leaves are used to represent narrative characters
(Photo: J. Green)

traces of sporadic or staccato movements are associated with dancing, while unbroken lines may be visual representations of journeys between locations that have been previously inscribed on the drawing space. These aspects of the semantic complex may be observed in real-time, or deduced from the traces that actions leave on the sand.

Another feature of lines drawn in sand stories is that they tend to be accurately configured in space, in what Wilkins has referred to as a 'geo-centred absolute frame of reference' (Wilkins 1997: 143). They convey correct spatial information, just as deictic or pointing gestures tend to do in Indigenous communities in Central Australia and in many other places in the world. So a line, particularly one representing motion, drawn in an east to west trajectory will be generally

*Figure 1.8.3*  An action results in a mark on the ground and ends with a deictic gesture in the air (video still, Eileen Pwerrerl Campbell, Ti Tree 2007) (Green 2014: 164)

interpreted as indicating motion on that cardinal axis, rather than in an arbitrary unspecified direction.

At the micro-analytic level, disassembling sand stories into a series of semantically coherent action units or 'moves' shows how units of action in sand stories are not amenable to bounded categories. The 'burden of information' (Levinson and Holler 2014: 1) may shift from one means of expression to another, and be distributed across different media. Let us follow a simple action that illustrates this point. The narrator's hand moves seamlessly across the soft surface of the ground, leaving a visible trace before moving into the air in a unitary action that crosses media – the earth and the air (Figure 1.8.3). This action disrupts pre-conceived notions of what the boundaries between gesture and drawing might be and demonstrates the semiotic possibilities of communicative actions that have graphic consequences. In the example shown in Figure 1.8.3, spatial information is distributed between the graphic traces on the ground and the pointing gesture in the air. This composite action that crosses media coheres as a semantic whole.

The types of vocal performance found in sand drawing are similarly complex, and without an interdisciplinary and collaborative approach that draws on insights from linguistics and musicology some of the richness of these vocal repertoires might be lost or overlooked. In addition to conventionalized or arbitrary aspects of spoken language that are amenable to traditional linguistic analyses, there are idiosyncratic vocal effects and poetic devices that add rhetorical flavour and texture to performances. Some sand stories include song, and repeated song texts may punctuate longer texts that are more speech-like. Some vocal phenomena fall on a continuum between sounds that are characterized as more like 'speech' and those that sound more like 'song'. Moving between vocal performances that are either more song-like or more like ordinary speech has the pragmatic effect of signalling degrees of formality of a sand story. An interdisciplinary methodology helps to investigate the relationship between various aspects of speech, such as pitch, intonation and rhythm, and features of music or song – beats, melody and musical rhythms. Delineating the similarities and differences between sand story songs and other song repertoires from Central Australia leads to a more sophisticated understanding of the ethnopoetics of the verbal arts from that region, and, more generally, of the diversity of verbal art forms world-wide.

In the Indigenous sand drawing tradition from Central Australia the integration of diverse communicative resources is complex and aesthetically appealing. Understanding how sand stories work provides an insight into the narrative traditions of an ancient culture. As a case study in complexity, enriched by an interdisciplinary approach, it contributes to the theory and analysis of multimodality in human communication, and shows how communicative messages that draw on multiple modalities are integrated. Developing tools for the analysis of multimodal narrative performances such as these challenges disciplinary boundaries and leads the study of 'language' in a new direction. This approach highlights similarities and differences between differing theoretical perspectives and research domains. More broadly it contributes to understandings of the human language capacity, its relationship to other aspects of cognition, and the role that various types of human action play in communication.

## References

Enfield, N. J. (2009). *The Anatomy of Meaning: Speech, Gesture, and Composite Utterances*. Cambridge, UK; New York: Cambridge University Press.

Goodwin, C. and LeBaron, C. (2011). *Embodied Interaction: Language and Body in the Material World*. Cambridge: Cambridge University Press.

Green, J. (2014). *Drawn from the Ground: Sound, Sign and Inscription in Central Australian Sand Stories*. Cambridge: Cambridge University Press.

Ingold, T. (2007). *Lines: A Brief History*. Abingdon: Routledge.

Jewitt, C., Bezemer, J. and O'Halloran, K. (2016). *Introducing Multimodality*. Abingdon: Routledge.

Kendon, A. (2004). *Gesture: Visible Action as Utterance*. Cambridge: Cambridge University Press.

Levinson, S. C. and Holler, J. (2014). The origin of human multi-modal communication. *Philosophical Transactions of the Royal Society B: Biological Sciences*, 369(1651): 20130302–20130302. doi:10.1098/rstb.2013.0302.

Munn, N. D. (1973). *Walbiri Iconography: Graphic Representation and Cultural Symbolism in a Central Australian Society*. Ithaca and London: Cornell University Press.

Nevile, M., Haddington, P., Heinemann, T. and Rauniomaa, M. (2014). *Interacting with Objects: Language, Materiality, and Social Activity*. Amsterdam: John Benjamins Publishing Company.

Seyfeddinipur, M. and Gullberg, M. (2014). *From Gesture in Conversation to Visible Action as Utterance*. Amsterdam: Benjamins.

Vigliocco, G., Perniss, P. and Vinson, D. (2014). Language as a multimodal phenomenon: implications for language learning, processing and evolution. *Philosophical Transactions of the Royal Society B: Biological Sciences*, 369(1651): 20130292–20130292. doi:10.1098/rstb.2013.0292.

Wilkins, D. P. (1997). Alternative representations of space: Arrernte narratives in sand. *Proceedings of the CLS Opening Academic Year*, 97(98): 133–164.

# 9

# Suspending

*Catherine Ayres and David Bissell*

---

> John shifts uncomfortably in his seat, tapping his coffee cup, and tells Catherine he can't do overnight hikes anymore, because his wife is ill. Catherine skips over this revelation with the callous speed that comes with research interview performance anxiety. She hastily directs John back to the 'research topic' at hand – national parks in Australia.

This interview encounter took place in 2013 as part of Catherine's doctoral research. Throughout the intervening years (and likely well into the future) Catherine regrets her anxious impatience to get back to a more 'relevant' line of discussion; this still feels like a missed opportunity to respond more caringly and attentively to such a delicate moment of vulnerability and trust. Many of us engaged in qualitative research have surely had similar experiences of regret. These intensities, however, are silenced, or at least muffled, in research outputs that omit these moments in favour of juicy narrative quotes that serve as evidence in support of arguments or findings. And yet these are the sticky moments that, although rarely acknowledged, slice into our research practices and into our lives.

We introduce the concept of 'suspending' here to highlight how the multiple durations that comprise interviews are a significant dimension of the research encounter that is often overlooked across a range of disciplines in the social sciences, with analytical attention instead devoted to the symbolic and rhetorical dimensions of what was said. Different durations resonate at different times, sometimes immediately, and sometimes years after the initial encounter. Following Ingold's (1993) observations about the multiple co-existent temporalities of landscapes, we want to suggest how the interview 'landscape' is steeped in the pasts and possible futures of researcher and researched alike, a site in which trajectories converge and transform. We want to revisit the interview event between Catherine and John to draw out 'suspending' as a methodological intervention filled with theoretical, practical and ethical possibilities for thinking empirical encounters.

In the context of qualitative interviews, researchers might feel compelled to adhere to core methodological tenets, such as generating 'valid' and 'relevant' data, and ensuring participants are informed and comfortable (Pitts and Miller-Day 2007). These are undoubtedly important considerations, but these, and other research conventions, may also inadvertently give rise to regrets such as the interview encounter with John. We argue that 'suspending' some assumptions to do with the performance of research interviews enables new research practices, new ways of sensing the multiple durations of interview encounters, and new forms of knowledge around the ethical considerations to which we researchers attend.

A common refrain in qualitative research literature is around the necessity to steer or guide unstructured or semi-structured interviews along the lines of preconceived research problems[1] and this ability to guide the research interview is often seen as a core research skill, where staying on track is the researcher's responsibility (Sarantakos 2013). Such persistent focus on the authority and skill of the researcher, however, reduces the importance of the singular twists and turns that might happen during the encounter itself. When we are steadfast in our notions of the research topic, defined by predefined research questions or problems, the illness of John's wife seems tangential, a disruption to the proper task of researching national parks. But suspending some of these assumptions of relevance to research topics enables more sensitive consideration of these little escapes. *Something* happened within John that moved attention in a different direction. Attending to what precipitated this change of direction calls into question the infinitesimal, imperceptible, or 'molecular' processes 'through which attention takes place' (McCormack 2007: 365). What transitions had occurred within John for him to deviate from discussion of national parks to such an intimate revelation? And crucially, how might attending to these molecular modulations enable new understandings of complex formations of identities, values or politics?

Paying attention to such molecular transitions might require different modes of communication than researchers have traditionally utilized in research encounters. Indeed, one reason for rushing over John's mention of his wife was Catherine's discomfort with such sudden intimacy that disrupted what she envisioned as a 'normal' mode of engagement between a researcher and a participant. Paul Harrison has gestured towards the significance of such instances in testimony, which 'confound, resist or simply withdraw from such engagement' (2010: 162). While we might usually see disruptive or unexpected instances as failures on the part of the researcher – for example, a failure of the researcher to engage the participant fully in the research topic – Harrison invites us to consider how these instances are, in fact, important constitutive elements of testimony. Paying attention to how and when such instances take place might, for example, require suspending the desire to adhere to norms that determine 'appropriate' modes of conversation that allow for smooth or easy communication.

Brian Massumi's writing on the energetics of events helps us to think about the multilayeredness of what is actually going on as we talk with someone in an interview. On one hand, there are the symbolic dimensions of expression. So, in our case, this would be the *content* of the talk that happens in interviews – the wordy sentences that end up as our interview transcripts, and the sorts of conventional meanings that often end up being ascribed to them. On the other hand, he points out that we are also affected at a much more immediate bodily level. This is the strength or duration of the *effect* of the expression of those spoken words on the bodies present. This is the dimension that can often be overlooked when we are trawling through the words in our transcripts, perhaps long after the interview itself took place. Massumi refers to this as its *intensity*.

Intensity has a felt dimension. What is so significant about this is that while there is a relationship between content and intensity, it is absolutely not predictable. The content, or *what* is being talked about, might amplify or dampen the intensity, for instance. But this felt intensity is much more unruly. As Massumi points out, this is because '[i]ntensity would seem to be associated with nonlinear processes: resonance and feedback which momentarily *suspend* the linear progress of the narrative present from past to future' (2002: 26, emphasis added). For instance, a long-forgotten memory might involuntarily cut in at an unexpected moment, perhaps ushered in by the precise words being spoken, or the manner in which they are said, adding something new to this present moment. As the intensity changes, our expectations about what might come next are destabilized.

The key point here is that it is not that we as researchers should be making a conscious choice to suspend the linear shape of a narrative. Suspending, in this sense, is not something that we force on a situation. Suspending is about being sensitive to the way that intensities can catch us off guard, surprising us, and changing the course of events. It is the intensity of the event itself through the precise playing out of talk as content and expression that, as Massumi says, creates a 'state of suspense, potentially of disruption' (2002: 26) from where it might be difficult to imagine what could happen next. So, if there is a skill here to be developed, we suggest that it is about cultivating our responsiveness to the singular moments that bead all encounters. Analytically, what this might mean is that rather than focusing on just the content of interviews in the vain hope of stitching together a coherent narrative, acknowledging the interview's intensities reminds us that such coherence is really just a fragile semblance made up of countless little suspensions.

To return to Catherine's interview encounter with John in this light demonstrates how heavily researchers rely on expectations of how a research interview can, or should, be performed. But adhering to habitual conventions of speech could reduce our openness to move with the subtle singularities each interview participant might offer. In this case, John offered something special, a complex signal that his body had moved from the topic at hand, revealing his own unique connection with the topic of national parks in Australia. Although Catherine tried to steer the discussion back to the topic, in doing so she could not help but feel she missed an opportunity to move *with* John, to be guided by him and his unique contribution as a research participant. Perhaps in this case Catherine might have remained silent, allowing John the option of elaborating further, or she could have suspended her own discomfort and directly inquired into the reasons why John connected the research topic and his wife, or offered an equally intimate connection. Or perhaps Catherine could have balanced her images of interview practices with an ethos of trying to notice 'different kinds of things that might be happening, or things that might be happening differently' (Coleman and Ringrose 2013: 4).

Catherine's reluctance to move with John was also shaped by an uncertainty over whether, indeed, John's revelation about his wife's illness should be treated as data. Of course, as Davies and Davies have put it, there are 'multiple possible trajectories in the tales that we, and our research participants, tell in the process of "generating data"' (Davies and Davies 2007: 1140). This particular trajectory, over time, became the richest and most profound moment in Catherine's doctoral research. John's mention of his wife's illness was pivotal in an argument around the complexities of how people connect to national parks. But Catherine had considerable qualms about utilizing this revelation in her research; after all, her information sheet and consent form – developed as part of a human research ethics application – said nothing about John's wife. To develop this moment into a full and rich empirical illustration, Catherine had to suspend her imagination of what, precisely, constitutes data and how it can be put to work. This moment also points to broader ethical concerns over what it is to conceptualize a person – with all their vulnerabilities and peculiarities – as a source of data for research outputs and how this utilitarian attitude can affect both participants and researchers in different and unforeseeable ways.

The temporal arcs of such ethical considerations can be unpredictable, with research encounters following us into our futures and leaping to mind, unbidden, with surprising intensity. Gail Lewis, for example, has outlined how her intense (yet secret) hatred of one of her interview participants has endured for 15 years, though her thinking around this event has transformed (Lewis 2010). These 'slow creep' (Bissell 2014) intensities of research encounters reveal the need for thinking about the multiple durations of research – the speeds, slownesses, and transformations through time. Deploying 'suspending' as a conceptual and methodological lure points towards the importance of these enduring capacities of empirical research to affect us in myriad ways far beyond the immediate interview, observation or ethnography. Suspending judgements on how and when research encounters are important requires us to be more open to uncertainties of research and perhaps calls for more nuanced evaluations of research ethics that accommodate the possibilities of such uncertainty.

The interview moment discussed here was pivotal in thinking about the complexity of people's relations to national parks. John revealed but one example of how these relations are steeped in complex, infinite assemblages of memories, ideas and practices that cannot (and, perhaps, *should* not) be easily reduced to the kind of ring-fenced ideas that Catherine was initially trying to explore in her research, which framed her initial sense of what 'relevant' lines of conversation would be for her interview with John. In this sense 'suspending' is not only a useful orientation for how to engage and interact in a research interview, but also a mode through which the multiple temporalities of such assemblages can be allowed to bloom. Suspending, we suggest, is therefore an orientation for attending to the multiple temporalities of research processes. Collecting, interpreting, analysing and presenting research materials, for example, each present opportunities for experimenting with 'suspending' as a methodological sensibility.

Introducing 'suspending' as part of a researcher's toolkit might enable radically different practices, politics and ethics of research. But doing so also, in some ways, demands *more* of researchers. Although the imperative of research ethics is around minimizing harm by increasing the comfort of participants at all costs, this risk-averse strategy risks closing off political possibilities afforded by moments of *discomfort* (Ahmed 2006). Catherine's encounter with John illustrates the importance of discomfort in research interviews. It was precisely this moment, this uncomfortable moment, which opened out into a fruitful illustration of the complexity of research encounters. The concern of research ethics in this case, then, is not to shut down or minimize the chance of discomfort, but rather, calls on researchers to employ strategies such as those we have suggested here in their navigation of these moments. Ethical decisions then become decisions about how to treat these moments with the care and consideration they deserve. But to move towards this mode of navigation, we must suspend our assumptions around how we can perform research interviews, and attend more carefully to how those encounters might endure in multiple ways.

## Note

1 Although some 'grounded theory' research claims to elide this convention, researchers in this area presumably do have at least a general field of inquiry in mind.

## References

Ahmed, S. (2006). *Queer Phenomenology*. Durham, NC: Duke University Press.
Bissell, D. (2014). Encountering stressed bodies: slow creep transformations and tipping points of commuting mobilities. *Geoforum*, 51: 191–201.
Coleman, R. and Ringrose, J. (2013). Introduction: Deleuze and research methodologies. In R. Coleman and J. Ringrose (Eds.) *Deleuze and Research Methodologies* (pp. 1–22). Edinburgh: Edinburgh University Press.
Davies, B. and Davies, C. (2007). Having, and being had by, 'experience': or, 'experience' in the social sciences after the discursive/poststructuralist turn. *Qualitative Inquiry*, 13(8): 1139–1159.
Harrison, P. (2010). Testimony and the truth of the other. In P. Harrison and B. Anderson (Eds.) *Taking-Place: Non-representational Theories and Geography* (pp. 161–183). London: Ashgate.

Ingold, T. (1993). The temporality of the landscape. *World Archaeology*, 25(2): 152–174.

Lewis, G. (2010). Animating hatreds: research encounters, organisational secrets, emotional truths. In R. Ryan-Flood and R. Gill (Eds.) *Secrecy and Silence in the Research Process: Feminist Reflections* (pp. 211–227). London: Routledge.

Massumi, B. (2002). *Parables for the Virtual: Movement, Affect, Sensation*. Durham, NC: Duke University Press.

McCormack, D. (2007). Molecular affects in human geographies. *Environment and Planning A*, 39(2): 359–377.

Pitts, M. and Miller-Day, M. (2007). Upward turning points and positive rapport-development across time in researcher–participant relationships. *Qualitative Research*, 7(2): 177–201.

Sarantakos, S. (2013). *Social Research* (4th ed.). New York: Palgrave Macmillan.

# Section 2
# Capturing and composing

# 1

# Capturing and composing

## Doing the epistemic and the ontic together

*Emma Uprichard*

This section brings together 12 contributions that collectively illustrate the many ways methods are continuously capturing and composing the world. In effect, capturing and composing are a set of what Celia Lury describes in the introduction to this handbook as 'compound methods': combinations of practices that are always interrupt*ing* (in) the now (and then) of the (historical and forthcoming) present. How capturing and composing interrupt is, in part, through the way they co-occur and co-act simultaneously. As we shall see, the recursion between the captured – the seized, taken, or recorded – and the composed is an important part of how the contributions in this section compound and interrupt knowing and being in the world. More specifically, it will be shown that capturing and composing bring together the epistemic and the ontic.

First, it is suggested that different methods capture and compose the world in particular ways. The fact that all methods compose their object of study epistemically is not a new idea and has been a key assumption across a range of authors for some time now (Collins and Pinch 1993; Latour and Woolgar 1986/1979; Law 2004; Mulkay 1990). But, as will be shown, the advantage of thinking of capturing and composing in this way is that it offers a foothold on the challenges and complexities of interdisciplinary methods. Second, the essays in this section, each in its own way, demonstrate a notion of *doing* method. The emphasis on *doing*, and specifically capturing and composing, brings with it a whole host of implications as to how methods are doing the world. A key consequence of doing method is that time and temporality are inscribed into method; that is, methods *move* things into being and becoming different to what they are. This leads into the concluding issue raised in this essay, namely, the ontic construction of what methods and things are together being and becoming. After all, it is not only that method shapes the world; it is also that the world recursively shapes method. By this I mean that methods are compounding the way the world is, was and can become as methods continually capture and compose.

## Methods of capturing and composing

It is fair to say, at least as a starting point, that in most other methods textbooks, capturing and composing might be thought of as collecting and analysing. And in many ways, collecting and analysing are what is meant by capturing and composing, except that capturing and composing are also interrupting and disrupting that which they capture and compose.

Capturing (from the Latin verb *caperer*) has many layers of meanings, among which are: to cause (data) to be stored in a computer; and from physics, capturing also refers to sub-atomic particle absorption. As these layers of meaning suggest, that which is captured is not necessarily freely giving itself – it is not data as given, but rather data as taken with or without 'will'. Capturing can be seen as 'seizing', forcing the storage or absorption of a thing. Alongside capturing the world, the entries in this section also implicitly describe the ways that methods are necessarily composing the world. Interestingly, the verb *to compose* is itself composed of many histories with multiple origins: from Middle English for 'put together' or 'construct'; from Old French for *composer* (to consist of, or to be drawn), and from Latin *componere* (forming the noun *component*) and *compositus* (to place). Note that composing can also involve *de*composing – breaking down components, rotting and putrefying that which is captured.

## Compounding capturing and composing

When multiple practices come together as compound methods (instead of multiple methods or mixed methods), we end up with multiple ensembles or assemblages of method – a mesh of methods – that are capturing and composing the world as it is continuously being and becoming. Capturing and composing take us beyond the singular or multiple; they have the capacity of holding, folding and twisting – that is, compounding – partials and wholes simultaneously. Interdisciplinary research, in particular, often demands compound methods capable of capturing and composing the world across contexts, in many ways and simultaneously.

We see an example of a compound method doing many things simultaneously in Leila Dawney's contribution Figurationing, where she notes that 'Figures are both material and semiotic: they convey meaning and have material substance and world-making effects'. Likewise, in her contribution on Abducting, Ana Teixeira de Melo draws on Pierce to propose the notion of abducting as a 'meta-methodological' practice. She writes: 'Abducting is considered as a meta-practice, a meta-method composed of different dimensions.' Among the things that go into the compound practice of abducting, de Melo argues, is the notion of surprise. Compound methods, like all methods, are both made in and through the world, and as such they are used to reflect – or rather as Leila Dawney argues – to 'diffract' the world. Diffracting is a way of representing the multiplicity of the world and acknowledging the ways in which every methodological 'cut' illuminates tangled in/visibilities. Similarly, Charles C. Ragin brings to light the way that combinations of quantitative and qualitative methods may be involved in 'casing the case'. He draws attention to the different and similar ways that qualitative and quantitative methods use and produce cases. For example, he argues, qualitative methods in the social sciences are mainly occupied with the 'patterns of difference' between cases whereas quantitative methods produce 'generalities' across cases. He suggests that in the humanities, casing is done differently again, since there the focus is typically on bringing out the specificities of one particular case.

## Doing methods

Put into practice as doings, methods enact and perform the world as they act and are acted on. This point is made most explicitly in Holger Pötzsch's contribution on iBorder/ing – the slash here being an important marker of the 'twin concept' he develops, which 'co-constitutes contingent subjectivities and practices'. He explains:

iBorder/ing in its composite form, thus, becomes conceivable as a fundamental cultural technique of in/exclusion that not so much identifies and processes already established practices and subjectivities, but becomes co-constitutive of them in and through technologically predisposed and procedurally framed everyday border work. In turn, an apparently clear-cut distinction between border technologies and institutions on the one hand and border subjects and practices on the other is problematized and replaced by a processual understanding of the border as a constantly emerging assemblage combining all the above phenomena.

Made visible as a doing or happening, iBordering activates the epistemologically crucial practice of inclusion/exclusion to make it available as a methodological resource for a method assemblage.

By explicitly approaching methods as doings, we can put ourselves into the method as (disciplinary and disciplined) subjects. We do methods and methods also do us. As has long been argued, classifying classifies the classified, and vice versa. We know the impact of methods being done only too well when working in interdisciplinary teams, where methods can quickly become the vehicles through which we dialogue, unite, oppress, negotiate, silence and bond with our colleagues. When methods are understood as doings, we can also account for the way that methods are acting as they are conducted, used and done. As Thomas and Hunt sum up:

> Just look at the vocabulary of methodologies: requirements, design, quality, communication, tests, deliverables – all good solid nouns, and not a verb in sight. Yet increasingly, [we] are coming to believe that these things, these nouns, aren't really that useful. Instead, we see that the real value lies in the processes that lead to the artifact's creation; the verbs are more valuable than the nouns.
>
> *Thomas and Hunt 2003: 82*

And once we understand methods as doings, we can also see how they act on themselves and each other recursively. For example, Google is an entity, a search engine, which continually reconstitutes itself as a method assemblage – in relation to a user or re-searcher – in the activity of Googling. In this way, as Alberto Corsín Jiménez suggests in his contribution, interdisciplinary methods are always producing knowledge that is 'forever in beta'. Indeed, methods themselves are also 'forever beta', since they too keep evolving, whether it be through the ways they are modified over time to measure or categorize differently or in relation to disciplinary vogues that come and go.

Consider, also, the method of visualizing as both capturing and composing the object being visualized. The object of a visualization is individuated in a recursive process: data that is visualized can show the object to be changing, that is to be growing or increasing and so on. Similarly, Carolin Gerlitz notes the capacity for 'retrieving' digital data by default generates other kinds of possibilities. She writes:

> The grammatization of a platform suggests that the data units it generates are comparable if not similar entities, while at the same time creating the conditions for third parties and users to fold heterogeneous interpretations of these grammars into the platform.

In Greg McInerny's essay, he makes the point that 'anything can be visualized' and he uses the notion of 'design space' to imagine 'the hyper-volume that contains all visualizations whether they have been made or not, and whether they are of use, or no use at all' Similarly, in Notating, Moritz Wedell says:

> Acts of notating do not only refer to static objects. Notations can efficiently represent movements: movements of mathematical operations, movements of bodies, of sound, of thought, etc. But acts of notating do not merely 'put' these movements 'into' a notational composition, or a choreography of things in order to make them re-accessible. Acts of notating detach these movements from the restrictions of the real world, and they allow us to proceed and explore movements in a virtual realm, a sphere only restricted by the rules of the notation. It is this freedom to carry out movements in an abstract realm of notational practice that is an important precondition for composition and invention both in the arts and in science.

In relation to these examples, we might think of how capturing and composing are made possible through and in relations with 'design spaces', where such spaces are dynamic, hyper-volumes, in which what is captured and composed is being and becoming in the doing of compound methods.

In summary, the focus on doing methods draws attention to the ways in which methods capture and compose epistemic things relative to *where, when, how, or for how long* the methods are carried out. Doing methods not only morphs the world epistemically (what methods make knowable), but also morphs the world ontically (what methods make to be and become). Methods act, and they act *intensively*; they act on and inflect the world in myriad ways that might or might not matter.

## Tensing method

As a doing, method itself can be and become. Doing method means activating time and temporality as well as space. Interviewing can capture experiences in the past as well as 'narratives of the future' (Uprichard 2011). Method-related words are, in fact, already riddled with the temporal. Oral history, forecasting methods, clustering or sequencing methods all indicate the bounded region of time that they will capture (the past, the present, the future), but they also make temporalities, that is, the relation between the past and the present, the present and the future. Different timings are etched into methodological activities such as recording, reporting, transcribing, modelling, simulating, forecasting, narrating, processing, reporting, interpreting, projecting, anticipating, explaining, predicting, often in the closely specified rules of specific disciplines, which assert their primacy to make (methodological) time.

As Luciana Duranti's chapter suggests, archiving as a form of 'documenting society' captures and composes society in a way that is, she notes, 'highly regulated by a discipline'. But, doing methods invites multiple timings, invites adverbs and negation (Fonteyn, De Smet and Heyvaert 2015). So, for example, we might imagine a doing method capable of capturing and composing one aspect of the world by slowing it down, while it captures and composes in other areas by speeding it up *at the same time*. Computational techniques, for example, have sped up the activities of capturing, retrieving, sorting, processing and visualizing data, sometimes allowing, sometimes requiring that they are done in parallel or in sequences of various kinds, allowing for different kinds of pausing, sometimes leading to various kinds of stoppage.

Barbara Adam's contribution on Timing is infused with tense and time. Her contribution is unique insofar as she captures and composes what she calls 'compressed' or 'distilled' theory in 'poetic shape' (see Adam 2004). 'Timing is,' she suggests, 'Doing time'. Emphasizing the doing of method 'challenges method and methodology' and 'troubles ontology-epistemology distinctions / Recognizing our complicity in generating data changes the nature of a fact':

> Time open, unbounded: imposing boundaries is our doing
> Holding time still to contemplate the social de-temporalizes
> Ossifies becoming; closes down processes; negates potentials
> All of past and future encoded in momentary slices of social life
> Temporal wholes expressed in moment-parts: present embedded
> Reworking temporal complexity fundamentally reconfigures theory
> Emphasizing process over product challenges method & methodology
> Embracing temporal extension reforms disciplines, categories & norms

As Adam puts it: 'Temporal wholes expressed in moment-parts' are always capturing and composing change and continuity. '[Q]uestions and answers become intermeshed / Once quests for interdisciplinary inquiry are seriously engaging process timing'. Time tenses method and this is turn moves the thing that is the object of the method.

Tensing calls into be*ing* when and where method is being and becoming. In much the same way as muscles are tensed to (not) move a part of the body, it is in the tensing of methods that parts of the world are (not) moved too. Tensing reveals the space of movement where and when methods become – or as Thrift (2004) might call it, tensing reveals the 'movement-space' that methods do, a movement-space which is 'relative rather than absolute'.

When method is tensed, then we can also imagine its object to be moved into a new space of being, making the object become something other than what it was, poised for action. For example, when I conduct an interview, the person I am speaking with is both subject and object, depending on the turn-taking of the conversation. Furthermore, the person being interviewed is always being and becoming (Uprichard 2008) as a person in the moment, at the time, before, during and after the present time of the interview itself. Both interviewer and interviewee are in action, co-constructing the emerging dialogue, which is being captured and composed dynamically through interaction. Similarly, when Google 'searches', the object of the search does not simply sit and wait for the algorithm to 'find' it. Instead, much of the work of Googling is to continuously capture and compose parts of the Internet data in a way that tenses it up, mov*ing* it into spaces where it can be matched, weighted, filtered, indexed, compared and ranked before appearing in a list. The data are kneaded like the baker's dough that folds whole and partial folds that are repeatedly refolding and unfolding the (un)foldings – iteratively capturing and composing data such that the end user has an object (a list of results) that might or might not be to his or her liking. Tensing the doing of method brings th-ings to 'life' by morphing them into something other than what or where or how they once were or might become.

## Ontic capturing and composing

Capturing and composing are epistemic constructs: they organize *knowledge in and about* the world; they are also ontic constructs and organize aspects *of* the world. After all, where individuals reflexively re-act to both actions and descriptions, there is a 'looping effect' – to use Hacking's (1995) term – as the world is re-made through the th-ings (concepts, actions, classifications) that have already gone into making it. The world keeps changing (and not changing) and so too do our modes of being, becoming and knowing the world. Interdisciplinary research is no different in this respect; it requires us to consider new possibilities for action under new kinds of description.

Once we interrogate how methods always compound the world in particular ways, then the question becomes less about what is and is not captured and composed (although this is certainly an important component in understanding how the world is compounded). Rather, the question becomes more about how we might (re)construct methods which expose the way methods compound. And vice versa, how does the world compound methods such that patterns of change and continuity are maintained?

Crucially, what we do when we approach methods in this way is what Alberto Corsín Jiménez refers to in his contribution as 'white-boxing': we expose sites of methodological praxis and 'disclose and re-source their capacities for agency and relation'. He explains:

> white-boxing allows for novel forms and new locations of social durability to emerge – new expressions of cultural, political and aesthetic materiality and critique. White-boxing finally comes full circle in helping us unpack and interrogate the very sources and resources through which we . . . take presence in the world. It invites us to reconsider the infrastructures, spaces and times that give shape to our research methodologies.

In other words, approaching method as compound methods not only invites the debates around 'politics of method' (May 2005) to re-turn to centre-stage, but also interrogates the ways in which methods in general compound disciplinary and interdisciplinary sites capture and de/re-compose each other.

Once we start thinking of ways in which methods are continually compounding particular epistemological and ontological spaces in the world, then interdisciplinarity and interdisciplinary methods take on new kinds of emancipatory possibilities. Indeed, how methods compound, de- and re-compose epistemic and ontic inter/disciplinary presents and futures arguably becomes a necessary (and potentially sufficient) part of any chance of disrupting and displacing the present (and future). As Hacking (2002: 100) puts it: 'Social change creates new categories of people, but the counting is no mere report of developments. It elaborately, often philanthropically, creates new ways for people to be.' Similarly, method creates new kinds of knowledge and interdisciplinary contexts often create new ways for knowledge to be and become also. Methods and knowledge make each other. They both conspire to create the world in the 'forever beta'. It is up to methodologists and interdisciplinary researchers alike to come together to 'white box' the ways in which both method and interdisciplinarity re-produce the capturings and composings that ultimately defy our individual and collective agency.

## Acknowledgements

I am grateful to Celia Lury for her comments and suggestions on earlier drafts, which have improved this discussion.

## References

Adam, B. (2004). *Time: Key Concepts*. Oxford: Polity Press.
Collins, H. and Pinch, T. (1993). *The Golem: What Everyone Should Know About Science*. Cambridge: Cambridge University Press.
Fonteyn, L., De Smet, H. and Heyvaert, L. (2015). What it means to verbalize: the changing discourse functions of the English gerund. *Journal of English Linguistics*, 43(1): 36–60.
Hacking, I. (1995). The looping effects of human kinds. In D. Sperber, D. Premack and A. J. Premack (Eds.) *Causal Cognition: A Multidisciplinary Debate* (pp. 351–383). Oxford: Oxford University Press.
Hacking, I. (2002). *Historical Ontology*. Cambridge, MA: Harvard University Press.

Latour, B. and Woolgar, S. (1986/1979). *Laboratory Life: The Construction of Scientific Facts* (2nd ed.). Princeton, NJ: Princeton University Press.

Law, J. (2004). *After Method: Mess in Social Science Research*. London: Routledge.

May, C. (2005). Methodological pluralism, British sociology and the evidence-based state: a reply to Payne. *Sociology*, 39 (3): 519–528.

Mulkay, M. (1990). *Sociology of Science*. Bloomington: Indiana University Press.

Thomas, D. and Hunt, A. (2003). Verbing the noun. *IEEE Software*, 20(4): 82–83.

Thrift, N. (2004). Movement-space: the changing domain of thinking resulting from the development of new kinds of spatial awareness. *Economy and Society*, 33(4): 582–604.

Uprichard, E. (2008). Children as 'being and becomings': children, childhood and temporality. *Children & Society*, 22(4): 303–313.

Uprichard, E. (2011). Narratives of the future. In M. Williams and P. Vogt (Eds.) *The SAGE Handbook of Innovation in Social Research Methods* (pp. 103–119). London: Sage.

# 2

# Abducting

*Ana Teixeira de Melo*

## Introduction

The world as we experience it, and particularly the social world, is constructed and portrayed in increasingly complex ways. The terrains of science are changing, in a close relationship with changes in the social world and dominant worldviews. The social world is assumed as complex, dynamical, governed by non-linear reciprocal causal processes and as showing unexpected and emergent properties. In parallel, there are richer and wider sources and varieties of information and more fluid definitions of what constitutes relevant data. New approaches to data exploration emerge and the boundaries between disciplines are progressively more fluid and flexible. The increasing dialogues between disciplines build patterns that bridge them. All these transformations invite scientists in general and social scientists in particular, to revise the internal organization and the predominant modes of operation of 'this thing we call science' (Chalmers 1982).

An increasingly complex world can only be understood and rendered manageable by an increasingly complex science. The complex is about particular organizations of relationships sustaining emergent phenomena. The question to be posed is: how can science organize its internal (and external) relationships so it can promote its own development through meaningful discoveries, propelled by doubt, uncertainty, novelty and surprise, as core ingredients of great scientific leaps?

The recursive nature of complex systems must be embraced by a complex science, capable of organizing its internal relationships to maximize its generative and creative potential. This includes understanding and exploring the nature of the interactions between different scientific disciplines and specific fields of inquiry, as well as the relationships between the methods and concepts within and between disciplines. A complex world is a dynamic, synergetic, creative world, supported by the non-linear nature of the interactions of its elements or components. So, should a complex science organize itself so that from its elements rich synergies can be created that surpass the capacities of the individual elements? The nature of the practices of relating concepts and methods within and between disciplines is constrained by modes of thinking, the researcher's preferred stances towards science itself as well as the practices and methods of enacting research and communicating its outcomes.

In the following sections, we explore abducting, considering it as meta-methodological practice and core integrated competency for twenty-first-century researchers, aiming to promote true scientific advance. We assume a multifaceted view of abducting as a foundational meta-practice for the enactment of an integrative science, which includes:

a   an ampliative and generative form of reasoning, as explored by Charles Sanders Peirce, associated with the formation of novel hypotheses;

b   a basic inquisitive stance or mind-set, facilitative of the abductive leap and characterized by openness, curiosity, exploration, humility and creativity;

c   a set of strategies and ways of relating between sciences that fosters rich and creative interdisciplinary interactions, promotes the development and expansion of each field and facilitates synergies, under the auspice of an integrative overarching complex science.

We will explore ways abducting can be used as an interdisciplinary-focused form of reasoning, when the researcher cultivates an inquisitive open mind-set and engages in rich interdisciplinary practices of relating that promote creativity and an expansion of each field through synergetic interactions.

## Abducting

Abduction was explored by Charles Peirce as a type of ampliative inference, where novel hypotheses are elaborated to explain surprising or unexpected facts. Peirce presented basic abductive inferences and other more extreme varieties of abduction in following the forms (CP 6.522–8, in Buchler 2014):

> A surprising fact C is observed,
> But if A were true, C would be a matter of course
> Hence, there is reason to suspect that A is true.
>
> *CP 5.189*

> A well-recognized kind of object M1, has for its ordinary predicates P1, P2, P3 etc., indistinctly recognized.
> The suggesting object, S, has these same predicates P1, P2, P3
> Hence, S of the kind M.
>
> *CP 5.542, 544–5 and James 1903, in Buchler 2014*

> M has, for example, the numerous marks P', P'', P''', etc.,
> S has the proportion r of the marks P', P'', P''', etc.
> Hence, probably, and approximately, S has an r-likeness to M.
>
> *CP, 2.694–7, in Buchler 1955*

In these forms of abduction, Peirce explores the nature of a given phenomenon by appealing to a novel principle that explains the differences and similarities between objects of attention.

Rozeboom (1997) supported Peirce's assertions regarding the abductive nature of true science. He talked about explanatory or ontological induction. In these forms of reasoning, in ways similar to abduction, inferences are made 'from observed patterns of data to the existence of theoretical entities . . . that could explain *why* the data are patterned' (Aogáin 2013). Peirce kept abductive reasoning close to data, but also attributed an important role to surprise (CP 7.202, in Fann 1970; Nubiola 2005). We believe abduction can be used more strategically and purposefully, attending also to the insufficiently explained, aiming to refine existing theories. Peirce suggested abduction as much as a process of attaining a hypothesis as the process of evaluating it (CP 5.171, 8.398, in Fann 1970). Abduction initiates a research cycle setting the stage for a sequence of inquiry where deduction and induction also play important roles (CP 15–59, in Fann 1970; CP 6.522–8, in Buchler 2014).

That said, abduction should be used strategically, as a goal-directed activity underlying the logic of discovery (Paavola 2004). As a set of strategies, and the practice of an attitude, abducting may be about activating or enhancing a creative potential (Gonzalez 2005), the 'flash of new suggestion' (CP 5.181, in Nubiola 2005), through available methods of data creation, collection, sense making and interpretation (Reichertz 2004; Timmermans and Tavory 2012). The nature of the relations between disciplines and fields can set more or less adequate stages for interdisciplinary abducting and creative enterprises.

## Interdisciplinary abducting

Abducting is considered, in this essay, as a meta-practice, a meta-method composed of different dimensions. It is essentially about the enactment of a set of practices of relating that we consider essential for the practice of a complex science such as:

a    practices that relate ideas (concepts and ideas about methods) in innovative ways;
b    practices of relating oneself with one's own discipline, science in general and the wider complex world with an open, inquisitive, curious and creative mind;
c    practices of relating to others to promote opportunities for the vivid and rich interaction of ideas, as well as practices pertaining to those ideas that create opportunities for abductive reasoning.

There is an underexplored potential for innovation and discovery lying between disciplines, that relies on the nature of their interactions (mediated, of course, by the interaction between the individual person of the researcher and collectivities of researchers), namely between their concepts and methods.

Interdisciplinary relationships may become strategies of abduction when:

d    questioning each other's assumptions, forcing clarification and revision of both central and peripheral concepts and methods;
e    amplifying each other's vulnerabilities and potentialities;
f    highlighting and pinpointing 'grey' or obscured areas to be explored or clarified;
g    illuminating the commonalities and differences and, thereby, suggesting underlying principles, mechanisms and processes to be collaboratively explored;
h    changing each other's landscape by creating different contexts, and offering such new perspectives or framings that the old concepts and the methods gain new meanings, are pushed to their limits, and experimented with variations and potentially transformations, including the emergence of new themes or lines of inquiry;
i    strategically incorporating aspects of other disciplines/fields that expand the possibilities of acting and reasoning, decrease the fragilities of a given network of concepts, or increase the congruence between them, the methods, and the research purposes.

Some forms of reasoning become strategies guiding a given type of relationship while eliciting and supporting an emergent hypothesis. Below, we present proposals for supporting interdisciplinary abductive reasoning, where A and B are different disciplines or fields of research.

A and B are similar but different.
If P were true,
Then both their differences and similarities would be a matter of fact,
Since they would share $Pc_{(common\ principle)}$,
while differing in $P_{A(specific\ principle\ for\ A)}$ and $P_{B\ (specific\ principle\ for\ B)}$

The well-recognized phenomenon A shows the complex features/underlying processes $C_1, C_2, C_3, C_i$.
The phenomenon B shows these features/processes appearing in similar form $C_1, C_2, C_3, C_i$, with the differences in the form of $C'_1, C'_2, C'_3, C'_i$,
B seems of the kind of A in regard to X,
And of a different kind of A in regard to Y.
But if $W_{(underlying\ processes/patterns\ that\ connect)}$ were true
The differences and similarities would be a matter of course

A and B are similar and different
$A_{concepts}\ (A_c)$ describe/explain A, and $A_{methods}\ (A_m)$ have been used
$B_{concepts}\ (B_c)$ describe/explain B, and $B_{methods}\ (B_m)$ have been used.
Given the differences between A and B,
For $B_{concepts}$ AND/OR $B_{methods}$ to provide meaningful insights regarding A,
Would not be surprising if
$B_{concepts} \rightarrow_{(map/translate\ into)} A_{concepts}$ AND/OR $B_{methods} \rightarrow_{(map/translate\ in\ X\ way\ into)} A_{methods}$,
OR $B_{concepts} \rightarrow_{(translate\ into/inform)} A_{methods}$ AND/OR $B_{methods\ (translate\ into/inform)} \rightarrow A_{concepts}$,
On the account of $X_{(underlying\ processes\ or\ patterns\ that\ connect)}$

Different disciplines might cooperate on constructing and cultivating these and other new ways of interdisciplinary relational forms of abducting.

## Conclusion

An integrative interdisciplinary relational practice of abducting might assist social sciences in increasing their own complexity and, with that, the possibilities of making more meaningful and pragmatically useful contributions for the construction, understanding and management of a world, where possibilities for positive action abound.

## References

Aogáin, E. M. (2013). An introduction to the work of William W. Rozeboom. In E. M. Aogáin *Scientific Inference: The Myth and the Reality. Selected Papers of William W. Rozeboom.* Dublin: Original Writing.
Buchler, J. (Ed.) (2014). *The Philosophical Writings of Peirce* [originally published 1940]. New York: Dover Publications, Inc.
Chalmers, A. F. (1982). *What Is This Thing Called Science?* (2nd ed.). Buckingham: Open University Press.
Fann, K. T. (1970). *Peirce's Theory of Abduction.* The Hague: Martinus Nijhoff.
Gonzalez, M. E. Q. (2005). Creativity: surprise and abductive reasoning. *Semiotica*, 153(1/4): 325–341.
Nubiola, J. (2005). Abduction or the logic of surprise. *Semiotica*, 153(1/4): 117–130.

Paavola, S. (2004). Abduction as a logic of discovery: the importance of strategies. *Foundations of Science*, 9(3): 267–283.

Reichertz, J. (2004). Abduction, deduction and induction in qualitative research. In U. Flick *et al.* (Eds.) *A Companion to Qualitative Research* (pp. 159–165). London: Sage.

Rozeboom, W. T. (1997). Good science is abductive, not hypothetico-deductive. First published in L. Harlow, S. A. Mulaik and J. H. Steiger (Eds.) *What If There Were No Significance Tests?* (pp. 366–391). New Jersey: Erlbaum.

Timmermans, S. and Tavory, I. (2012). Theory construction in qualitative research: from grounded theory to abductive analysis. *Sociological Theory*, 30: 167–186.

# 3

# Archiving

*Luciana Duranti*

## Introduction

The verb 'to archive', contrary to what many believe, does not find its origin in archival science but in data management, in the context of which it means moving data from a high-cost primary storage to a low-cost high-capacity storage for long-term retention. However, starting in the 1990s, when electronic records began to replace paper records on a large scale, the verb 'to archive' used by computer scientists to refer to saving the records to a system began to become popular in the world of archives to refer to two distinct ideas: the idea of transfer of organizational records to an archival repository for long-term preservation, and the idea of creating accumulations of documentary materials related to people's lives or social events. Both ideas are part of much broader activities, respectively, permanent preservation for cultural or evidentiary purposes of the records produced by public and private bodies in the course of their activities, and documentation of society. Both ideas involve research, though not necessarily the same research methods.

## Archiving for permanent preservation

Permanent preservation of the written record of society is the realm of archival science, a field of knowledge that originated as an autonomous discipline in Europe in the sixteenth century but whose practices are rooted in the Sumerian culture and whose body of concepts goes back to Roman law. Though the word 'archives' derives from the Greek ἀρχεῖον (*arkheion*), meaning public records, or office of chief magistrates, ἀρχή (*arkhē*) (from the verb ἄρχω (*arkhō*), to rule, to govern), it is referred to in the Justinian Code as *locus publicus in quo instrumenta deponuntur* (the public place where deeds are deposited), *quatenus incorrupta maneant* (so that they remain uncorrupted), *fidem faciant* (provide trustworthy evidence), and *perpetua rei memoria sit* (be continuing memory of that to which they attest). In the ancient world, an archives (note that in UK English the term is singular, while in North American English it is plural) was a place of preservation under the jurisdiction of a public authority. This public place endowed the documents that passed its threshold with trustworthiness, thereby giving them the capacity of serving as evidence and continuing memory of facts and acts. Over the centuries, archives have acquired additional functions, among which is that of preserving the records of society at large, including those of individuals and private organizations; and they have taken different shape and forms, as national institutions have been joined by archives of lower administrations (for example, state, region, city, province), and business archives (for example, banks, industries, legal firms, insurances) have become as rich and complex as those of universities, churches and other traditional institutions. As a consequence, the body of knowledge that governed the archival field grew as well and, from a primarily legal and historical discipline, became a science.

The archival *discipline*, Trevor Livelton writes, 'denotes a form of study with a distinct methodology used to gain knowledge. A discipline encompasses both a way of gaining knowledge – rules of procedure that discipline the scholar's search – and the resulting knowledge itself' (Livelton 1996: 44). Archival *science* can be regarded as a system inclusive of theory, methodology, practice and scholarship, which owes its integrity to its logical cohesion and to the existence of a clear purpose that rules it from the outside, determining the boundaries in which the system is designed to operate. Archival theory – the whole of the ideas about what archives are, and archival methodology – the whole of the ideas about how to treat them, govern the entire system of archival science. Thus, archival methodology controls the practice and scholarship of permanent preservation on the basis of the fundamental concepts of archival theory, which include: (a) archives hold in trust the archival documents of society and guarantee their continuing reliability as witnesses of action; (b) archival material is the natural, authentic, impartial, interrelated and unique by-product of human activities; (c) antiquity provides archival documents with the highest authority; and (d) unbroken custody ensures documents' authenticity. As a consequence, archival methodology has been based on the ideas of respect for context, provenance, relationships, order, structure and form.

Archival methodology encompasses several functions, all of which aim to achieve the permanent preservation of an authentic record of society. They are commonly identified as Appraisal (for Selection and Acquisition), Arrangement and Description, Retention and Preservation (this function specifically deals with the physical and technological stabilization of the material and protection of its intellectual content), Management and Administration, and Reference and Access. With the advent of the digital era, permanent preservation has acquired an additional key function, that of Control on Records Creation and Maintenance. 'Preservation starts at creation' is the motto of modern day digital archivists. The fulfilment of each and every one of these functions requires research, which is carried out using *the archival method*. In this context, it can be stated that 'archiving' is the use of the archival method to ensure the permanent preservation of the world documentary heritage.

The archival method is a sort of meta-methodology governing all methods used to carry out the archival functions mentioned above. It was developed by borrowing from different fields, thus, it is from its origin interdisciplinary: the nineteenth-century debate between historians and archivists led to the affirmation of the historical method of analysis for arrangement and description; the twentieth-century strong relationship between archival and library science affected the conduct of research on retrieval and access; the introduction of mechanization in public administration changed the method for the development of documentation processes and workflows; the twenty-first-century alliance with information science added understanding of a new technological context to the research on digital records preservation; social sciences methods have guided the questioning of user behaviours, creators' processes, and archival policies; and post-modernism has spurred debates about archival identity and purposes of appraisal.

Regardless, at the core of 'archiving' is the belief held by archivists that the subject of their research is not determined by personal interest, but by the nature of the material for which they are responsible, the archival institution's or programme's acquisition and description priorities, and the needs of the many services archivists carry out to support the work of both records creators and records users. For this reason, archivists, in fulfilling their professional functions, cast their research questions in a juridical and administrative as well as an historical framework. They analyse 'the phenomena and structure of records and record aggregations, which are not examined for their content, but for the meaning of their characteristics, form, organization, and administrative, functional, procedural and documentary context, as archival theory dictates'. 'Research guided by the archival method is inferential in nature, because it occurs within the interpretive framework of archival theory, according to which proof, truth, and evidence are extra-textual' (Duranti and Michetti 2016: 88, 90).

Thus, although archival science is nourished by methodologies from other fields, 'in a process that continuously broadens and refines its core without altering it', such methodologies are used to foster useful transfers to the archival field 'in emerging areas of endeavor and investigation, to eliminate the duplication of theoretical efforts in different fields, and to promote consistency of scientific knowledge' (Duranti and Michetti 2016: 83). However, when archivists use other methodologies, they confront them with archival theory, methods, practice and scholarship,

> subject them to a feedback process, and insert them into the fundamental structure of the system. Only in this way will they be able to maintain the integrity and continuity of their discipline while at the same time fostering its enrichment and growth [through an inter-disciplinary process of borrowing from other disciplines and assimilating their concepts and methods], integrity and continuity that are vital to their ability to preserve all records, regardless of medium.
>
> *Duranti and Michetti 2016: 83*

And this archival belief in the need to maintain the integrity of the archival discipline (Duranti 1994) leads to a brief discussion of the other idea linked to the term 'archiving', that of documenting society.

## Archiving for documenting society

For about three decades now, archivists and researchers, primarily in North America and the UK, have been lamenting the absence from the holdings of archives of the personal documents of individuals and minorities, as well as documentation of events of historical impact from a non-governmental perspective. The term 'silence' has been used many times in relation to the voices of marginalized groups and peoples, and filling such silence with those voices has been one of the major endeavours of several archival institutions as well as of a substantial component of the archival profession, to the point that the traditional view of the archivist as an objective and impartial professional conducting research that supports the control, understanding, preservation and dissemination of the existing records rather than anyone's specific interest has begun to be regarded as negative.

This is not the place for a discussion of the role of archival institutions and archivists in society, as the focus of this essay is on the 'archiving method'. But, if archiving is documenting society, what is the method for documenting society? Certainly, there has to be a method for deciding what deserves to be documented and where, and then a variety of methods for documenting the chosen subject, depending on whether written documents about it exist, who holds them, etc. Clearly, each topic, in each context, and for each purpose will require a different methodology, or a combination of methodologies, which is impossible to envision a priori.

Many archival authors have written about the development of documentation strategies and plans without ever converging on one 'archiving' approach (see the many entries on these matters in Duranti and Franks (2015)), and archivists are increasingly faced with the challenge of 'de-colonizing' archives by freeing archiving from the imperfections (and perfections) of our established models and practices in relation to what we see in the real world. Digital technologies are both fuelling the need for change and providing the principal means of effecting change, but the new generation of archivists still maintains that decolonization is fundamentally about attitudes and ideas, both technical and social, as new technologies are themselves historically and culturally contingent. They clearly see the need for a new approach to archiving, one based on activism, plurality and social responsibility (see, for example, *Archivaria* 80).

## Conclusion

The term 'archiving' is today commonly used to refer to two different ideas: permanently preserving the records of our society and documenting society. Though both ideas relate to the formation of a documentary heritage and the protection of the sources for historical memory, they differ in that the former derives from a juridical and administrative obligation to accountability, transparency, and proof, while the latter results from a social need to understand and preserve cultural identity. As a consequence, archiving intended as permanent preservation is highly regulated by a discipline, archival science, developed through several millennia and universal in nature, regardless of technological changes, while archiving intended as documentation of society is highly contextualized, based on the needs of specific groups and peoples, as well as their traditions and beliefs, and variously enabled by technological developments. Most recently, both ideas have been challenged as entrenched in a 'colonial' mindset, and a third idea of archiving is surfacing, one that, regarding as creators and owners of archives those about whom the documents talk, gives them the right to decide what information will be documented, retained, preserved and made accessible, how, and where. No methodology has been developed yet to carry out this third idea of archiving, but it is very likely that, being culturally motivated, it will be highly interdisciplinary.

## References

*Archivaria* 80 (2015). http://archivaria.ca/index.php/archivaria/issue/view/463

Duranti, L. (1994). The concept of appraisal and archival theory. *American Archivist*, 57(Spring): 328–345.

Duranti, L. and Franks, P. (Eds.) (2015). *Encyclopedia of Archival Science*. Lanham, MD: Rowman & Littlefield Publishing Group. Charles Harmon, Executive Editor.

Duranti, L. and Michetti, G. (2016). The archival method: rediscovering a research tradition. In A. J. Gilliland, S. McKemmish and A. Lau (Eds.) *Research in the Archival Multiverse* (pp. 74–95). Melbourne: Monash Publishing.

Justinian (529–565). *Corpus Juris Civilis, Novella 15* 'De Defensoribus civitatum', 'Et a defensoribus', *Digestum48, no. 19* 'De Poenis', *Codex I, no. 4* 'De episcopali audientia'.

Livelton, T. (1996). *Archival Theory, Records, and the Public*. Lanham, MD and London: The Society of American Archivists and The Scarecrow Press, Inc.

# 4

# iBorder/ing

*Holger Pötzsch*

## Introduction

This chapter presents a terminological trajectory of border-related concepts. It also highlights the methodological implications of a move from the descriptions of allegedly static border technologies to an assessment of the mundane practices through which these technologies are activated, re/appropriated, or subverted. The slash both connecting and separating iBorder and its verb iBordering in the title of the chapter is indicative of this methodological and theoretical double-move, described in more detail below.

Starting with a critical interrogation of interconnections between states and borders, the essay moves on to introduce conceptual advances that aim at grasping the increasing dislocation of borders that spread from the fringes of nation states to ubiquity in everyday life. Connected to this often technologically driven development is a shift in attention from static border locations and institutions to contingent practices of bordering at the level of day-to-day performances. Subsequently, I introduce the twin concept of iBorder/iBordering to account for the impacts of new technologies of dataveillance, biometrics, algorithmic analytics and human–machine coordination on these processes, before, finally, I argue that iBorder/ing constitutes a fundamental cultural technique that not so much processes given subjectivities and practices, but rather co-constitutes them.

## From borders to bordering: dislocating state borders

Contemporary border studies perceive borders as dynamic, multidimensional entities that constantly change and shift, and that function across a variety of registers (Rumford 2012; Brambilla 2014; Pötzsch 2015). Borders both divide and enable contact and exchange, they are resources providing orientation in contradictory terrains and designate zones of exception that allow for extraordinary measures to be taken against non-normative subjects. In recognition of this ambivalence, border research broadens its outlook, moving beyond a confinement to state borders and sovereignty. As Rumford has claimed with reference to what he terms 'multiperspectival border studies' (2012: 888), 'the state does not exhaust the meaning of the border' (2012: 894).

Attention to micro-processes of bordering at the heart of the sovereign exception, however, does not render states insignificant for border research. As O'Dowd (2010) has noted, even in times of globalization and increasing cross-border flows and connections state borders deserve continued attention. Arguing against an overexpansion of the border concept, he claims that '[c]ontemporary border studies . . . risk seeing nation states and state borders simply as fixed ideological constructs or ideas, rather than as territorial projections of infrastructural power' (2010: 1044). According to him, the concrete, economic and coercive power of states still has significant impacts on the lives and wellbeing of subjects and merits critical scholarly attention.

Acknowledging O'Dowd's position, I argue for the necessity of adopting multidisciplinary perspectives not so much with the objective of reducing the significance of states as actors in processes of bordering, but so as to enable a clearer understanding of exactly how states (and other actors) today project power – including forms of power extending to the level of mundane, day-to-day practices. As such, rather than dismissing the state, border research should direct critical attention to the multiple and dynamic scales and dimensions that predispose and frame varying forms of governance and sovereignty. Such an extended research trajectory will enable a productive exchange between traditional, political science-based approaches to inter-national relations and border studies on the one hand, and disciplines that enable a bottom-up phenomenological outlook on border processes, practices, and institutions on the other.

Every border is, by necessity, enacted from below at the level of day-to-day practices and can only be realized through everyday performances carried out by situated subjectivities (Hall 2012; Rumford 2012; Bigo 2014; Côté-Boucher, Infantino and Salter 2014). The term bordering is often used to describe this mundane level of constrained agency, the 'lived meaning, expression, contestation and reproduction of . . . boundaries and hierarchies in . . . everyday routines' (Hall 2012: 2) that underlie even the most pervasive and apparently efficient border regime. In their works, both Hall (2012) and Bigo (2014) employ qualitative methods inspired by anthropological approaches to highlight micro-processes of bordering at specific localities and in this way bring vernacular bottom-up perspectives to traditional state-based and systems-oriented border research.

This chapter highlights the interrelation between technologies of tracking, tracing and profil-ing, and mundane practices of in/exclusion at the contemporary dislocated and increasingly ubiquitous border. In doing so, I follow Côté-Boucher et al.'s (2014) call for a practice-based agenda in border research and weight a description of particular socio-technical potentials for management and control against the various practices and performances through which such potentials are actualized, negotiated or subverted. The term iBorder/iBordering will enable a conceptualizing of some of the implications of such an interdisciplinary methodological trajectory.

## iBorder/iBordering: bringing technology into border research

Vukov and Sheller have noted a transformation of borders 'away from static demarcators of hard territorial boundaries toward much more sophisticated, flexible, and mobile devices of tracking, filtration, and exclusion' (2013: 225) that enlist everyday life and practices in technologically enhanced processes of in/exclusion. Bringing critical approaches to technology into dialogue with border research, I have coined the twin concept of iBorder/iBordering to enable an understanding of the possible impacts of such a fusion of new technologies and border-related processes (Pötzsch 2015).

iBorder/iBordering allows for a critical interrogation of state-driven[1] bordering practices in complex socio-technical environments. The first segment – iBorder – enables a mapping of technological potentials for dispersed governance, i.e. new means of surveying, accessing, and analysing global communication flows as well as new techniques of identifying, tracking and tracing individual subjects and abstracted patterns of life. The second element – iBordering – draws attention to the (often messy) realization and/or subversion of these potentials at the level of everyday practices.

## The socio-technical apparatus of iBorder

The socio-technical apparatus of iBorder consists of new technologies of biometric identification, digital tracking, and algorithmic mapping that afford both an individualizing and a massifying trajectory. Biometric techniques such as facial recognition, iris scans, fingerprint and sentiments analysis, as well as gait or keystroke pattern recognition are combined with remotely accessible RFID-equipped passports and interoperable databases to enable an increasingly comprehensive identification and tracking of individuals. At the same time, new surveillance programmes directed at mobile and Internet-based communications make possible a largely automated assembling and assessment of population-level content, movement, and connection data that enables an identification of potentially threatening, abstracted patterns of life. For instance, as Edward Snowden has revealed, by acquiring access to key servers, fibre-optic cables and Internet exchange points, and by gathering phone records as well as geolocation and connection data, state agencies such as the NSA (US) or GCHQ (UK) have managed to survey a significant part of global communications over an extended period of time.[2]

The amount of data gathered by these agencies is too vast to be processed by humans. Algorithms are therefore used to find correlations and identify significant deviations from implied norms. The machine-generated actionable information resulting from these processes then selectively informs human decision-making cycles with probabilistic assessments leading to what Amoore (2013: 5) has termed a *politics of possibility* – a 'governing of emergent, uncertain, possible futures'. For instance, as signature strikes in drone warfare illustrate, this pre-emptive form of politics densely intertwines human cognition and agency with complex and dynamic socio-technical systems – an interaction that often entails deadly consequences.

As a result of these developments, today 'borders as bounded topographical locations or zones recede and reemerge as iBorder – an ephemeral, technologically afforded aura that attaches itself to the individual' (Pötzsch 2015: 111) and follows the individual wherever he or she might move. Biological and behavioural markers stored in increasingly interoperable databases, RFID-equipped passports and ID cards, as well as the almost constant accessibility of movement and connection data through wearable technical interfaces ubiquitously exposes subjects to the gaze of a de-territorialized border apparatus. This sticky everywhere-border is inherently uncrossable, denies non-normative subjectivities refuge from this condition, and can lead to detention or ultimately death. The present description of the technological potentials for management, control and coercion, however, can only provide a partial account of the complex processes of contemporary technologically facilitated bordering.

## The contingent practice of iBordering

A transition from iBorder to iBordering entails a shift in focus from overarching technological frames to 'technological work' (Walters 2011: 58) – the myriad minute daily endeavours through which the potentials for management and control inherent in these frames are activated, negotiated or subverted. This move also implies a change in methodology toward a phenomeno-logical inquiry into the micro-physics of power – the everyday practices and life-worlds of the subjectivities constitutively intertwined with contemporary technologically facilitated processes of bordering.

In accordance with this trajectory, Brambilla has recently advocated a 'need to humanise borders' by recovering their 'phenomenological dimension' (2014: 27). She suggests a combination of methods drawn from ethnography, anthropology, cultural analysis and visual cultural studies to account for the 'ontological multidimensionality of borders' (2014: 26) stretching from daily routines to geopolitics, and including state-driven procedures of management and control as well as counter-hegemonic articulations and performances.

I list several examples pointing to the limits of, and resistances to, the apparatus of iBorder at the level of day-to-day practices and experiences (Pötzsch 2015). The counter-practices range from the random exchange of sim-cards by Afghan insurgents to trick targeting protocols of US drones, via the unintended incentives created by the EU's Dublin accord to misrepresent migration flows to avoid national responsibilities, to the tricky business of establishing standards that bridge analogue and digital biometric registration practices. Based on these examples, I argue for the imminent necessity to separate 'ambitions of comprehensive surveillance, management, and control' articulated by a global security apparatus 'from the often messy realities of their incremental day-to-day implementation' (Pötzsch 2015: 112).

iBorder/iBordering shows how technology changes and partially enhances the capacity of states to exert power over its subjects and how these capacities are increasingly detached from distinct territorial locations towards an inherently boundless global space. However, as the examples described above attest, state governance is executed, limited and framed by the everyday practices of situated subjectivities. A proper understanding of contemporary borders and processes of bordering, thus, also requires qualitative methods that provide access to the life-worlds and mundane experiences of individual border subjects.

## Conclusion: iBorder/ing as a cultural technique

Besides the fissures and limits of its technological apparatus, there is another issue at stake in connection to the concept of iBorder/iBordering: the question of possible performative qualities of the involved technologies and practices. Does the socio-technical apparatus of iBorder process and manage or does it actively constitute contingent subjectivities and practices? Can iBorder be separated at all from the performance of iBordering, or does the pair form a mutually constitutive whole – iBorder/ing?

Introducing the term cultural technique from the context of recent German media theory, Winthrop-Young (2013) argues for the mutually constitutive nature of human subjectivities, technical objects and the procedures interconnecting the two. He argues that the practice of writing, the form of pencil and paper, as well as the subjectivity of a writer can only heuristically be divided. In reality, the entities involved constantly shape and mould one another in a perpetuated dynamic process of feedback and adaptation. Similarly, I argue that the socio-technical apparatus of iBorder, its procedures and protocols, as well as the subjectivities of those operating within its frame can be seen as mutually constitutive. iBorder/ing in its composite form, thus, becomes conceivable as a fundamental cultural technique of in/exclusion that not so much identifies and processes already established practices and subjectivities, but becomes co-constitutive of them in and through technologically predisposed and procedurally framed everyday border work. In turn, an apparently clear-cut distinction between border technologies and institutions on the one hand and border subjects and practices on the other is problematized and replaced by a processual understanding of the border as a constantly emerging assemblage combining all the above phenomena.

In iBorder/ing, automatically assembled and assessed hypothetical 'life signatures' (Amoore 2013: 81) spread through inter-operative databases and frame the everyday practices of police and border guards who encounter the marked emergent subjectivities. This way, virtual profiles pointing to mere possibilities of future actions and intentions entail actual consequences for certain non-normative subjects. The critical question then becomes whether, or not, such negative consequences at least in part lead to the actualization of the very intentions these technologies claim to identify and avert. Seen from this vantage point, iBorder/ing might be conceived as producing the very threats and aberrations it allegedly identifies and prevents.

iBorder/ing as a cultural technique connects the concrete subjectivities of situated individuals to biometrically and algorithmically determined data doubles populating databases and spread sheets. Practices and technologies of iBorder/ing, as such, selectively activate and operationalize various contingent identity-potentials inherent in these doubles that then feed back into the lives of these individuals entailing concrete physical and embodied effects. In this perspective, categories such as trusted traveller or terrorist threat emerge not as a priori givens justifying particular state conduct, but as the contingent results of technologically facilitated projections of state power pointing to possible futures. The concept of iBorder/ing enables border research to engage with these complexities of contemporary border regimes and practices by productively combining methods and insights from traditional top-down approaches with the phenomenological and critical frameworks providing alternative accounts from below.

## Notes

1 'State-driven' includes activities delegated to transnational or private actors, which then implement measures on behalf of states.
2 For an overview of the documents leaked by Snowden see the online archive established by the *Canadian Journalists for Free Expression* at https://snowdenarchive.cjfe.org/greenstone/cgi-bin/library.cgi and the documents available via *The Intercept* at https://theintercept.com/documents/.

## References

Amoore, L. (2013). *The Politics of Possibility: Risk and Security beyond Probability*. London: Routledge.
Bigo, D. (2014). The (in)securitization practices of the three universes of EU border control: military/navy – border guards/police – database analysts. *Security Dialogue*, 45(8): 209–225.
Brambilla, C. (2014). Exploring the critical potential of the borderscapes concept. *Geopolitics*, 20(1): 14–34.
Côté-Boucher, K., Infantino, F. and Salter, M. B. (2014). Border security as practice: an agenda for research. *Security Dialogue*, 45(3): 195–208.
Hall, A. (2012). *Border Watch: Cultures of Immigration, Detention and Control*. London: Pluto Press.
O'Dowd, L. (2010). From a 'borderless world' to a 'world of borders': bringing history back in. *Environment & Planning D: Society & Space*, 28(6): 1031–1050.
Pötzsch, H. (2015). The emergence of iBorder: bordering bodies, networks, and machines. *Environment & Planning D: Society & Space*, 33(1): 101–118.
Rumford, C. (2012). Towards a multiperspectival study of borders. *Geopolitics*, 17(4): 887–902.
Vukov, T. and Sheller, M. (2013). Borderwork: surveillant assemblages, virtual fences, and tactical countermedia. *Social Semiotics*, 23(2): 225–241.
Walters, W. (2011). Rezoning the global: technological zones, technological work and the (un-)making of biometric borders. In V. Squire (Ed.) *The Contested Politics of Mobility: Borderzones and Irregularity* (pp. 51–73). London: Routledge.
Winthrop-Young, G. (2013). Cultural techniques: preliminary remarks. *Theory, Culture & Society*, 30(6): 3–19.

# 5

# Casing

*Charles C. Ragin*

## Introduction

> Everything that happens once can never happen again. But everything that happens twice will surely happen a third time.
>
> *Paulo Coelho,* The Alchemist, *p. 157*

Paraphrasing Coelho: Every circumscribable instance is unique in its specificity; when we see two instances as more or less 'the same', we typically must invoke some sort of categorization, which, in turn, opens up the possibility of multiple instances. In social science, the most fundamental and consequential categorization is the 'case'. The willingness of social scientists to invoke cases and to focus on patterns across cases as the key to generalizing about them distinguishes much of social science from much of the humanities, where the focus is often on the challenge of representing the specificity of each case. Casing is fundamental to the practice of social science; it could be argued that casing is the quintessential social scientific research act. Researchers 'case' their evidence to bring closure to difficult issues in conceptualization and research design and thereby allow cross-case analysis to proceed. Once evidence is packaged in the form of multiple cases, they can be compared and contrasted, which in turn enables empirical generalization – a key goal of social scientific inquiry (Ragin 1992).

Empirical evidence is infinite in its complexity, specificity and contextuality. Casing focuses attention on specific, limited aspects of that infinity, highlighting some aspects as relevant and obscuring many others. In short, casing provides much-needed blinders, making it possible for researchers to see through, or past, complexity. Different casings provide different blinders, different findings and different connections to theory, research literatures and research communities. Casing locates research in the vast domain of social science, linking it to the efforts of some researchers and severing its ties with others (Ragin 2009).

Quantitative and qualitative researchers alike invoke cases. However, the two discourses are very different, despite several formal similarities. The remainder of this essay sketches these different invocations, with a focus on the distinctiveness of each approach.

## Casing in qualitative research

Qualitative research often begins with an interest in specific phenomena, outcomes or settings. At first, the casing of the phenomenon is fluid and open to revision and reformulation. The usual expectation is that the casing of the phenomenon will become more completely specified as more is learned, usually through in-depth research at the case level. Thus, the initial focus is often on 'good' instances of the phenomenon in question, and there is a back-and-forth between the identification of 'good' instances and the specification of the casing of the phenomenon.

Qualitative researchers construct diverse casings. At a formal level, the research focus is often on a specific category of phenomena, its constitutive features, and relevant antecedent conditions and processes. In other words, after establishing 'what it is', researchers focus on 'how does it come about?' Similarities across instances of the phenomenon in question are a key focus in research of this type. If the search for similarities proves to be unproductive, researchers may 're-case' their instances, often with an eye toward differentiating types of cases (George and Bennett 2005).

From the perspective of conventional quantitative research, the qualitative approach just sketched might seem ludicrous. First of all, the explanandum is more or less the same across all instances. Thus, the 'dependent variable' does not vary substantially and, accordingly, there is little 'variation' to explain. Second, because the qualitative researcher has selected cases that have a limited range of values on the outcome (that is, on the 'dependent variable'), correlations between antecedent conditions and the outcome are necessarily attenuated (see King, Keohane and Verba 1994), which leads, in turn, to abundant type II errors (that is, accepting the null hypothesis and concluding erroneously that antecedent conditions are irrelevant to the outcome). Third, and more generally, because of its focus on in-depth case analysis, the qualitative approach is necessarily limited to small Ns, which, from the perspective of conventional quantitative research, poses real obstacles to the use of probabilistic assessment and to the utilization of sophisticated multivariate techniques – both essential inferential tools.

In response, the qualitative researcher would reply that there are many analytical tools available to social scientists in addition to those based on correlation, and that there are many modes of empirical analysis that are not focused on the problem of accounting for variation in a dependent variable via some form of correlational analysis. For example, identifying an antecedent condition shared by instances of an outcome might signal the existence of a necessary condition for that outcome. Likewise, an absence of shared antecedent conditions could signal that there are different outcome types and that the researcher's next step should be to 're-case' the evidence, based on the identification of key differences between types.

## Casing in quantitative research

Quantitative research tends to be more deductive than qualitative research. The back-and-forth between cases and concepts that is central to qualitative inquiry is almost completely foreign to quantitative research, with its emphasis on theory testing. For quantitative research to proceed, cases must be abundant, independent of each other, and homogeneous with respect to the operation of causal variables. Additionally, they should be drawn from a well-defined population. The populations of quantitative social science are often given or taken for granted. The key is that the population of observations (that is, cases) must be circumscribable; otherwise, sampling bias cannot be evaluated.

Often, the definition of the relevant population in quantitative research is contestable. Consider research on the causes of mass protest in Third World countries against austerity measures mandated by the International Monetary Fund (IMF) as conditions for debt restructuring. While it is possible to identify positive cases (that is, countries with protest), the set of relevant negative cases is somewhat arbitrary. Should the study include all less-developed countries as candidates for IMF protest? Less-developed countries with high levels of debt? Less-developed, debtor countries with recent debt negotiations? Less-developed, debtor countries subjected to severe IMF conditionality? Each narrowing of the set of relevant cases reduces the N of cases available for quantitative analysis, which in turn undermines the possible utilization of inferential technique. Understandably, quantitative researchers generally avoid narrowly circumscribed populations. When Ns are small, standard errors are large, and it is more difficult to generate findings that are statistically significant. For this reason, quantitative researchers tend to err on the side of being over-inclusive. In the example just presented, for instance, the typical solution might be to use all less-developed countries and to include debt level and extent of IMF renegotiations as 'independent' variables.

While this solution seems plausible, at least on the surface, there is a world of difference between using debt level and extent of IMF renegotiations as independent variables, on the one hand, and using these same variables to circumscribe the population of relevant candidates for protest against the IMF, on the other. Not only are these two uses very different from a mathematical perspective, they are also very different from a casing perspective. Using them as independent variables entails a casing that embraces all less-developed countries as candidates for IMF protest; using them to circumscribe the relevant population shifts the casing to a relatively small but clearly delineated subset of less-developed countries – those that satisfy certain antecedent conditions.

It is not widely recognized that boosting the N of cases carries with it an increased danger of type I errors – erroneously rejecting the null hypothesis of no relationship. If the N of cases is artificially enlarged by including irrelevant negative cases (that is, cases that are not plausible candidates for the outcome in question), then the correlations between causal and outcome variables are likely to be spuriously inflated (Mahoney and Goertz 2004). This artificial inflation occurs because irrelevant negative cases are very likely to have low scores on both the independent variables and the outcome and thus will appear to be theory-confirming, when in fact they are simply irrelevant. Correlational analysis is completely symmetrical in its calculation; therefore, a case with low (or null) values on both the causal and outcome variables is just as theory-confirming as a case with high values on both. It is important to note as well that an artificially inflated N of cases also increases the danger of type I errors by reducing the size of estimated standard errors, which, in turn, makes statistical significance easier to achieve.

The quantitative researcher would respond by arguing that a central goal of social science is general knowledge and that studying wider, more inclusive populations serves this goal more directly than studying narrowly circumscribed populations. Furthermore, the concern about type I errors can be addressed by using a more stringent significance level or by applying other technical fixes that safeguard against spuriousness. Finally, models incorporating statistical interaction can be used to address causally relevant conditions that enable the impact of other causal conditions.

The fact that qualitative and quantitative researchers alike invoke cases is an important commonality, uniting different approaches to research under the banner of social science. In both arenas, casings enable analysis. Still, the differences between the two discourses are striking, and the opportunities for miscommunication and misunderstanding are many.

# References

Coelho, P. (1993). *The Alchemist* (trans. Alan R. Clarke). New York, NY: HarperPerennial.

George, A. and Bennett, A. (2005). *Case Studies and Theory Development*. Cambridge, MA: MIT Press.

King, G., Keohane, R. and Verba, S. (1994). *Designing Social Inquiry: Scientific Inference in Qualitative Research*. Princeton, NJ: Princeton University Press.

Mahoney, J. and Goertz, G. (2004). The possibility principle: choosing negative cases in comparative research. *American Political Science Review*, 98: 653–669.

Ragin, C. (1992). Casing and the process of social inquiry. In C. Ragin and H. Becker (Eds.) *What Is a Case?* (pp. 217–226). New York, NY: Cambridge University Press.

Ragin, C. (2009). Reflections on casing and case-oriented research. In D. Byrne and C. Ragin (Eds.) *The Sage Handbook of Case-based Methods* (pp. 522–534). Los Angeles, CA: Sage Publications.

# 6

# Diffracting

*Leila Dawney*

## Introduction

Against a logic of data 'collection' that assumes that data pre-exists its production through processes of investigation and research, 'diffracting' offers us a take on methodology that pays attention to the researcher as *composer*: as active participant in the making of worlds and objects. The term 'diffracting' as an approach to scholarly research was originally developed by Donna Haraway and later elaborated by Karen Barad. For Barad, diffraction is an 'ethico-onto-epistemological matter' (Barad 2007: 381): through acknowledging that as researchers we are part of the making of the world, we then have an ethical responsibility in how we orient to that world. It is about making visible the histories of objects and ideas, about letting the complexities of these histories speak without reducing them to a single narrative, and about intervening in the production of knowledge.

## What is diffraction?

'Diffraction' in physics is the change in direction of waves as they encounter an obstruction or overlap with other waves. Water waves, for example, travel around corners, around obstacles and through openings, as can be seen in the changing patterns of waves as they come up against boats and walls in a harbour. Diffraction apparatuses are instruments that study the effects of such interference and difference. In the physics classroom, these apparatuses often shine a laser through slits onto a screen to make visible the patterns that the light waves produce as they diffract. Diffraction apparatuses were central to the development of quantum physics since they demonstrated that matter, in certain circumstances, behaves like a wave insofar as it can demonstrate diffraction patterns. If we take this concept into the world of interdisciplinary methodologies, we can understand diffracting as a way of attending to and experimenting with interference patterns, as an ethos of embracing and, indeed, playing with contamination and entanglement to see what happens, and also as a means of exposing the complexity of the world. Diffraction apparatuses are used to produce knowledge about both the object being passed through the apparatus, but also about the apparatus itself. In this way, diffracting can be understood as a practice that acknowledges its participation in world-making, and which makes visible its own interference and the material effects of this interference.

## Diffraction in Haraway and Barad

Donna Haraway discusses diffraction as a research approach developed in contradistinction to reflection and reflexivity. Against reflection, but using a similarly optical metaphor, diffraction challenges the drive to undertake research by representing an object in a different form elsewhere, to make a new picture of the research object, instead focusing on the patterns of difference, movement and entanglement that come into play when making visible the histories of entangled objects and ideas: '[d]iffraction patterns record the history of interaction, interference, reinforcement, difference. Diffraction is about heterogeneous history, not originals' (Haraway and Randolf 1997: 273). In other words, the diffraction apparatus that makes interference visible becomes a metaphor for a particular approach to knowledge production. Diffraction in Haraway's work becomes a two-fold practice: first, a making-visible of the histories of the production of an object; a recording of the passage of the material histories of its ongoing formation, and the registering of that process, and second, a distinctly critical practice for making a difference in the world – diffraction makes it impossible for us to be a voyeur, but asks us to engage critically, to interfere (Haraway and Goodeve 2000). Haraway's collaborations with the artist Lynn Randolf are a good example of this – their arguments participate in the world and alter its flow (Haraway and Randolf 1997). Their work reconfigures the human through the figure of the cyborg: they bring the 'naturecultures' that separate scientific knowledge from cultural knowledge to visibility.

Karen Barad further elaborates the concept of diffraction as a scholarly approach, arguing for a diffractive method that makes visible the entanglements of scientific practices with the social. Barad discusses how diffraction apparatuses 'measure the effects of difference, [and] even more profoundly, they highlight, exhibit, and make evident the entangled structure of the changing and contingent ontology of the world' (Barad 2007: 73). Her book develops the concept comprehensively as a method and approach to research ontology and epistemology. She discusses ways to build a diffraction apparatus in order to study the effects of entanglements and interferences of the natural and the social, of the discursive and the material. Central to this argument is the refusal to hold anything still; the apparatus plays each aspect against each other, encouraging the visibilities produced through their entanglements to question their integrity and highlight their messiness. Diffraction also involves experimenting and intervening in the world: as Barad puts it: 'diffraction is a material practice for making a difference, for topologically reconfiguring connections' (Barad 2007: 381). Diffracting is about 'how differences matter' (Barad 2007: 378), about asking which differences matter, and how they are made to matter.

## Diffracting as social science methodology

Arguing against using mixed methods research as a means of simply validating or reinforcing one 'dominant' method, Emma Uprichard and I have argued for diffracting as an approach to social science methodology that refuses to take as given any pre-existing research object (Uprichard and Dawney 2016). Building on Barad and Haraway, we are interested in how a diffractive approach precludes the possibility of using one set of data to illustrate, enrich or verify another, since that would involve the 'holding still' of one and a refusal to see it as a messy, processual entity. The methods that we choose when conducting research enact 'cuts' through the world that make certain aspects of research phenomena visible, and these cuts are, in themselves, part of the ongoing production of knowledge-making that produces particular phenomena as objects.

Diffracting as an approach to research might involve a consideration of the ways in which 'cuts', when read through their interaction with each other, and with other objects, such as philosophical or literary texts, can produce different objects and even call into question the ontological stability of the object in question. It is a project of making 'manifest the extraordinary liveliness of the world' (Barad 2007: 91) that refuses to reduce or reflect, or to hold anything still. It is about experimentation: putting things together and seeing what happens; or putting ourselves in particular situations; or recovering histories and genealogies that became lost as objects were being made into objects. Diffracting involves thinking with disjuncture; thinking about where data rubs up against data and what that exposes about how subjects and objects of research are made through the research. As an approach to the study of philosophical and cultural texts, diffracting involves reading them against each other, without situating one as a fixed frame of reference, and seeing what happens. It unsettles and mixes, producing disturbances to watch how they pan out. It delves into forgotten material histories, making visible and asking what caused them to be forgotten: it exposes the world in its complexity and messiness.

## From diffraction to diffracting

What would it mean to change a noun to a verb? What does it mean for us to 'do' diffracting? Diffracting is about doing research as participating, experimenting and inserting ourselves into the world in ways that have particular effects, and paying attention to those effects; it is about acknowledging how we may be the diffraction apparatus, the object and the screen all at once. Diffracting in research is about intervening and world-making: it is a provocation to a status quo that assumes the stability of object and/or researcher. In refusing to tie anything into a single narrative, it allows social entities to exist as multiple, tangled ontologies, tracing their multiplicity and entanglement. The final section elaborates this by pointing to three things that diffracting can *do* to research practice.

1 *Diffracting troubles the case and messes the categories.* Diffracting involves getting rid of the assumption that there is an unproblematic 'object': it is a practice of attending to relationality, process and messiness in the always-incomplete object. It welcomes the emergence of disjunctures, things that go against the grain, lacunae, difference and diversion as a means of troubling the research case as a bounded, isolated unit and revealing the ways in which processes of objectification, the making of the research object, take place.

2 *Diffracting lets the world be messy and complex.* It allows histories of objects to participate in their making; it makes visible the complexity, messiness and instability of research objects, and their excess to knowing. It lets data non-cohere and disintegrate: composing and recomposing objects, cases and phenomena; questioning how we remake cases methodologically; confusing and clarifying.

3 *Diffracting acknowledges how researchers participate in world- and knowledge-making – as composers of data and as diffraction apparatuses.* Diffracting involves thinking about how researchers participate in the making of objects, knowledges and worlds. Diffracting acknowledges how we represent the world, through language, creativity, social scientific and natural scientific practices. It refuses representationalism and operates in a performative mode. It 'does not concern homologies but attends to specific material entanglements' (Barad 2007: 88). When we approach problems, we bring some aspects of them to visibility, and in doing so alter their histories and contribute to their making. Research norms and practices make objects; they substantiate them through composing them as such.

Overall, diffracting opens up opportunities for researchers across the disciplines to approach research practices a little differently from those approaches more usually written into social research methods textbooks. It provides a framework for a performative politics of social research practices and objects, acknowledging the role of the research process in the 'making' of objects and worlds.

## References

Barad, K. (2007). *Meeting the Universe Halfway: Quantum Physics and the Entanglement of Matter and Meaning.* Durham, NC and London: Duke University Press.

Haraway, D. J. and Randolf, L. M. (1997). *Modest_witness@second_millennium: femaleman_meets_oncomouse: feminism and technoscience.* London: Routledge.

Haraway, D. J. and Goodeve, T. N. (2000). *How Like a Leaf: An Interview with Thyrza Nichols Goodeve.* Psychology Press.

Uprichard, E. and Dawney, L. (20 October 2016). Data diffraction: challenging data integration in mixed methods research. *Journal of Mixed Methods Research.* Online first. https://doi.org/10.1177/1558689816674650

# 7

# Figurationing

*Leila Dawney*

A figure is more than a representation. It is an emergent accretion of images, ideas and association that resonates within each of its iterations. At once material body, media image and cultural signifier, it is both subject and object, indeterminate container for meaning, and affective conduit. It can testify, attract, alienate, revulse and captivate. It may also feel, and weep and laugh. A specific body might be a figure (for example, Princess Diana or the First World War veteran Harry Patch), but so might a more generalized category of person (such as the figure of the soldier and the mother). Figuration is the iterative process through which particular figures are given substance and become visible, accruing meaning and political-affective force. Figures are thus understood here as humanoid effects of processes of figuration, that 'body forth' into the world (Castaneda 2002: 3). I have argued that the figure can thus be seen as a 'way of thinking about the relations engendered by bodies or categories of bodies in their social and cultural specificity, and through their iterations and differential repetitions: what we might think about as *encoded corporeality*' (Dawney 2013: 30). Figures are both material and semiotic: they convey meaning and have material substance and world-making effects.

The concept of the figure has diverse histories and genealogies, in literary theory, sociology, philosophy and science studies. Elias' 'figurational sociology' proposed an understanding of social reality based on interdependencies and process rather than atomized individuals, emphasizing figurations as objects of analysis rather than society or individual (Elias and Jephcott 1982). The historian Erich Auerbach approaches figuration as a mimetic practice in Western literature, describing how figures 'establish . . . a connection between two events or persons in such a way that the first signifies not only itself but also the second, while the second involves or fulfils the first' (Auerbach and Said 2013: 73). Auerbach shows how the figure in literature emerges from a discursive tradition of figural realism where the story of Christ is read into Old Testament Scripture, emphasizing narratives of salvation, sacrifice and progress (Auerbach 1984; Auerbach and Said 2013). In science and technology studies, Suchman uses the term 'configuration' to expand the material-discursive relation of the figure beyond the human towards socio-technical assemblages (Suchman 2012).

Figurationing is a mimetic practice that maps our world, producing stories to which subjects can attach themselves, or can gain purchase on different forms of power and life. Haraway writes of figuration as a 'contaminated practice' which throws figures into the world, drawing on stories and associations, on allegories and material histories in the production of ideas and imaginaries (Haraway and Randolf 1997: 8). Figures, then, shape our world and understandings, providing a means through which power is given form and made material and meaningful. To study the figure and processes of figuration is not only to study the various representations of the figure, but the means by which figures travel – in art and literature, across social media, in gossip, conversations in the pub, dreams and daydreams. Drawing on these diverse genealogies of the figure, figurational critique involves understanding and unravelling the work that figures do, and/or mobilizing them in new ways. It attends to figures in two ways: as forms and embodiments of power, and as positions from which or through which to rethink power. As a methodological approach broadly allied to cultural studies, it has clear potential for moving beyond these disciplinary boundaries to shed light on formations of social, political, economic and material life.

Foucault's *The History of Sexuality* (1978), is an example of the use of the figure as a form or embodiment of power. In this work, he identifies four figures of the nineteenth century that both signalled and stood for the emergence of biopolitical power: the hysterical woman, the masturbating child, the Malthusian couple and the sexual deviant (Foucault 1978). As the 'simultaneously material and semiotic effect of specific practices', he suggests that such figures have a double force – as 'constitutive effect and generative circulation' (Foucault 1978: 3). Foucault's figures, then, embody, materialize and reproduce biopolitical power and as such are central to its workings. More recently, Elizabeth Povinelli has posited 'four figures of the Anthropocene' as representing what she calls 'geontopower' (Povinelli 2015). As such, figures can stand for articulations of power and counter-power, providing a means of substantializing ideas as they take shape. Social movements, for example, may coalesce around particular figures – such as the miner as a figure of labour that comes to represent all workers – and figures may be invoked in the manipulation of public sentiment, for example in the scapegoating of welfare claimants and asylum seekers (Tyler 2013).

But figures are also sites of possibility and hope: the figures of the cyborg (Haraway 1991; Haraway and Randolf 1997), the stranger (Simmel 1950) and the nomad (Braidotti 1994) provide alternative positions – they open up possibilities, allowing us to think differently, and 'construct geometric possibilities in the cracks of the matrices of domination', providing 'performative image[s] of the future' (Kember 1996: 264). Thus the work that figures do can be channelled towards a politics of hope and transformation. There is also critical work to be done in inhabiting figures differently, as Haraway and Randolf suggest, referring to the problematic links between the concept of the figure and Christian redemption narratives discussed above:

> [W]e inhabit and are inhabited by such figures that map universes of knowledge, practice and power. To read such maps with mixed and differential literacies and without the totality, appropriations, apocalyptic disasters, comedic resolutions, and salvation histories of secularised Christian realism is the task of the mutated modest witness.
>
> *Haraway and Randolf 1997: 11*

It is one task of figurational critique, then, to make a difference: to unravel the processes of figuration that tell particular stories, and tell other stories in the process.

The figure, for Haraway, has two main facets. First, it operates as a trope: 'figures do not have to be representational and mimetic, but they do have to be tropic; that is, they cannot be literal and self-identical. Figures must involve at least some kind of displacement that can trouble identifications and certainties' (Haraway and Randolf 1997: 10). In other words, figures always refer outside of themselves and become a locus for cultural stories, myths, sentiments and hopes. Haraway's entities – cyborg, primate, oncomouse, are figurations 'involved in a kind of narrative interpellation into ways of living in the world' (Haraway and Randolf 1997: 140). Haraway describes such figures as 'performative images that can be inhabited', drawing on an understanding of contemporary forms and logics of life as an 'implosion of bodies, texts and property' (Haraway and Randolf 1997: 7–11).

Thinking about figures, and practices of figuration, involves dealing with images and texts from a materialist perspective. It involves recognizing that these images and texts are more than representations: they have material substance and effects, and participate in material worlds through which sense and meaning are produced. In my own work on the figure of Harry Patch, the last First World War veteran in the United Kingdom, I argued that Patch operated as a figure of a form of experiential authority through his material embodiment: his frailty and working-class masculinity, and his experience of pain and suffering. His figuration across media and in public imaginations could not be understood in terms of an analysis of one broadcast or one image. Figurational critique involves an understanding of the affective and material impact that these constellations of images, broadcasts and iterations in the public sphere and in individual worlds have.

How might we approach the work that figures do, and think about how they circulate and participate in the production of the social? Innovative theoretical-methodological practices are needed that bridge the space between body, text and world where this work happens. Castaneda, in her work on the figure of the child, adopts a figurative method in order to 'describe in some detail the constellation of practices, materialities, and knowledges through which a particular figuration occurs, and in turn, to identify the significance of that figuration for making wider cultural claims' (Castaneda 2002: 8). As I argue elsewhere, invoking the 'figure' makes possible, a 'way of conceptualizing the affective capacities that are held by figures that are both material and symbolic, that are produced by and produce the social' (Dawney 2013: 43). Like Foucauldian genealogy, figurative methods track figures across and through their figuration; across media forms and other forms of representation; and through everyday practices. They pay attention to the material-semiotic apparatuses of which they are part, and in which they participate.

Like genealogy, figuration is interested in differences and displacements; in the movement of figures in an economy of truth production; in their indeterminacy and in the power that such indeterminacy wields. Figurative methods need to draw attention to how figures participate in the making of worlds, tracing their various articulations but also the archetypes and cultural narratives that they draw into their storytelling. However, a genealogical analysis of sites of figuration alone does not necessarily help us understand the mechanisms through which figures work through bodies: what gives them grip, and captivates people. Figures operate, in part, through their encounters with bodies, and their operation is dependent on those bodies' material histories of association and experience: their inscription as subjects. It is for this reason, then, that we need to consider encounters between bodies and figures: what fears and fascinations do they rub against? What histories and material traces of those histories participate in the encounter? We need to sit alongside such figures, allow them to do their work, to play upon and bring to the surface the cultural myths, the fears, hopes, desires and negations that they call into being in their tropic resonances. In other words, figurational critiquing asks us to manifest the spectres that haunt and lurk behind the figure. Castaneda's work pushes at these questions in her discussion of how the mythological ghost of 'La Llorena' (the weeping woman) in Hispanic America draws on fears about the selling of Guatemalan children to wealthy childless couples from the USA. The power of the figure lies in its ability to establish such connections on the level of affect: to tap into the histories of subjectivation that emerge in bodies in their encounters with texts.

If we understand figures as a locus for affective forces that coalesce around them, perhaps we can see more clearly how they might grip or captivate, or inspire us to disgust or to hate.

In being captivated ourselves, we can follow the lines of association to the material and historical conditions that make it possible for our bodies to respond to these figures in such ways. We can investigate the traces of history and experience in our own bodies that they rub up against: the desires, anxieties, material insecurities and existential fears that are already in process that are triggered through these figurings in a way that both augments what is there already and contributes to the ongoing formation of the subject and, by association, the social. This involves attention to the affective register of bodies, through which these relations take hold and the extent to which these figures augment and intensify affective resonances in bodies. A discussion of the circulation of stories and images is thus supplemented with thinking about how they work in and through affective bodies and the political effects of such circulation. If we are to follow this methodological challenge through, we need to develop tools for reading figures not only through texts but also through the affective micropolitics of daily life: for thinking about moving train carriages to avoid someone who looks like a 'terrorist', a smile at a pregnant woman, weeping while watching Mo Farah winning a gold medal in the 2012 Olympics, or the authority of Malala Yousafzai. Figures not involve only the tropic association of texts and representations, but are chains of constellations of images and associations and memories that gain meaning in their affective encounters with bodies, held together by repetition, association and the memory of the effect of related images and ideas.

Figurational critique, then, needs to involve sitting with the figure: allowing ourselves to be taken along with the figure, letting it do its work and affect us, opening ourselves up to its various referrals and deferrals, associations and dissociations, attachments and alienations. It involves an unravelling, but also a remaking, a retelling of figural stories.

## References

Auerbach, E. (1984). *Scenes from the Drama of European Literature*. Minneapolis: University of Minnesota Press.

Auerbach, E. and Said, E. W. (2013). *Mimesis: The Representation of Reality in Western Literature*. Princeton, NJ: Princeton University Press.

Braidotti, R. (1994). *Nomadic Subjects: Embodiment and Sexual Difference in Contemporary Feminist Theory*. New York: Columbia University Press.

Castaneda, C. (2002). *Figurations: Child, Bodies, Worlds*. Durham and London: Duke University Press.

Dawney, L. (2013). The figure of authority: the affective biopolitics of the mother and the dying man. *Journal of Political Power*, 6(1): 29–47.

Elias, N. and Jephcott, E. (1982). *The Civilizing Process*. Oxford: Blackwell.

Foucault, M. (1978). *The History of Sexuality, Volume 1: An Introduction*. New York: Pantheon Books.

Haraway, D. J. (1991). *Simians, Cyborgs, and Women: The Reinvention of Nature*. London: Free Association.

Haraway, D. J. and Randolf, L. M. (1997). *Modest_witness@second_millennium: femaleman_meets_oncomouse: feminism and technoscience*. London: Routledge.

Kember, S. (1996). Feminist figuration and the question of origin. In G. Robertson *et al.* (Eds.) *FutureNatural: Nature/Science/Culture* (pp. 256–269). London and New York: Routledge.

Povinelli, E. A. (2015). Transgender creeks and the three figures of power in late liberalism. *differences*, 26(1): 168–187.

Simmel, G. (1950). The stranger. In K. H. Wolff (Ed. and trans.) *The Sociology of Georg Simmel* (pp. 402–408). New York: The Free Press.

Suchman, L. (2012). Configuration. In C. Lury and N. Wakeford (Eds.) *Inventive Methods: The Happening of the Social* (pp. 48–60). London: Routledge.

Tyler, I. (2013). *Revolting Subjects: Social Abjection and Resistance in Neoliberal Britain*. London, Zed Books.

# 8

# Notating

*Moritz Wedell*

Notating has a vast and multifaceted history that contrasts strongly with a limited theory of notation. In this essay, I trace the epistemic productivity of notational acts, particularly those aspects that cannot be grasped within an analytical scope derived from the investigation of notational systems.

In Western culture, the earliest traces of notating appear in language. The etymological roots of specific words relating to communication and number point to a primitive, though fundamental, type of notating. Words, such as the English *to tell*, *to count*, the German *zählen*, *erzählen*, and the French *conter*, *compter* etymologically refer to scenes in which notches are cut into a given material. Notating in its historical core is to invest pieces of wood or bones with notches: primitive series of basic uniform symbols, which however help both to recall a number and an account of what happened. Tally sticks work numerically and semantically. In some rural areas their use remained a dominant practice until the beginnings of the twentieth century; occasionally, it was *the* procedure of notating when people could not or did not want to write alphabetically. Later, the Middle Ages, building on ancient traditions, witnessed an explosion of specialized notations beyond alphabetic language. These helped to organize books, to record musical practice, to articulate and foster scientific progress (Wedell 2015: 1209–1211). With modernity, new developments emerged, such as the handling of moveable types, the evolution of mathematical sign languages, the elaboration of the modern musical score. From the twentieth and twenty-first centuries, researchers document a broad spectrum of individual notations, such as Mary Wigman's choreographic sketches, or John Cage's experimental graphic scores. On the technical side, punch-cards, the Turing machine, and digital code can also count as notational practices (further examples can be found in Grube, Kogge and Krämer 2005). Notating, in an iconic and performative perspective, overlaps broadly with writing as a cultural technique: a notational practice that unfolds in the interplay of three dimensions – a graphic (or spatial), a symbolic (or referential) and an operational (or performative) aspect – and is expressly not limited to vocal language (Krämer 2003; Grube *et al.* 2005).

Despite the fact that a historical phenomenology of notating identifies such a wide panorama of practices, the only established theory of notation cuts down this multitude to next to nothing. Nelson Goodman's aim was to define what notations *really* are. For Goodman, notating meant the strict application of specific rules, rules that define well-formed symbol systems and their unequivocal relations to fields of reference. Only a very few practices comply with these conditions: a musical score, for example, does, but alphabetic writing does not (Goodman 1976: 127–173). What we learn from Goodman are the requirements for notations, which will be free of any trouble caused by ambiguity. It is most useful to adapt Goodman's specific concept of a 'notational system' for its precise terminological discrimination. But it is most instructive, too, to study what he omitted: studying *acts of notating* – instead of *notational systems* – leads to the investigation of myriad practices, the social implications and epistemological potentials that emerge specifically from ambiguity.

★★★

The realm of social interaction is crucial to the study of acts of notating in two ways, both affecting the interplay of storage and retrieval: As to the first – shared know-how – it is important to note that no notational practice emerges as a universal. All ways to notate were practised first, and sometimes only, individually or in small communities. This is even true for the prehistory of our basic mathematical language. In the premodern period, the Hindu-Arabic numerals underwent radical transformations, and most of their local forms and rules of use disappeared eventually. Similarly, most coding languages of the late twentieth century have not survived. Against their specific requirements many of these notations showed a fine performance. But every notational system must be reliable beyond individual use. It is only collective recognition and application that makes it run.

As to the second – shared situational knowledge – even the most limited notational practice is not necessarily doomed to fail. The inventory of signs of a primitive tally stick is limited to just one, the simple notch. Nevertheless, tally sticks may refer to economically complex situations. The notches do not capture complex information per se, as a system of notation, but the notational procedure includes also the agents involved in the transaction. The more a milieu is organized by brain memory, the more its notations will work as memorial aids. When notational practices are under-determined in terms of formal distinction and reference, they might require social memory to kick in. Conversely, the more sophisticated a notational procedure is in terms of formal and referential distinction, the more we can expect that captured information will be retrievable without additional information, and the broader is the realm of possible addressees.

★★★

The epistemic dynamics of notating, however, is not limited to the interplay of storage and retrieval. Beyond its documentary function, every notational procedure displays an inherent potential for exploration. One motor is the reassignment of symbol schemes to new fields of reference. When the Hindu-Arabic numerals were first received in the West, two scribes of the tenth century just listed them in an encyclopedia (Isidor of Seville's *Etymologies*) as a supplement to the names of the Roman numerals, without any specified function. In Toledo, monks of the tenth through twelfth centuries used them as a notation for quantity when they marked pebbles of an abacus with the new symbols, thus simplifying the manual procedure of calculation. Other schools of the early Middle Ages used them as representations of extension and proportion to understand problems of geometry and land surveying. In yet another realm of reception, clerics of the twelfth and thirteenth centuries used the numerals to explore the arithmetic potential of the decimal system. In the early fourteenth century Leonardo Fibonacci linked these arithmetic procedures to real-world problems, namely the processing of economic transactions. All this happened long before mathematical operators and logical refinements were introduced and established, making number the paragon of scientific rationality (Wedell 2015: 1212–1215, 1233–1243).

An important contemporary example of notational reassignment is the digitization of printed texts. Here, alphabetic writing is encoded in digital data, ready to be both processed computationally and represented as readable writing on a screen. As a form of notating that operates on two levels, digitized corpora do not just multiply the resources for our actual reading. Giving access to both hermeneutics and statistics, they open fresh ways to investigate established texts on unexplored levels. This is not an issue of retrieval, because it is about information that no one deliberately stored. Rather it is an issue of shaping what we search, of exploring new patterns of knowledge.

A method that takes advantage of literature as digital data is 'distant reading' or 'computational criticism', as opposed to 'close reading'. The underlying process, 'operationalizing' as Franco Moretti, founder of the University of Stanford's Literature Lab calls it, is itself derived from the sciences. It is a 'process whereby concepts are transformed into a series of operations – which, in their turn, allow to measure all sorts of objects' (Moretti 2013b: 1). In the field of literary studies, this interdisciplinary approach may be applied to hermeneutical concepts, such as *the protagonist* or *style*. Operationalizing transforms them from 'concepts' into 'instruments' to capture formal aspects that no human reader can detect (2013b: 9). The parameters of computational analysis might seem banal in some cases, such as the frequency and distribution of functional words such as 'the' in a text. But they help. On a small scale, they can contribute to define parameters such as the stylometric fingerprint of an author or a specific genre. Hence they help, for example, to identify unspecified pieces of writing (think of anonymous fragments in historical sources, or of plagiarism). On a large scale, researchers expect them to capture no less than the very nature of 'world literature', outside the normative power of national canons, beyond the biased claims of individual literary historians caught in their local trends and tastes (Moretti 2013a: 43–62).

Not every digital corpus, however, is reliable and transparent, nor is every available tool. Take Google Books, for example. The sheer mass of included texts promises to facilitate work with big data in the humanities. But the corpus is unexplained regarding its composition and unreliable in the matter of text recognition. Further, Google's reading aid, the 'n gram viewer', while easy to use, is opaque at the level of programming. It is virtually impossible to know what it actually represents. Thomas Weitin, founder of the University of Konstanz's Literature Lab, therefore suggests we turn to smaller but more controllable corpora (such as AntConc, Voyant or CATMA). He advocates 'scalable reading' of smart data, interweaving hermeneutics and statistics (Weitin 2015).

How smart data are, however, depends on the work of appropriate encoding, which presents notating as a task of interdisciplinary collaboration. The case of classical literature clarifies what is at stake: ancient texts might have come down to us in fragments scattered over more than a millennium. They might have been notated originally on materials such as stone, papyrus rolls or parchment codices. The sources could have survived in several countries, maybe on different continents. Therefore, different fields of knowledge – papyrology, codicology, paleography, epigraphy, linguistics, philology and information technology, among others – will be required to assemble the digital document, to provide an appropriate annotation with markups and to define the processes of operationalizing (Revellio 2015). Ancient literatures might not be the most prominent field of research today. But it is most illuminating to study what it needs, on a level of digital notating, to produce a smart corpus: to provide documents effective for an accountable reader and for a counting machine.

The early use of tally sticks involved an integrated complex of numerical and semantic dimensions. This essential double grip of notating returns in a sophisticated interplay of statistics and hermeneutics when we process re-encoded notated materials on a digital basis.

★★★

Acts of notating do not only refer to static objects. Notations can most efficiently represent movements: movements of mathematical operations, movements of bodies, of sound, of thought. But acts of notating do not merely 'put' these movements 'into' a notational composition, or a choreography of things to make them reaccessible. Acts of notating detach these movements from the restrictions of the real world, and in doing so they allow researchers to explore movements in a virtual realm, a sphere only restricted by the rules of the notation. It is this freedom to carry out movements in an abstract realm of notational practice that is an important precondition for composition and invention, both in the arts and in science. Recent interdisciplinary research that exemplifies this approach includes the fields of the technical image and the epistemology of models.

The act of 'putting into a notation', however, is not unidirectional. It unfolds in a flexible double bound setting and opens up another dimension of epistemic exploration. The example of writing down music in a classic musical score shows how radically a notational system limits the scope of what we *think music is*. But the limits given by notation do not only contrast music against mere sound. The discrimination will most probably affect what we *perceive as music*. To notate, in this sense, is a practice, which *informs* what we note. Paradoxically, one might also argue the other way around that to notate is primarily the opposite, to *represent* what we note: the example of processing notes by pen and paper illustrates that we are highly skilled and free to use notating as a way to represent what we perceive, imagine or process mentally (Kogge, in Grube *et al.* 2005: 162–167). The paradox makes clear that notating is neither only a way to *represent* nor only to *inform* what we note. Notating embraces both, representation and information of mental activity, reciprocally. In a positive view, notating is, then, a means to constitute cultural and scientific stability; in a pessimistic view, epistemological standstill.

In a reflective move, we can also understand how this notational complex is an instrument that fosters epistemological progress. When notating means to put on display, to stage, *what* we note, then it allows us at the same time to explore and to investigate into *how* we note: notating appears to be useful to evaluate our perceptions, imaginations, mental operations *with respect to the ways we mirror these activities notationally*, and that may set the notational complex in motion.

Grammar, for example, shapes our understanding of language and its parts. *What* we note is: grammar stages (represents) and determines (informs) our perception of language, and thus stabilizes our understanding of what is correct and what is wrong. At the same time, grammar's notational presence determines *how* we note: in a scriptural culture, we cannot see language but through an ensemble of normative routines. The fact, however, that grammar is fixed notationally – both documented as a set of written rules, and incorporated in the structures of written texts – puts forward an opportunity of examination at a second level, focusing on grammar itself. When our (notated) rules fail to correspond with our (practical) routines, a tension may arise within the notational complex. This is what makes the history of grammar exciting. Historically, our conventional grammatical terms and rules derived from the study of classical languages. Eventually, when the tension between contemporary language use and these rules became problematic, philologists broke up with the absolute supremacy of classical grammar (a process of considerable political explosiveness in some periods, for example, in the seventeenth and eighteenth centuries). One of the epistemological effects of this dynamic was that it encouraged linguists to discriminate between different models of grammar (classical school grammars,

grammars of national languages, descriptive grammars of spoken language, generative grammar, and so on), each of which has a different take on how language is meaningful, and hence multiplies what we note/know about language and how we note/know it. It is an ongoing process. Some of the basic questions raised by Martianus Capella in the sixth century are still being discussed in disciplinary and interdisciplinary research (Stockhammer 2014).

★★★

In a broader perspective, this interplay between the notional and the notational resonates with the 'looping effects', which the philosopher of science Ian Hacking described to explain processes of social construction (Hacking 1999). Social categories represent (or label) distinctive features of social groups. They associate with specific experiences and expectations of how people in this group will act. Therefore, a specific category might also loop back and inform what people understand as their very personal room of opportunity for being themselves. It is not only on a metaphorical level of 'inscribing identity' that these observations relate to notating. Looping effects might well extend to actual writing about oneself, say in a diary or in letters (think of Foucault's 'techniques of the self'), or on social media. Facebook's timeline, among other functions, appears as a tool for notational identity formation in this sense. Notating, here, adopts the function of a socially relevant form of self-reflection. As an analytical concept, notating can help to understand developments – unfolding or stagnation – of the social performance of individuals and groups.

When the philosopher Alva Noë took up the metaphor of looping effects, he shifted the focus to practices that induce looping effects intentionally. While Hacking assumes more generally that looping effects are everywhere, Noë demonstrates that they are specifically at the core of art (Noë 2015: 29–48, and passim). In his approach, art is, like philosophy, a practice that produces knowledge. Art is a tool that brings into focus our unquestioned practical routines and intellectual attitudes, or, as Noë puts it, our ways *to be organized*, to live by *organized activities*. Art, by staging patterns of our organized activities presents us with the means for *reorganizing ourselves*. Even though not all forms of art are notational, Noë sees in writing – notating language – the very model of how art works. An experiment of thought clarifies the starting point of Noë's argument: imagine speech as an activity untouched by the power of representation. Against this scenario, imagine the transformative power of writing: writing not only as the fundamental act to create and unfold a graphical image for speech (a notation that will loop back on the way people speak, and then again affect the routines of writing, and so forth); but first and foremost as 'a technique for thinking about whatever domain it is, we are writing about' (Noë 2015: 40). The context of art demonstrates how the function of notating can shift from documenting information to producing knowledge, and from representing what we think to reorganizing who we are.

★★★

When we count, it is inevitable that we answer the question '*what* counts?' too. When we notate, the act of notating necessarily responds to an inquiry into '*Why* is it worth noting?' The 'why' changes, and hence notational practices evolve. When conventional forms of notation resist interventions of artistic or scientific 'rewriting' they may become subject to reorganization in their turn. Avant-garde art and poetry of the early twentieth century are only the most striking examples of the reorganization of writing schemes, perhaps more shining and accessible than the invention of (equally effective) scientific models, such as Gottlob Frege's '*Begriffsschrift*'. When the available routines, schemes, or systems of notation cease to represent our ways to perceive, to imagine and to think, the tension within the notational complex will generate a need to break up and re-form notational practices, even to set up new notational systems in Goodman's sense. These progressive forms of notating, again, will eventually vanish or establish new patterns to capture and to compose the world we live in.

## References

Goodman, N. (1976). *Languages of Art: An Approach to a Theory of Symbols*. Indianapolis, IN: Hacket Publishing Company.

Grube, G., Kogge, W. and Krämer, S. (2005). *Schrift. Kulturtechnik zwischen Auge, Hand und Maschine*. Munich: Wilhelm Fink Verlag.

Hacking, I. (1999). *The Social Construction of What?* Cambridge, MA: Harvard University Press.

Krämer, S. (2003). Notational iconicity, calculus: on writing as a cultural technique. *Modern Language Notes*, 118: 518–537.

Moretti, F. (2013a). *Distant Reading*. London and New York: Verso.

Moretti, F. (2013b). Operationalizing: or, the function of measurement in modern literary theory. *Stanford Literary Lab. Pamphlet* 6: 1–13.

Noë, A. (2015). *Strange Tools: Art and Human Nature*. New York, NY: Hill and Wang.

Revellio, M. (2015). Classics and the digital age: advantages and limitations of digital text analysis in classical philology. *Konstanz LitLingLab. Pamphlet* 2: 1–16. Retrieved from: http://nbn-resolving.de/urn:nbn:de:bsz:352-0-320377

Stockhammer, R. (2014). *Grammatik – Wissen und Macht in der Geschichte einer sprachlichen Institution*. Berlin: Suhrkamp.

Wedell, M. (2015). Numbers. In A. Classen (Ed.) *Handbook of Medieval Culture: Fundamental Aspects and Conditions of the European Middle Ages* (vol. 2) (pp. 1205–1260). Berlin and Boston: De Gruyter.

Weitin, T. (2015). Thinking slowly: reading literature in the aftermath of big data. *Konstanz LitLingLab. Pamphlet* 1: 1–18. Retrieved from: http://nbn-resolving.de/urn:nbn:de:bsz:352-0-285900. English version at: www.digitalhumanitiescooperation.de/en/pamphlete/pamphlet-1-thinking-slowly/

# 9

# Prototyping

*Alberto Corsín Jiménez*

## Re-sourcing

The website and digital database of the Spanish guerrilla architectural platform *Inteligencia Colectiva* (IC) showcases hundreds of designs of do-it-yourself, retrofitted, community-driven, grassroots technological and architectural solutions and adaptations from all over the world.[1] These designs, which include diagrams and architectural sketches, 3D renders, text, photographs and even videos, are all licensed freely with Creative Commons licences. They are open-source prototypes, as IC likes to think of them: designs that are at once *gratis* and free to access, but also technical templates that simultaneously 'free' the very practice of design by enabling third parties to extend, modify, adapt or simply share the original source code. They are prototypes that liberate design as both a technical and cultural practice.

The prototypes that IC documents in their website offer an example of how the philosophy of free and open-source (F/OS) software has taken root in domains of practice that extend well beyond the digital realm. Architects, artists, designers, engineers, activists and grassroots organizations have found in the philosophy F/OS a toolkit for reimagining the material, aesthetic and environmental affordances of their work. Every component and aspect of their work, whether it be the infrastructural systems that support it or the aesthetic languages and registers that frame it, whether the dimensions of its material exchanges or its collaborative dynamics, are held up to examination as both *resources* and *re-sources*: that is, for their capacity to work as foundations and support-structures but also as springs and openings for future extensions and modifications. In this sense, we can think of the concept of *open-source prototyping* as a figure of complexity that is 'less than one and more than many': a prototype is always less than itself, in the sense that the source code enables expansions and bifurcations to the original design. Therefore, the design never reaches closure, it is never 'one' properly speaking. At the same time, the very possibility of having new additions and modifications defines the infra-ontology of prototyping as 'more than many', for the source code remains forever open to 'more' designs being added to the 'many' already in existence (Corsín Jiménez 2014).[2]

In this guise, the language and praxis of open-source prototyping provides an instance of complex, adaptive, self-organized systems whose heuristic and metaphoric vectors are not drawn from biology, chemistry or physics (cf. Barad 2007; Stengers 2010). These are emergent socio-infrastructural assemblages that re-source themselves through an inventive and ongoing exploration of how human and nonhuman sources and resources – including legal licences, technical specifications, organic and inorganic processes, or social relations – work on each other. They delineate the contours of ecologies that are forever 'in beta' (Corsín Jiménez and Estalella 2017).

Such an understanding of how open-source prototypes work differs in important and significant ways from how the concept of prototypes has historically been used in design and engineering contexts. As the standard dictionary definition has it, a prototype is an original on which something is modelled; that is, it is an exemplar or first model. In the history of science and technology, the process of prototyping has traditionally been marginalized as a backstage operation. Prototypes were hidden inside the 'black boxes' of engineering and design processes (Winner 1993): the prototyping phase of a design project indexed that stage where decisions regarding the use of components, materials, protocols or standards were thought of as residual and of secondary importance; processes whose very subjection to ongoing revision and reassembling, whose inhabiting a space of messiness and uncertainty invited no interest or curiosity. Prototypes were thought to be little more than drafts, provisional templates whose bodies functioned as test-grounds for trials and errors, groundwork and experimental assays whose efforts pointed somewhat unremarkably to things-that-were-not-quite-objects-yet.

## White-boxing

Recent interest in the material semiotics and biographies of modelling practices has allowed us to unpack and recuperate the prototype from its relegation within the black boxes of the history of science and technology (Chadarevian and Hopwood 2004; Daston 2000). In particular, scholarship on critical making and open-source design invites us to reconsider prototyping from a *white-boxing* perspective, where the emphasis is on how socio-technical practices constantly exfoliate, disclose and re-source their capacities for agency and relation (Corsín Jiménez, Estalella and Zoohaus Collective 2014).

As both method and epistemic orientation, white-boxing opens up novel vistas for social theory. First, it shows us how the work of prototyping has traditionally been tensed against the proprietorial boundaries of technology. It makes evident the role that the opacity of black boxes has played in keeping industrial designs secret. Prototypes were only allowed to upgrade from the 'proto' phase to the final 'type' stage once properly locked into an intellectual and industrial property regime (Biagioli 2011). Theirs was therefore an existence *before* and/or *in suspension of property*: where the relations between persons and things remained as yet unmediated by the entanglements of ownership or authorship. Similarly, black boxes contributed towards keeping knowledge 'in places' (in boxes, so to speak), such that it became meaningful to speak of the places and geographies of technoscience: university and industry laboratories, science parks, etc. (Smith and Agar 1998). In their place, from the perspective of white-boxing, we can ask instead about the recursive processes through which the 'proto' and the 'type' have been parenthesized with respect to each other: What tensions and practices traverse such states of suspension? Why are there parentheses to start with? What is it about a prototype that makes us imagine it as being 'stuck' in a permanent state of anteriority? And, finally, how does form finally emerge out from the parentheses, what drives the extraction of the proto-type?

White-boxing further allows us to understand how prototyping has demarcated a political economy for the circulation of expert knowledge. Prototypes have signalled to the umbra separating experts from lay people, scientists from amateurs, insiders and outsiders to the technoscientific black box. In this sense, a perspective focused on white-boxing challenges how expert knowledge travels and circulates in society by destabilizing the epistemic status of technology itself as a sociological vector. The sources and resources most often associated with technoscientific expertise – laboratory equipment, scientific authority and credentials, university settings, intellectual property – are seen, when looked at from a white-boxing perspective, to be constantly unfolding, hacked and re-sourced. If 'technology is society made durable', as Latour famously put it (1991), prototyping and white-boxing would therefore seem to make technology un-durable, fragmented and dispersed. Or perhaps it is 'the social' that we should be looking at unboxing and destabilizing? Chris Kelty (2008) speaks for instance of how F/OS software 'modulates' anew the relation between power and knowledge: free software programmers work on themselves as a social and political community by working on (writing code, editing, patching, compiling, improving) the infrastructure that enables their coming into being in the first place. As an expression of a prototyping culture, such 'recursive publics', as Kelty calls them, would indeed seem to point to how white-boxing allows for novel forms and new locations of social durability to emerge – new expressions of cultural, political and aesthetic materiality and critique.

White-boxing finally comes full circle in helping us unpack and interrogate the very sources and resources through which we, as social scientists, take presence in the world. It invites us to reconsider the infrastructures, spaces and times that give shape to our research methodologies: What would social scientific inquiry look like if it were an open-source design, if it were modulated recursively as it responded and accommodated to the disturbances, challenges and co-inventions of the fieldwork process? How does research wireframe itself into suspension? What would method look like as a prototype?

I have pointed at three re-sources through which the prototyping of method might take shape. First, by reconsidering the proprietary and, more amply, legal economy of research. In other words, making sure that we open access (consultation, readership, editing) to our work. There are a number of ways in which this can be done: licensing our work with Creative Commons licences, avoiding publication in subscription journals, promoting the use of Open Access institutional repositories, as well as helping to define the terms of discoverability (the algorithms, the metrics) that will shape the future public sphere of scholarly communication.

Second, the method of prototyping invites us to reconsider how the process of research separates 'researchers' from 'informants', experts from amateurs, knowledge-makers from knowledge-users. For instance, we might wish to ask about the role that methods play in the design of inquiry and analysis. What technologies, spaces, materials and social relations are bundled together in the production of method? Who gets to question our methods, when and in what terms, and do our methods have the capacity to incorporate such voices? In other words, is there scope for white-boxing our methods, making the people with whom we work party to our methodological agenda and design?

Finally, this explicit self-grounding of the problem of method as something that is constantly 'prototyping' itself calls for re-examining the multiple and criss-crossing 'trajectories of apprenticeship' through which research comes into being (Pignarre and Stengers 2011). We prototype by designing an ecology for every method: one where we take stock of our mutual competences and constraints, our respective obligations and requirements, the resources and the conditions through which we each, collectively and individually, operate. We prototype every method when we come to the realization that the organization of research calls for nothing less than the design of an *infrastructure of apprenticeships*.

## Notes

1 Inteligencias Colectiva, www.inteligenciascolectivas.org/
2 A playful take on the 'more than one and less than many' that Marilyn Strathern (2004) suggested characterizes fractality as a figure of contemporary complexity.

# References

Barad, K. (2007). *Meeting the Universe Halfway: Quantum Physics and the Entanglement of Matter and Meaning*. Durham, NC: Duke University Press.

Biagioli, M. (2011). Patent specification and political representation. In M. Biagioli, P. Jaszi and M. Woodmansee (Eds.) *Making and Unmaking Intellectual Property: Creative Production in Legal and Cultural Perspective* (pp. 25–39). Chicago, IL and London: The University of Chicago Press.

Chadarevian, S. de and Hopwood, N. (Eds.) (2004). *Models: The Third Dimension of Science*. Stanford, CA: Stanford University Press.

Corsín Jiménez, A. (2014). Introduction. The prototype: more than many and less than one. *Journal of Cultural Economy*, 7(4): 381–398.

Corsín Jiménez, A. and Estalella, A. (2017). Ecologies in beta: the city as infrastructure of apprenticeships. In P. Harvey, C. B. Jensen and A. Morita (Eds.) *Infrastructures and Social Complexity* (pp. 141–156). London and New York, NY: Routledge.

Corsín Jiménez, A., Estalella, A. and Zoohaus Collective (2014). The interior design of [free] knowledge. *Journal of Cultural Economy*, 7(4): 493–515.

Daston, L. (2000). *Biographies of Scientific Objects*. Chicago: University of Chicago Press.

Kelty, C. M. (2008). *Two Bits: The Cultural Significance of Free Software*. Durham, NC and London: Duke University Press.

Latour, B. (1991). Technology is society made durable. In J. Law (Ed.) *A Sociology of Monsters: Essays on Power, Technology and Domination* (pp. 103–131). London: Routledge.

Pignarre, P. and Stengers, I. (2011). *Capitalist Sorcery: Breaking the Spell*. London and New York, NY: Palgrave Macmillan.

Smith, C. and Agar, J. (Eds.) (1998). *Making Space for Science: Territorial Themes in the Shaping of Knowledge*. Basingstoke: Macmillan Press.

Stengers, I. (2010). *Cosmopolitics I*. Minneapolis, MN and London: University of Minnesota Press.

Strathern, M. (2004). *Partial Connections*. Walnut Creek, CA: AltaMira Press.

Winner, L. (1993). Upon opening the black box and finding it empty: social constructivism and the philosophy of technology. *Science, Technology, & Human Values*, 18(3): 362–378.

# 10

# Retrieving

*Carolin Gerlitz*

## Introduction

Digital and social media have opened up new avenues for data collection about social, cultural and political life. Since the advent of digital online media, their data have been of interest to a range of disciplines which have approached digital data with their respective questions and methodologies. Within sociology, online and social media data have initially led to the hope that these new data formats may not have been 'contaminated' by the interferences of researchers and their methodologies (Savage and Burrows 2009) – as opposed to traditional sociological methods such as interviews or questionnaires. Engaging with the making and technicity of digital and social media data, however, soon confronted researchers with multiple inscriptions of media, methods and their tools (Ruppert, Law and Savage 2013). Various methodological approaches to access such data preformatted by media have emerged in the context of media and communication studies, sociology and Science and Technology Studies (STS). In the field of digital research methods, digital media are treated as research devices, capable of structured data production (Rogers 2013; Weltevrede 2015). STS have embraced digital data as defined by the actors themselves rather than researchers (Callon 2006) and contributions from the field of Actor Network Theory (Latour *et al.* 2012) have pointed out that digital data offer both a granular view on individual actions and an aggregated overview, allowing the possibility to cut across the micro/macro distinction that has been so central to sociological debates.

Across disciplines, different ways of accessing data from online and social media have emerged, most notably scraping, that is the extraction of preformatted data from user interfaces (Marres and Weltevrede 2013), but also retrieval, the extraction of data via application programming interfaces (APIs) offered by digital and social media platforms. APIs are software interfaces that enable researchers and other third parties to connect to associated databases in order to produce content for, or extract data from, platforms. This access is usually highly structured, standardized and regulated by the associated platform, offering data access to developers, business partners and researchers among others. As scraping is limited by what can be extracted from user interfaces, retrieving has gained increasing relevance in the context of ever-growing volumes of data.

Interdisciplinary methodological debates have drawn attention to the various inscriptions at stake when working with digital and social media data and have attended to possible bias built into the data by the media that pre-structure them for their own purposes (Marres and Gerlitz 2016). Twitter data, for instance, may well be used to study public debates, but has originally been structured alongside Twitter's own valuation logic which largely focuses on identifying popularity and trending topics. Data retrieval through APIs has emerged as an interdisciplinary methodological approach which enables access to such actor defined data, allowing researchers to attune their methods to their research object. However, when tracing the inscriptions of retrieved data, it becomes apparent that retrieval not only confronts researchers with the self-categorization of the medium that provides APIs in the first place, but with a larger cascade of inscriptions as the data accessible from one platform might not necessarily have its origin in that same platform. Retrieval, this chapter suggests, poses particular challenges to inventive methods that seek to account for the ongoing happening of social life (Lury and Wakeford 2012) while attending to the mutual inscription of method and problem. In the context of complex data ecologies and interoperability between platforms, the question emerges as to what data retrieval makes accessible in the first place and to which inscription or bias methods need to attune. To account for the inventiveness in retrieving, we need to attend to its technicity, namely APIs, first.

## Application programming interfaces and the grammatization of data

Many digital and social media platforms offer (a variety of) APIs to build upon, produce content or extract data, enabling structured access to platform databases. In the case of Twitter, for example, the social media platform offers a so-called REST API for discrete queries, a Streaming API for continued data-capture in real time and an Advertising API.[1] The data available via APIs can be considered pre-structured on many levels. On a first level, it is the result of standardized platform activities – or grammars of action to draw on the work of Philip Agre (1994) – which enable users to act in particular ways and platforms to instantly capture data about these actions in standardized form. In the context of Twitter, these grammars focus on organizing user relations (friending, following, muting), comments, likes, retweets, status updates and posts among others, as well as their metadata. The grammatization of user action in the front-end is met with another layer of grammatization in the back-end in the form of API commands and regulations that determine which data can be accessed by whom in what quantities. To remain with the case of Twitter, API access is organized through OAuth,[1] a personalized access token for third-party API access. Once access is granted, input to or retrieval from the database are pre-structured through the platform's developer facing grammars. In the case of Twitter's REST API grammars are organized alongside query-related GET commands which retrieve data and activity-focused POST commands, that allow the API user to post content. These commands largely mirror the grammars of front ends, but also offer additional data and are policed through extensive documentation, good practice cases and Twitter's rules of conduct.

Looking at data retrieval through the lens of grammatization brings to attention that organizing data into categories and units is increasingly distributed, at least between the researcher, users realizing these grammars and the media, as platforms define what data formats users can generate and retrieve through APIs. This redistribution of ordering capacities away from the researcher towards media platforms was initially perceived as the promise of transactional digital data resulting from the direct capture of user activities. Following Callon, 'One way of testing the relevance and robustness of a proposed categorization is to allow the entities studied to participate in the enterprise of classification' (2006: 8). Indeed, the grammatization of API data suggests that researchers should attune their methods to the categorizations, inferences and objections made *by the medium*.

## Retrieval and realism

However, attending only to the categorizations suggested by a medium, data research enabled through API retrieval runs the risk of re-enacting a specific form of realism. If we follow Alain Desrosieres (2001), there are different degrees of data realism: metrological realism, which assumes an unproblematic relationship between the world and its measure; accounting realism, which establishes the trustworthiness of metrics through standardized practices; and proof-in-use realism, in which realities are defined by the databases that promise to describe them, while paying only little attention to how the data are made, captured and animated.

> For users in this third group, 'reality' is nothing more than the database to which they have access. Normally, such users do not want to (or cannot) know what happened before the data entered the base. They want to be able to trust the 'source' (here the database) as blindly as possible to make their arguments.
>
> *Desrosieres 2001: 346*

Such trust in a data source is akin to the hopes and phantasies characteristic of those early advocates of big data debates who claimed digital transactional data would be more 'raw' than qualitative research data. Despite considerable caution being expressed about this approach, such proof-in-use realism may still inform today's investment in data as pre-structured by or specific to a medium. Desrosieres' analysis suggests that while the increasing valorization of transactional and actor categorized data in the context of social media research might initially have been driven by an interest in attuning methods to objects, it might yet end up contributing to a proof-in-use realism if it remains inattentive to the question of how the grammars of platform databases are actually counting, capturing and composing in the first place. To understand what digital data is animated by and which other actors participate in its categorization, it is relevant to attend to the wider infrastructures in which retrieval operates.

## The ecosystem of data retrieval

APIs not only lend themselves to data retrieval, but also enable a variety of third parties to build on top and produce content for respective platforms, allowing a variety of actors to participate in data production. Let us return to the case of Twitter. Since its launch in 2006, Twitter has offered APIs for developers to extract and input data. The POST commands have resulted in a proliferating ecosystem of third-party Twitter clients, sources and access points, which allow Twitter users to engage with the content and grammars of Twitter via alternative interfaces (Gerlitz and Rieder 2014). Among these access points are Twitter specific clients who are concerned with the de- and re-composition of topical streams by offering multiple timelines, professional clients focused on team tweeting, follower growth, journalistic or marketing practices such as Hootsuite, as well as automators such as If This Then That and cross-syndication apps that allow the sharing of content of one platform with another. Each of these clients are built on Twitter's front-end and back-end grammars, but they also extend them, as they are informed by different ideas of 'being on Twitter' (Gerlitz and Rieder 2014). Such clients not only provide alternative interfaces to Twitter grammars, they also might come with a re-interpretation or expansion of these grammars. Take the example of Twitter's previous favourite and current like button: While some third-party apps interpreted favourites as a means to bookmark and save tweets into collections, others treated them as signs of appreciation and collected favourites received into rankings of popularity (Passmann and Gerlitz 2014).

Such divergent interpretations of the same action enabled even more activities to fold into the same grammar and thus same data point. Doing so, third-party clients contributed to realize the 'interpretative flexibility' of platform grammars (Bijker, Hughes and Pinch 1987), which may come fixed in form, but offer users a certain flexibility to define what a tweet, a favourite/like or a @reply stands for.

In addition, data retrieval has to face a third layer of grammatization, as clients are not only able to reinterpret grammars, but are also able to fold the data of one platform into the grammars of another platform. In the case of cross-platform syndication, that is the automatic posting of content from one platform to another, the grammars of one platform (hashtags/posts/images on Instagram or Facebook) are transposed into the grammars of another platform (Twitter). The data researchers retrieve from the Twitter API might seem comparable and countable through their standardized forms, but might not even have been created for Twitter in the first place, but cross-syndicated from Facebook, transformed from RSS feeds or automatically created from news postings. Is a hashtag produced within Twitter's web interface comparable to a hashtag cross-syndicated from Instagram or a hashtag automatically selected through software? The grammatization of a platform suggests that the data units it generates are comparable if not similar entities, while at the same time creating the conditions for third parties and users to fold heterogeneous interpretations of these grammars into the platform.

## Lively metrics

Looking at the proliferating client-ecosystem thus challenges a proof-in-use realism of retrieved data by leading us to ask: what do we actually count when retrieving data through APIs? The activity of categorizing is not only distributed between the researcher and the platform, but is realized through users, their practices and interpretations, third-party clients and cross-platform syndication. What API data retrieval gives access to, therefore, are 'lively metrics', that is data categories that are internally dynamic, situated, localized and alive. Their liveliness – as opposed to mere currency or liveness (Marres and Weltevrede 2013) refers to the multiple ways in which platform grammars can be realized and interpreted. It is hence not only the platform that categorizes and grammatizes the data that can be retrieved, the lively metrics available via APIs are animated by the entire ecosystem of users, practices and clients. Hence, the moment researchers retrieve data through APIs, it has already been pre-composed in dynamic, local and distributed ways. Or, put the other way around, expanding Agre's work in the context of social media platforms, grammatization not only enables capture, but establishes avenues for new, dynamic and thus lively forms of data composition, which are often made invisible through standardized data and its retrieval infrastructures.

This opens up new avenues for inventive methods that seek to let objects pose their own problems (Lury and Wakeford 2012).

1   Lively metrics refuse a single interpretation. Aggregated data units provided through API data retrieval are not comparable from the outset, but need to be made comparable through additional interpretation of the wider ecosystem of actors and practices in which the data are produced. The retrieval of pre-structured data is thus not a discrete process but invites an attentiveness as to how the capture and composition of data are entangled on many levels.

2   Retrieving data from a single platform means working with data from a multiplicity of sources. The proliferation of clients and cross-platform syndication allows the grammar of one medium to fold into the grammar of another. API retrieval thus gives access to data

formats that are themselves already composed as distributed accomplishments. To attune methods to the data retrieved requires the researcher to move beyond a single medium perspective and advance the notion of medium-specificity (Rogers 2013) to include the distributed ecologies of platforms.

3   Retrieving lively data confronts researchers with the insight that it is not only social life that can be regarded as 'happening' (Lury and Wakeford 2012), but data, their capture and composition, are equally subject to such happening. The process of retrieval contributes to the happening of the categorization of data, as it creates specially composed samples.

4   The liveliness of metrics should not be addressed as a matter of data cleaning but be part of the quest to engage with the messiness and internal heterogeneity of data. Rather than seeking to retain only data formats that are fully comparable and rely on the same interpretation of grammars, data retrieval asks us to attend to the wider dynamics through which data formats are animated. Many platform APIs offer cues for such approaches as they allow to retrieve the source or client from which platform data was produced in the first place (Gerlitz and Rieder 2014).

As data retrieval through APIs is enabled by platforms that seek to open themselves to multiple stakeholders, retrieval should be expanded to capture these relations, foldings and distributed accomplishment of grammars. Engaging with the assembly of capture and composition at stake in platform data suggests that data retrieval has more in common with established methods such as interviews and survey than with big data hopes for raw and unfiltered transactional data, as the data comes with cascades of inscriptions. On the other hand, data retrieval confronts researchers with an even more distributed process of categorization, objection and inscription, which is made partly invisible by data infrastructures and which expands beyond the data source and the researcher to involve all of a platform's stakeholders. In the context of platform media, data categorization itself can be understood as happening and inventive retrieval infrastructures can be called to account for its liveliness.

## Note

1   https://dev.twitter.com/overview/documentation

## References

Agre, P. E. (1994). Surveillance and capture: two models of privacy. *The Information Society*, 10(2): 101–127.

Bijker, W. E., Hughes, T. P. and Pinch, T. J. (1987). *The Social Construction of Technological Systems: New Directions in the Sociology and History of Technology*. Cambridge, MA: MIT Press.

Callon, M. (2006). Can methods for analysing large numbers organize a productive dialogue with the actors they study? *European Management Review*, 3(1): 7–16.

Desrosieres, A. (2001). How real are statistics? Four possible attitudes. *Social Research*, 68(2): 339–355.

Gerlitz, C. and Rieder, B. (2014). Tweets Are Not Created Equal. Intersecting Devices in the 1% Sample. Presentation at the AoIR conference, Daegu, South Korea.

Latour, B. *et al.* (2012). 'The whole is always smaller than its parts': a digital test of Gabriel Tardes' monads. *The British Journal of Sociology*, 63(4): 590–615.

Lury, C. and Wakeford, N. (2012). *Inventive Methods: The Happening of the Social*. London: Routledge.

Marres, N. and Gerlitz, C. (2016). Interface methods: renegotiating relations between digital social research, STS and sociology. *The Sociological Review*, 64(1): 21–46.

Marres, N. and Weltevrede, E. (2013). Scraping the social? *Journal of Cultural Economy*, 6(3): 313–335.

Passmann, J. and Gerlitz, C. (2014). 'Good' platform political reasons for 'bad' platform data. Zur sozio-technischen Geschichte der Plattformaktivitäten Fav, Retweet und Like. *Datenkritik*. Retrieved

March 2018 from: www.medialekontrolle.de/wp-content/uploads/2014/09/Passmann-Johannes-Gerlitz-Carolin-2014-03-01.pdf

Rogers, R. (2013). *Digital Methods*. Cambridge, MA: The MIT Press.

Ruppert, E., Law, J. and Savage, M. (2013). Reassembling social science methods: the challenge of digital devices. *Theory, Culture & Society*, 30(4): 22–46.

Savage, M. and Burrows, R. (2009). Some further reflections on the coming crisis of empirical sociology. *Sociology*, 43(4): 762–772.

Weltevrede, E. (2015). *Repurposing Digital Methods: The Research Affordances of Platforms and Engines*, Amsterdam: PhD dissertation.

# 11
# Timing

*Barbara Adam*

Timing is
Doing time
Synchronizing
Temporalizing all
Capturing invisibility
Composing with fluidity
Knowing interdependency
Depending on right moments
Seeing through a temporal lens
Embracing radical indeterminacy
Extending perspective to open past
Presencing all of time in this moment
Encompassing unknowable open futures
Consciously framing temporal boundaries
Knowing that these choices impact on results
Equally appreciating externally imposed frames
Collecting as doing, frozen in outcomes of collection
Needs re-animating arrested time: product to process
Re-tracing the social to very first material (inter)actions
Time open, unbounded: imposing boundaries is our doing
Holding time still to contemplate the social de-temporalizes
Ossifies becoming; closes down processes; negates potentials
All of past and future encoded in momentary slices of social life
Temporal wholes expressed in moment-parts: present embedded
Reworking temporal complexity fundamentally reconfigures theory
Emphasizing process over product challenges method & methodology
Embracing temporal extension reforms disciplines, categories & norms
Interaction as original social troubles ontology-epistemology distinctions
Recognizing our complicity in generating data changes the nature of a fact
Explicitly acknowledging use of theoretical and methodological technologies
On conceptual path of no return, questions and answers become intermeshed
Once quests for interdisciplinary inquiry are seriously engaging process timing

# 12

# Visualizing data

## A view from design space

*Greg McInerny*

## The promise

Anything can be visualized – whether it is financial trends, phylogenetic relationships, partisanship in senate voting patterns or even the concept of evolution (Figure 2.12.1). Visualizing data brings quantities, forms and relationships into view when the subject matter is minuscule or distant, abstract or intangible, transient or multiscale.

Visualizing data can be essential to making sense of data by enabling discoveries and increased understanding. Visualizations can also facilitate education and enjoyment, and have even become cultural icons. As a new 'photojournalism' (Stefaner 2014), visualizing data can reveal unseen issues; such as when a humble chart catalysed the creation of the Bill and Melinda Gates' foundation – 'that rotavirus slice in the pie chart set us on fire' (Gates 2013).

Visualization tools make businesses intelligent, allowing us to 'Answer questions as quickly as you can think of them' (Tableau™ 2010). This might be unsurprising. When we visualize, we wire data into our cognition via advanced graphics technologies and the highly evolved human visual system. More information is consumed 'through vision than through all of the other senses combined' (Ware 2012), so why consume information any other way? As Peter Hall (2008)

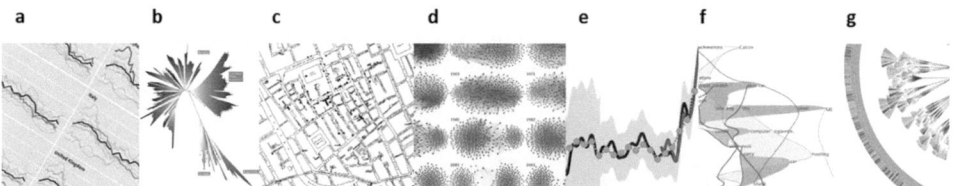

*Figure 2.12.1*   (a) Visualizing data from financial trends; (b) the tree of life; (c) cholera incidence; (d) partisanship in senate voting patterns; (e) climate change trends; (f) poems; (g) the evolution of *On the Origin of Species* by Charles Darwin.

Image credits: (a) OECD (2017), Inflation (CPI) (indicator). doi: 10.1787/eee82e6e-en; (b) www.nature.com/nrmicro/journal/v14/n6/full/nrmicro.2016.63.html; (c) https://en.wikipedia.org/wiki/John_Snow#/media/File:Snow-cholera-map-1.jpg; (d) www.mamartino.com/projects/rise_of_partisanship/; (e) https://en.wikipedia.org/wiki/File:T_comp_61-90.pdf; (f) www.sci.utah.edu/~nmccurdy/Poemage/; (g) http://moma.org/interactives/exhibitions/2011/talktome/assets/TTM_124-large.jpg

suggests, as data and information inundate our lives, 'diagrams, maps, and visualisation tools offer a means to filter and make sense of it'.

## Do visualizations visualize?

But that is only the promise. Peter Hall (2008) closed the statement above with a caveat – 'to visualise it is to understand it, or so we hope'.

When data are visualized ineffectively, the invisible is not necessarily made visible. No matter what level of promise the information industries bestow on visualization, '*to visualize*' is not always a guarantee that discoveries or sense will be made. Rendering data in visual objects is not always sufficient. Not all pictures are worth a thousand words. Not all visualizations 'visualize'.

Consider Anscombe's Quartet (Anscombe 1973), a quintessential example of visualization where inspecting the data or a simple statistical investigation reveals little information (Figure 2.12.2). Yet when the Quartet is 'visualized' patterns immediately pop out and ideas spark (Figure 2.12.3).

Anscombe's demonstration has, however, been staged. It depends on a particular graphical representation, where the categories are separated across multiple graphs that have common scales. This staging can easily be undermined (Figure 2.12.4) by encoding data in ways our perception does not instinctively decode, or that our cognition cannot translate. If our perception and cognition fail, does a visualization actually visualize?

$X_I = \{10,8,13,9,11,14,6,4,12,7,5\}$
$Y_I = \{8.04, 6.95, 7.58, 8.81,8.33,9.96,7.24,4.26,10.84,4.82,5.68\}$
$X_{II} = \{10, 8,13, 9, 11,14, 6,4,12,7,5\}$
$Y_{II} = \{9.14,8.14, 8.74, 8.77,9.26,8.1,6.13,3.1,9.13,7.26,4.74\}$
$X_{III} = \{10,8,13,9,11,14,6,4,12,7,5\}$
$Y_{III} = \{7.46,6.77,12.74,7.11,7.81,8.84,6.08,5.39,8.15,6.42,5.73\}$
$X_{IIII} = \{8,8,8,8,8,8,8,19,8,8,8\}$
$Y_{IIII} = \{6.58,5.76,7.71,8.84,8.47,7.04,5.25,12.5,5.56,7.91,6.89\}$

```
mean(x)=9
var(x) = 11
mean(y) = 7.5
var(y) = 4.1
cor(x,y) = 0.816
Y = 0.5x+3
```

*Figure 2.12.2*   Anscombe's Quartet, as seen as raw data (left) and simple exploratory statistics (right) (after Anscombe 1973).

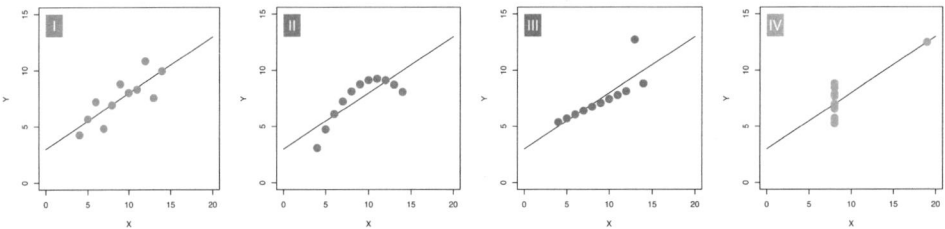

*Figure 2.12.3*   Visualizing Anscombe's Quartet, an example that features in very many books and teaching materials (after Anscombe 1973).

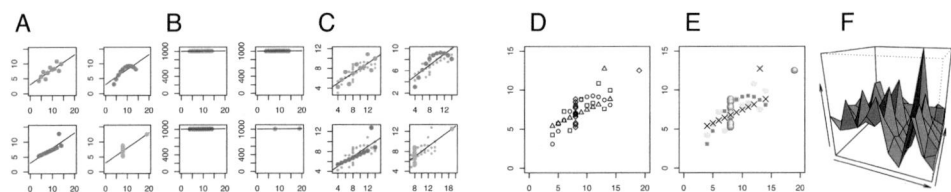

*Figure 2.12.4*   Reducing the effectiveness of Anscombe's Quartet. (A) The original rendition; (B) with data increased by 1,000; (C) individual panels scaled to focal variables; (D) one plot with overlaid symbols; (E) and then perceptually informed symbols (see Ware 2012); and finally (F) a 3D surface.

## Design Space

Visualizations are often described by the constraints from which they were designed, such as the resources to be used (e.g. data, tools, media), the tasks to be enabled (e.g. to locate, compare, reflect) and the context of use (e.g. users, situation, device). For example, various renditions of Anscombe's Quartet (e.g. Anscombe 1973; Figure 2.12.5) share the analytical goal of visualizing patterns and trends, with the specified data. Most of these renditions aim to demonstrate the power of 'visualization' and a few aim to examine frailties in the generality of that claim (Figures 2.12.4 and 2.12.5).

Any single design is just one realization of the design constraints, with alternative designs arising when the data (Figure 2.12.4B) or media change (Figure 2.12.4F), or when non-standard plots are required (Figure 2.12.5). Divergent designs might arise from the same design constraints (Figure 2.12.6), and differing design constraints might produce convergent designs. A discussion of 'visualization' – whether as objects, a set of methods or a subject – could be hindered by our predisposition to viewing visualization through the design constraints that can also define function and how function is evaluated.

To shed some of these inbuilt values and perspectives, we could consider visualizations as collections of visual objects and nothing more, and consider all design possibilities, even those that have not been made material. We can use this 'Design Space' as a shorthand term for the infinite variation of visualizations. Some areas in this space will have a use, or multiple uses, and some will have no conceivable purpose. Design Space is envisaged as a hyper-volume of *all* possible visualization designs with as many dimensions as there are ways to visualize data using coordinate and mapping systems, visual encodings and formatting, scales and sizing, sampling and aggregation methods, etc.

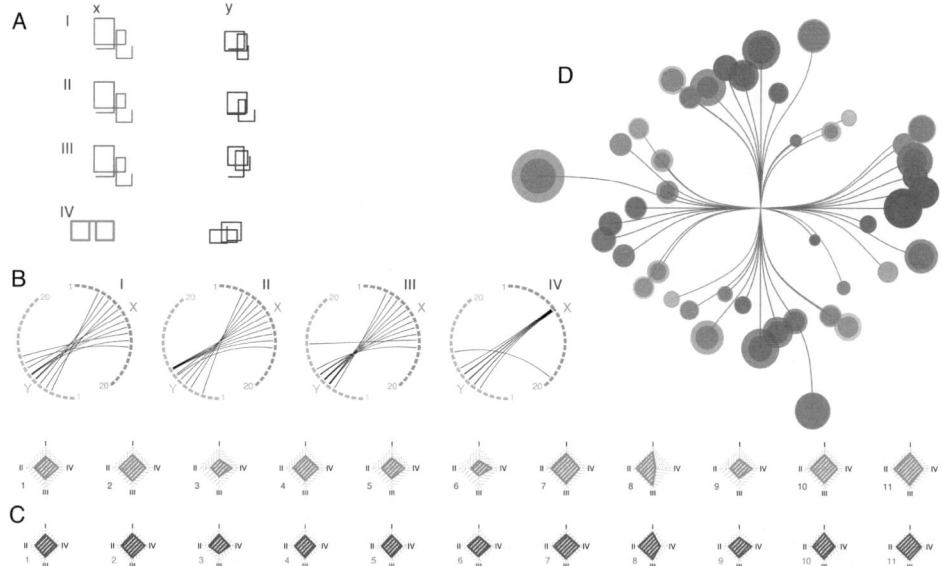

*Figure 2.12.5*   Further variations on Anscombe's Quartet: (A) A tribute to George Nees' 1964 computer drawing '23-ecken' (note the discovery of spectacles in the X variable of category 4); (B) chord diagrams; (C) radar plots comparing the row values; and (D) with inspiration from Jan Willem Tulp's 'Ghost Counties' but designed to reveal nothing in particular.

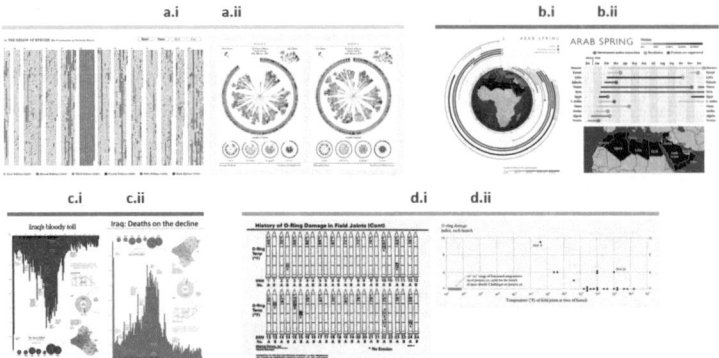

*Figure 2.12.6*　Paired examples of the same data visualized in different areas of Design Space – (a.i–ii) visualizations of the evolution of *On the Origin of Species* by Charles Darwin; (b.i–ii) timelines of Arab Spring events; (c.i–ii) death toll in Iraq during the American occupancy; (d.i–ii) O-ring damage during space shuttle launches.

Image credits: (a.i) https://fathom.info/traces/; (a.ii) www.moma.org/interactives/exhibitions/2011/talktome/
objects/145525/; (b.i) www.informationisbeautifulawards.com/showcase/113-arab-spring; (b.ii) www.thefunctional
art.com/2015/02/redesigning-circular-timeline.html; (c.i) www.scmp.com/infographics/article/1284683/iraqs-
bloody-toll; (c.ii) www.youtube.com/watch?v=Ybwh4lejYO4; (d.i) and (d.ii) Reprinted by Permission, from Visual
Explanations, Edward Tufte, Graphics Press.

There are issues with this definition of Design Space, but its vagueness forces us to reflect on how we define, evaluate and interpret visualizations. Different disciplines can impose highly specific views onto qualities such as 'effectiveness' or 'beauty', and how visualizations might be used and created. In what follows, we will explore topics such as function, technology, aesthetics and our approaches to studying visualization. To start, let us consider if Design Space might be charted, and what parts of this n-dimensional space 'work'?

## Lost in Design Space

Many books and blogs assist the craft of visualizing data, by suggesting how to visualize data effectively using different coordinate systems, visual encodings, patterns of emphasis and data manipulations. Each perspective, however, will at some point fail. Visualization 'rules' are often drawn from experimental evaluations that compare simplified, tractable compartments of Design Space. As visualization science lacks a wholly predictive theory (Kindlmann and Scheidegger 2014), the science accumulates contingent rules to understand and compare the relative suitability of designs given specific data types or tasks. Rather than providing a reliable rule-based mapping between designs and their properties, these studies instead point to the unavoidable difficulties of a predictive theory as there are instabilities in Design Space where the properties of a design depend on the data. In Design Space, contingency reigns.

For example, even simple datasets can experience conflicts between the 'rules of thumb' that should assist us when visualizing data, such as when different categories conflict in their demands for a truncated axis (an axis not starting at zero) or demand differing aspect ratios (the relative dimensions of the plot) (Figure 2.12.7). When visualizing data, it might be inevitable that we hide some patterns as we reveal others. Patterns can be a composite of features that might be optimally revealed in different kinds of charts and not viewable in any single graph.

Even in a simple chart, the accuracy of comparing different data combinations may vary widely (Figure 2.12.8) suggesting that information cannot always be reliably retrieved. We could

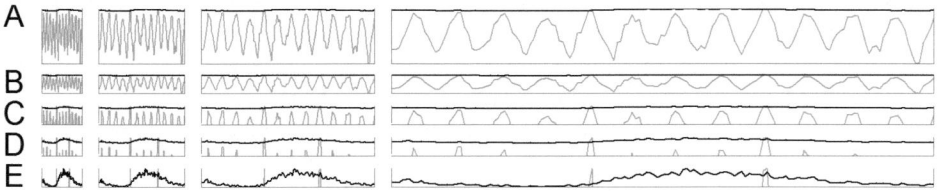

*Figure 2.12.7*  Visualizing two data series with contrasting demands in one graph. In the top
row (A), the details of the grey oscillation are revealed by stretching the graph.
More detail is seen by squashing the *y*-axis (B), which increases the aspect
ratio further, showing the different rates of increase and decline. However, the
pattern in the black data becomes increasingly hidden. Each stretch, and each
squash, flattens the black pattern. More could be seen of the black data in the
thinnest and, relatively, tallest plot (left hand side of A) where the grey data
was least visible. By zooming in the detail of trend, and fine scale oscillations
around that trend, are shown for the black data (C–E), but at the expense of
the grey data. In (E) we have contravened what some might call a golden rule
by truncating the *y*-axis. The format of a graph might not always suit all the
patterns it contains. Arbitrary data selections were downloaded and modified
from www.sidc.be/silso/datafiles and for the Waddington data station http://
data.giss.nasa.gov/ gistemp/stdata/.

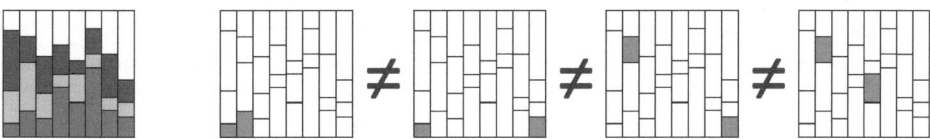

*Figure 2.12.8*  As shown by visualization studies (e.g. Talbot, Setlur and Anand 2014), making
comparisons within a single chart is subject to a variety of position dependent
errors. Comparisons of size are not equally accurate for the data highlighted in
each chart. Where the data are categorical a designer would select the
ordering, or enforce an arbitrary ordering such as alphabetical order, but each
design will affect what is seen in different ways.

then say, with some confidence, that visualizations are intrinsically biased, each design having its
own hallucinatory or jumbling effects (e.g. Kindlmann and Scheidegger 2014) that are specific
to the visualization technique or the data, or both. Despite the wealth of visualization research,
it is easy to get lost in Design Space.

## Degrees of freedom

Design Space might initially appear to be small for simple data sets. For instance, a scatterplot
might seem the only choice when visualizing two vectors of continuous data, such as for a
category in Anscombe's Quartet (e.g. $X_I$, $Y_I$ in Figure 2.12.2). Yet the axes of a scatterplot can
be aligned to produce a parallel-coordinates plot, then bent to simulate a chord diagram or hive
plot, or the values can be summed for a stacked bar chart, which can be bowed into a pie chart
and then punctured to produce a donut plot. Each jump in Design Space can modify the
meaning and information content (Figures 2.12.3, 2.12.4 and 2.12.5), even when the symbols,
shapes and scales are unchanged.

More and more degrees of freedom are presented to the designer as data increase in their
dimensionality or become dispersed in complex ways, or if the data contain multiscale attributes

or lack intrinsic ordering. As data become complicated there are more and more ways to sample and aggregate, and then to lay out, arrange and format the visual encodings. A designer can simultaneously reduce and increase the dimensionality of data by aggregating and framing the data, and introducing emphasis and interpretations that do not appear in the 'raw' data. Any particular design can, then, be considered as both arbitrary and specific.

## Designed by defaults

Without armies of avid draughtspersons, software are essential for mapping data within visual objects. Nonetheless, software coerces as it enables by suggesting, or insisting on, idioms that represent small portions of Design Space. These idioms need not neatly map onto designers' goals, or the data's structure, or even the best practices proposed by visualization research. For example, defaults do not always enable the 'small multiples' (Tufte 1990) rendition of Anscombe's Quartet without some tinkering. For more bespoke or artistic visualizations – that often inspire us to visualize (e.g. Figure 2.12.1(f)) – software might lack the templates and analytical functions necessary to emulate these works.

Different software reveal and optimize different design possibilities, determining what can be defined and manipulated programmatically, or otherwise. For instance, data manipulation and analysis might be easier in some software (*R*; www.r-project.org/), whereas control over the form of shapes and interactivity is easier in others (*Processing*; https://processing.org) and interactive web applications might be more naturally created elsewhere (*P5*; https://p5js.org/ and *D3*; https://d3js.org/). Each software offers a different view of Design Space. Spreadsheet applications can launch users towards apparently polished forms, but designing beyond the defaults requires flexibility within software, and the facilities to create new templates and functions in code, or by other means. Design Space is too vast, and its contingencies too many, to be entirely contained within defaults.

## Points of view

In his book review entitled 'Pretty vacant', Kevin Walker (2014) critiques the apparent hollowness of some visualizations which substitute function and precision with frivolity and fun. Despite being more likely to reside in coffee table books than to inform system-critical decisions, these 'vacant visualizations' can face incredibly strong criticisms that have included censorship campaigns. Neither art nor science, these visual stories do not necessarily claim any grand discoveries or offer experiential epiphanies. The vigorous critique is often aimed at explorations in Design Space that go beyond the software defaults.

However, the differences in this 'infographic' genre are not always recognized when it is critiqued. Vacant visualizations may use 'fun' illustrations and pictograms that aid memorability and recall (Borkin *et al.* 2015) and so improve understanding in ways that pared-back graphs cannot. This does not stop purists being concerned with the use of chart junk, in what they might already consider to be junk charts. Other approaches that use 'arbitrary encodings' (those that must be learnt through the visualization itself (Ware 2012)) are more readily accepted due to their apparent aesthetic qualities. Some propose that arbitrariness might stimulate deliberative reasoning which could benefit comprehension (Hullman, Adar and Shah 2011). This strategy might only work when aesthetics seduce the reader sufficiently for them to invest in decoding the images, though the seduction need not lead to anything more.

For the designer Georgia Lupi (2012), her approach is to use 'non-linear story telling' where multiple layers of information are overlaid in what others might deem to be ornamentation or complication of a design. Lupi says this method is 'just ours' (Lupi 2012), with no claims to advance visualization techniques in general. Rather than aiming for dimension reduction to optimize information retrieval, this artistic approach adds dimensionality to optimize interest. By intertwining what other disciplines would remove or separate into multiple panels, a more sublime, aesthetic appeal can be produced in designs that are cryptic despite using familiar forms of visual encoding. Not all areas of Design Space can be understood by studies that seek to minimize response times. Design Space can be understood in many opposing ways. Without stepping back from our own point of view there is a lot we can miss.

## 20/20 visualization

How we see the subject of visualization and see Design Space can be determined by a critical condition; a 'visualization myopia' that develops when research predominantly focuses on the design and interpretation of visualizations up to and including cognition. Perhaps this visualization myopia originates from an academic community rooted in computer graphics, and that reaches into perceptual and cognitive sciences. Then, Design Space is understood in most detail through functional errors revealed by experiments and through user-study reports on designing with-science, for-science.

Other perspectives obviously exist, each with its own characteristic focus and reciprocal myopias – such as an artist focused on unmapped techniques, or digital methods researchers focused on new translations. When visualizing data we are confronted with complementary, but distinctive approaches that have alternative views of how Design Space might be mapped (if it has to be), and how dimensions such as functionality, aesthetics and technology might be positioned in relation to Design Space. Design Space is, then, a key pivot point for all areas of study: a shared space within and onto which multiple values and perspectives can be imposed. Without looking out from Design Space on to visualization we could miss a 20/20 view of what happens when we visualize data.

These myopias share a lack of understanding of what happens post-cognition, after the visual is seen. What impact do visualizations actually have? And those that do not really visualize? Do visualizations reconfigure and feed back into cultural and socio-technical systems? And how do they do that? After all, the legend where the discovery of cholera transmission hinged on a visualization ignores that map's purpose and other forms of information and knowledge upon which decisions were made (Brody, Rip, Vinten-Johansen, Paneth and Rachman 2000; and see Figure 2.12.9). Similarly, a graph might not have averted the space shuttle disaster on its own (Robison, Boisjoly, Hoeker and Young 2012). In the wilds beyond the interfaces of engineers or economists or experimentalists, we engage with and respond to visualizations among a host of other data, information and knowledge.

## The visible spectrum

Right now we are exposed to the broadest spectrum of visualization expertise, literacy, use, tools and interest that has ever existed. At one set of extremes we have the purer reflections of the promise, where inspirational bespoke interactive visuals have inspired large changes in the prac-tices, structure and audiences of influential organizations. At the other extremes are people who do not know what visualizations are and do not use digital technologies.

This spectrum offers diverse opportunities to develop lenses to see beyond our myopias and beyond the myopic brouhahas of data-ink ratios. Probing this spectrum could help define and reconcile how Design Space might be mapped to concepts beyond functionality and

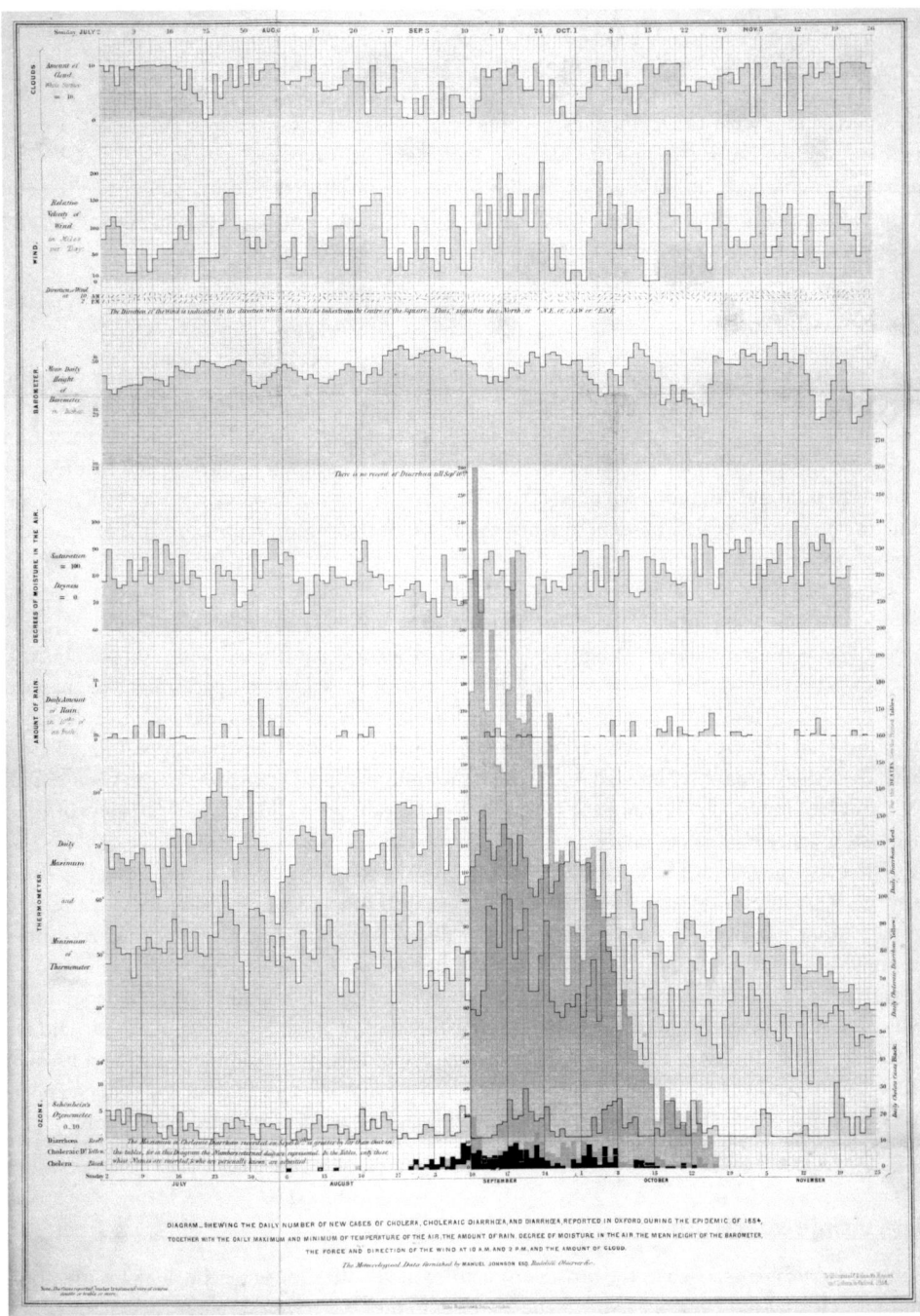

*Figure 2.12.9* Visualizing this data in this way did not help the discovery of cholera transmission during an outbreak in Oxford during 1854. Cholera incidence is visualized against many variables, but not necessarily variables that were relevant to the transmission mechanism. Image credit: http://libweb5. princeton.edu/visual_materials/maps/websites/thematic-maps/quantitative/ medicine/acland-chart.jpg

aesthetics, such as visual cultures, power and the construction of knowledge. At present, we often navigate these issues with speculation and opinions that do not challenge how our myopias determine what we see.

We know that data are, by definition cooked, maps are not the territory, and that visualizations are biased propositions. These issues of representation are not always revealed when inferences about data and systems are made using a visualization, nor in our analyses of how we visualize data. The question is, then, how visualizations elicit change as a joint function of data, design, technology, perception, cognition, cultures and socio-technical systems. Novel interdisciplinary approaches are undoubtedly needed to look out from Design Space, into this spectrum, with the aim of developing a 2020 view of visualizing data.

# References

Anscombe, F. J. (1973). Graphs in statistical analysis. *American Statistician*, 27(1): 17–21.

Borkin, M., Bylinskii, Z., Kim, N., Bainbridge, C., Yeh, C., Borkin, D., Pfister, H. and Oliva, A. (2015). Beyond memorability: visualization recognition and recall. *IEEE Transactions on Visualization and Computer Graphics*, 22(1): 519–528.

Brody, H., Rip, M. R., Vinten-Johansen, P., Paneth, N. and Rachman, S. (2010). Map-making and myth-making in Broad Street: The London cholera epidemic, 1854. *The Lancet*, 356: 64–68.

Gates, B. (2013). *Bill Gates: Dimbleby lecture.* [online] Retrieved 6 July 2016 from: www.gatesfoundation. org/media-center/speeches/2013/01/bill-gates-dimbleby-lecture

Hall, P. (2008). Critical visualization. In P. Antonelli (Ed.) *Design and the Elastic Mind* (pp. 122–131). New York, NY: Museum of Modern Art, Harrison.

Hullman, J., Adar, E. and Shah, P. (2011). Benefitting InfoVis with visual difficulties. *IEEE Transactions on Visualization and Computer Graphics*, 17(12): 2213–2222.

Kindlmann, G. and Scheidegger, C. (2014). An algebraic process for visualization design. *IEEE Transactions on Visualization and Computer Graphics*, 20(12): 2181–2190.

Lupi, G. (2012). Non-linear storytelling: journalism through 'Info-spatial' compositions. *Parsons Journal for Information Mapping*, IV(4): 1–11.

Robison, W., Boisjoly, R., Hoeker, D. and Young, S. (2002). Representation and misrepresentation: Tufte and the Morton Thiokol engineers on the Challenger. *Science and Engineering Ethics*, 8(1): 59–81.

Stefaner, M. (2014). *Worlds, not stories.* [online] Retrieved 6 July 2016 from: http://well-formed-data.net/archives/1027/worlds-not-stories

Tableau (2016). Answer questions as fast as you can think of them. [Online] Retrieved 29 April 2016 from: http://get.tableau.com/trial/p3group.html?width=300&height=300&inline=true

Talbot, J., Setlur, V. and Anand, A. (2014). Four experiments on the perception of bar charts. *IEEE Transactions on Visualization and Computer Graphics*, 20(12): 2152–2160.

Tufte, E. R. (1990). *Visual Explanations: Images and Quantities, Evidence and Narrative.* Cheshire, CT: Graphics Press.

Walker, K. (2014). Pretty vacant: what we're not seeing in graphics today. *New Scientist.* [Online] Retrieved 6 July 2016 from: www.newscientist.com/article/mg22429991-700-pretty-vacant-what-were-not-seeing-in-graphics-today/

Ware, C. (2012). *Information Visualisation: Perception for Design* (3rd ed.). Burlington, MA: Morgan Kauffman.

# Section 3
# Engaging and distributing

# 1

# Engaging and distributing

*Sybille Lammes*

The entries grouped under 'Engaging and distributing' share an interest in how engagement and distribution can be understood as important parts of methodological processes. The theme of *engaging* prompts us to think how we are involved with our research material, requiring us to consider how different actors, including both ourselves and our research subjects, are implicated in the process of doing research. Indeed, it spurs us to think how such involvements can be managed to provide new methodological opportunities. *Distributing* is inevitably linked to engaging, as when we acknowledge a multifaceted engagement in our approach, it leads us to consider knowledge production as dispersed and disseminated. And similarly, new modes of distribution can become an opportunity for methodological innovation rather than being the mere dissemination of already produced knowledge.

Engaging and distributing have always been of methodological concern. However, we are drawing attention to them here because of the new opportunities afforded by changes in research infrastructures including the rise of ubiquitous computing, social media, digital games, changes in user interfaces, initiatives relating to open data and open access, the growing opportunities and political drive towards participation and collaboration and increases in computational power. Some authors in this section discuss processes of engagement and distribution by focusing on a specific case, while others show us more general strategies. All work from the principle that engagement gives rise to processes in which positions, shapes and networks of knowledge production become distributed, transformable and mobile. They show how this rethinking has crystallized by considering: media archaeology (Parikka); mobile media (Büscher); net-research on political movements (Rogers); sensing atmospheres (Engelmann and McCormack); collaboration in data-sprints (Venturini, Munk and Meunier); phenomenological film analysis (Marks); and ethical approaches in play studies (Sicart). Hailing from diverse inter- and trans-disciplinary backgrounds, all authors ask questions pertinent to how methods can be sensitive to the ways in which we as researchers engage with our research materials, be it in the field of media studies, play studies, STS, design studies or critical geography. They shed light on how such engagements can give rise to, and inter-relate with, new modes of distribution and knowledge production.

## Hybridization: rethinking methods

The reflections in this section relate to an ongoing re-evaluation of how we as researchers can engage with and recognize distributed relations between ourselves and 'our' research materials in a constellation that Sasha Engelmann and Derek McCormack call 'the materiality of spacetimes' (that is, the spatio-temporal relations brought into existence in the bringing together of things, other people, the social, technical and natural environment). As we engage with research, we might recognize that the shape of our research field shifts and disperses through this engagement. Even more, our research also continually invites us as researchers to shift, shape and reconsider our position as researcher. If we accept this invitation, our methods will emerge in a reciprocal process of hybridization, in which our position as researcher is co-produced through what we engage with and (re-)distribute.

If, for example, we use a method like Digging, as Jussi Parikka proposes in his contribution, we will make use of the affordances of the instruments that we use for digging into the ground – a spade, a trowel, a pick – engaging with them and acknowledging their agency and affordances in how they contribute to what we want to achieve. Yet as they serve us, we serve them, hence hybridizing instruments, bodies, objects and subjects. Because of this fluidity or 'flux' (Hayles 2002), we cannot understand a method as a fully stable or a priori established thing. Instead, the capacities and qualities – or character – of a method are brought into being through the uses it invites us to employ and through the ground we want to cover by using 'it'.

## Modes of engaging: reflexivity and situatedness

One well-established way to recognize the co-production of methods, things and fields has been to pay attention to how we as researchers are materially situated within our research networks and to reflect consciously on how our modes of engagements unfold while doing research. What was learned from the reflexive turn in this respect is that there is much to be ascertained by situating ourselves in our research contexts, whether it concerns research about fan-culture, games, a design company, a political movement or the use of technologies in daily life.

Ways of making our situatedness visible to ourselves and others through reflexivity have been advocated in anthropology, gender studies, STS and other fields in the social sciences and humanities (Alvesson and Sköldberg 2009; Ateljevic, Harris, Wilson and Collins 2005; Davies 1999; England 1994; Woolgar and Ashmore 1988; Hughes and Lury 2013). These approaches underscore that researchers should pay methodological heed to processes of becoming involved with the research 'materials' that they engage. This recognition especially comes to the fore when engaging *in vivo* with research fields and real-time experiments 'in the wild', either virtual or physical (Alvesson and Sköldberg 2009; Ateljevic *et al.* 2005; Davies 1999; England 1994). After all, 'live' situations ask the researcher to constantly, consciously and directly adapt to how events unfold. It, therefore, becomes even more important that one constantly adapts to new research circumstances and that new connections with actors and other material come into being during the process.

Methods are tangibly dependent on what we encounter – and how that invites us to develop new strategies. We need to think on our feet, constantly translating methods from where we stand and what we (unpredictably) touch and engage with. In critical geography, for example, experimental projects require methods that account for our involvement in such experiments and the alternative knowledge produced by untried – and often interdisciplinary – encounters (Last 2012; McCormack 2013; Thompson 2015). These projects can be as diverse as the experimental architectural artistic practices discussed by Engelmann and McCormack, experimental work on sensing (Gabrys 2012), dance (McCormack 2008), art-labs (Jellis 2015) or art exhibitions (Paglen 2008).

More generally, over the last 50 years situatedness and reflexivity have been widely acknowledged in different disciplines as hybridizing modes of methodological engagement for what Celia Lury calls compound methods. Yet, as Karen Barad observes (1998), situatedness and reflexivity do not always lead to critical reflection. They can, for example, be used to tie knowledge production to problem solving and to a restricted notion of accountability to society, without acknowledging the hybridity of research and considering our partial position as researchers in the society to which we are accountable. Preferring refraction or diffraction to reflexivity to emphasize a more critical approach to research, feminist scholars such as Karen Barad (2014) and Donna Haraway (1988) argue that it is important to show how we as researchers are situated within our research in a strategically political way. Such an approach counters hegemonic epistemologies that use a sharp distinction between the researcher as active 'agent' and research subjects as passive and 'raw' material which have been described as positions linked to masculine and white ideologies of domination (Haraway 1988) and based on false dichotomies like culture/nature. The basic starting-point for these alternative critical approaches is to acknowledge our positionality as researcher and pay attention to how it shapes our approach and outcomes, as well as how methods come into being through an engagement with our research field. When researchers engage with their research materials, a process unfolds in which delineations between researchers, methods and others, can become increasingly untenable, or at least open to change and differential distributions. The essays in this section of the book present us with ways to acknowledge such processes and offer new approaches to make this a productive part of research methods.

## Modes of distributing: movements

Understanding methods as processual, allows us to explicate how the knowledges that methods can produce emerge in and through their distribution. Or as Monika Büscher maintains in her chapter, it enables us to underscore that methods are always (in) movement. With a background in mobility studies, Büscher draws attention to how mobile methods implicate movement on many levels, both in terms of participation and distribution. She maintains that it would be a mistake to understand mobile or moving methods solely as a situation in which researchers look for new approaches to research mobile media, such as car-navigation (Brown and Laurier 2005), social media apps such as Grindr (Licoppe, Rivière and Morel 2016) and many more mobile technologies. Moving methods are, for Büscher, rooted in a deeper understanding of reality as processual and 'becoming'. Furthermore, and maybe as a consequence, when a researcher engages with methods from this perspective it means they acknowledge that their approach can be moved itself, sometimes even pushed into making interdisciplinary translations with and via newly produced domains.

Such 'moving' attitudes towards research are of course intimately connected to how research methods foreground processuality by being transformative and dependent on the engagement of researchers and other actors. This engagement can also include the multisensory affects that are felt and performed when doing research. We might be bodily moved, to paraphrase Büscher, and this process includes more senses than just the eyes that we use for reflection, observation and visualization. Accounting for the multisensory as part of our method is a rather novel perspective and very little work has been done on this so far, in part because senses such as touch or smell (McLean 2012; Quercia, Schifanella, Aiello and McLean 2015) often go beyond words, yet can be made intrinsic to our methods. It is precisely these possibilities that

Engelmann and McCormack address in their chapter, describing experiments of sensing the atmospheric as a 'complex relationship between experience, technique and technology'. Also, film scholar Laura U. Marks engages with the multi-dimensionality of the senses and affect when she proposes in her contribution a phenomenologically informed method for film analyses, describing this as involving an examination of the bodily and non-cognitive dynamic engagements of researchers with the films they investigate.

Movement can also be linked to play, as play is not only a kind of movement (as, for example, in the play in a piece of machinery), but also moves us into other kinds of dynamic approaches to research. In his contribution, 'Playing with ethics', Miguel Angel Sicart maintains that play, which he approaches as dynamic and relational, is fundamental to how we understand change-able relations between play, 'playthings' (Sicart 2014: 70), players and researchers. Sicart argues that this relationality has consequences for how methods of playing address the fluid connections between the 'game' and the engaged position of the player. He draws attention to the ethical dimensions of this approach. After all, being playful means that we embrace risk and danger as part of our methodology. In Sicart's view, when we want to study play we should always aim to 'have access to the reflective work that players do when engaging with moral challenges' and make this part of the dynamics of methods for play.

The contribution from Tommaso Venturini, Anders Munk and Axel Meunier 'Data-sprinting' is a substantiation of Büscher's argument that moving methods need not only pertain to the physical mobility of researcher and research, but might also involve a deeper understanding of how research moves reality – in their case political reality. Rooted in Actor-Network-Theory (ANT) (Callon 1999, 1986; Latour 1996, 2005; Callon, Lascoumes and Barthe 2009; Law 2009), the method of controversy mapping embraces 'radical uncertainty' and an interdisciplinary approach in which the public are engaged as actors, echoing Sicart's advocacy of the value of engaging players in research. The data-sprints they describe are a method in which researchers engage with research 'subjects' and vice versa; they are intensive gatherings in which public controversies are addressed, situations *par excellence* where social 'realities are in the making and capricious'. The authors explore the ways in which positions between researcher and researched hybridize. Relatedly, Richard Rogers' chapter shows how digital methods enable researchers to engage within distributed networks, in his case, the Internet. In his essay, 'Issuecrawling', the ongoing engagement of the researcher with its field is recognized in a detailed 'manual' on the moves, methods and steps Rogers took when researching relations between right-wing populist websites.

## Engaging and distributing in the digital age

Many authors in this section share a keen interest in how research questions play out in our contemporary mediatized and often digitalized culture. This emphasis on the digital is in no way a coincidence, as such new digitalized 'networks of control' (Galloway 2004) have made possible new forms of engagement and distribution. Described as interactive (Coleman 2010), processual (Berry 2015), playful (Sicart 2014), accelerated (Cubitt 1999), or hybrid and conver-gent (Jenkins 2006; Chadwick 2013), such shifting networks or assemblages invite researchers to acknowledge the significance of diverse form of distribution for moving methods.

This is not to say that prior to the digital age no such distributed approaches existed or were needed, only that contemporary media have furthered the need for the development of (situated and) distributed methods. In a kind of methodological recursion, researchers are increasingly drawn to the use or doing of digital media to study them, and this doubled reflexive engagement has produced new research methods that emphasize the significance of the processual and incorporate the networked possibilities of research methods.

The subfield of digital methods (see also Gerlitz, this volume) is an influential example of how the digital era has given rise to new networked approaches. Broadly speaking, digital methods are a series of methods that are digital in themselves – as opposed to being digitized – with a strong emphasis on web-based techniques to follow digital traces so as to allow the capturing of the action of distributed networks (Rogers 2009). An example is the crawling technique used by Rogers in his chapter, the method he uses to map distributed networks of political connections on the Internet. Other methods that have been developed in this subfield include scraping (Marres and Weltevrede 2013), controversy analyses (Marres 2015; Venturini 2012) and diverse methods for analysing the multiplicities of 'real-timeness' (Weltevrede, Helmond and Gerlitz 2014). Yet other methods have also emerged that might not neatly fall under the denominator of digital methods, but are still intimately related to new digital phenomena, including mobile methods that make use of digital audio-visual devices for more and other purposes than what Rogers would call a 'digitized' recording method (2009), for example digital ethnographies (Pink *et al.* 2015) that include participant-observations of the use of mobile phones and navigation devices (Wilmott 2016; Hjorth and Pink 2014), or (auto)ethnographic methods for researching digital games (Witkowski 2015; Cuttell 2015; Taylor 2009). All these methods share an interest in accounting for the engaged and distributed relations between researchers and digital research materials and methods.

## Conclusion: the -ings of things

Engaging and distributing as tropes in research not only draw attention to how researchers participate in and are involved with their research materials and shape and share their research objectives through them, they also prompt us to acknowledge the relational dynamics of digital and non-digital things and assemblages. Engagement nudges us to develop adaptive and reflexive methods for producing and visualizing results. It also stimulates us to recognize the manifoldness of methods, asking us to consider knowledge production as a co-production, emerging in relations with other actors in the assemblage as well as ourselves, such as players, citizens, and also other things and species (Haraway 2003, 2013).

As a consequence, our research can often seem 'messy' (Law 2004). Methods only come into being as part of complex and dynamic assemblages that can at times confuse us. In the networks that we spin, we have to deal with the play in and of translations, or in Latourian terms with hybridizations that are constantly shifting, undergoing redistribution, depending on what we do, where we stand, and where we wish to go. In fields such as ANT, cultural studies (Alvesson and Sköldberg 2009; Ateljevic *et al.* 2005; Davies 1999), (digital) media studies (Hess 2001; Pink 2008) and critical geography (Wittel 2000), we have seen a proliferation of new approaches that try to recognize such dynamics. The chapters in this section seek to recognize this complexity by working from an understanding of researcher and researched as not fixed but in flux. In these fields, engaging and distributing are key to how we consider and acknowledge the translations of knowledge that come into being through a critical involvement between researchers, other actors, methods and things.

## References

Alvesson, M. and Sköldberg, K. (2009). *Reflexive Methodology: New Vistas for Qualitative Research*. London, Thousand Oaks, New Delhi: Sage Publication.

Ateljevic, I., Harris, C., Wilson, E. and Collins, F. L. (2005). Getting 'entangled': reflexivity and the 'critical turn' in tourism studies. *Tourism Recreation Research*, 30(2): 9–21.

Barad, K. (1998). Getting real: technoscientific practices and the materialization of reality. *Differences: A Journal of Feminist Cultural Studies*, 10(2): 87–91.

Barad, K. (2014). Diffracting diffraction: cutting together-apart. *Parallax*, 20(3): 168–187.

Berry, D. M. (2015). *Critical Theory and the Digital*. New York, NY: Bloomsbury Publishing USA.

Brown, B. and Laurier, E. (2005). Maps and journeys: an ethno-methodological investigation. *Cartographica: The International Journal for Geographic Information and Geovisualization*, 40(3): 17–33.

Callon, M. (1986). The sociology of an actor-network: the case of the electric vehicle. In M. Callon, A. Rip and J. Law (Eds.) *Mapping the Dynamics of Science and Technology* (pp. 19–34). London: Macmillan.

Callon, M. (1999). Actor-network theory: the market test. *The Sociological Review*, 47(S1): 181–195.

Callon, M., Lascoumes, P. and Barthe, Y. (2009). *Acting in an Uncertain World: An Essay on Technical Democracy* (G. Burchell, Trans.). Cambridge, MA: MIT Press.

Chadwick, A. (2013). *The Hybrid Media System: Politics and Power*. Oxford: Oxford University Press.

Coleman, E. Gabriella (2010). Ethnographic approaches to digital media. *Annual Review of Anthropology*, 39: 487–505.

Cubitt, S. (1999). Virilio and new media. *Theory, Culture & Society*, 16(5–6): 127–142.

Cuttell, J. (2015). Arguing for an immersive method: reflexive meaning-making, the visible researcher, and moral responses to gameplay. *Journal of Comparative Research in Anthropology and Sociology*, 6(1): 55–75.

Davies, C. A. 1999. *Reflexive Ethnography: A Guide to Researching Selves and Others*. London, New York: Routledge.

England, K. (1994). Getting personal: reflexivity, positionality, and feminist research. *The Professional Geographer*, 46(1): 80–89.

Gabrys, J. (2012). Sensing an experimental forest: processing environments and distributing relations. *Computational Culture*, 2: n.p.

Galloway, A. R. (2004). *Protocol: How Control Exists after Decentralization*. Cambridge, MA: The MIT Press.

Haraway, D. J. (1988). Situated knowledges: the science question in feminism and the privilege of partial perspective. *Feminist Studies*, 14(3): 575–599.

Haraway, D. J. (2003). *The Companion Species Manifesto: Dogs, People, and Significant Otherness*. Vol. 1. Chicago, IL: Prickly Paradigm Press.

Haraway, D. (2013). *Simians, Cyborgs, and Women: The Reinvention of Nature*. New York, NY: Routledge.

Hayles, K. N. (2002). Flesh and metal: reconfiguring the mindbody in virtual environments. *Configurations*, 10(2): 297–320.

Hess, D. (2001). Ethnography and the development of science and technology studies. In P. Atkinson, A. Coffey, S. Delamont, J. Lofland and L. Lofland (Eds.) *Handbook of Ethnography* (pp. 234–245). London: SAGE.

Hjorth, L. and Pink, S. (2014). New visualities and the digital wayfarer: reconceptualizing camera phone photography and locative media. *Mobile Media & Communication*, 2(1): 40–57.

Hughes, C. and Lury, C. (2013). Re-turning feminist methodologies: from a social to an ecological epistemology. *Gender and Education*, 25(6): 786–799.

Jellis, T. (2015). Spatial experiments: art, geography, pedagogy. *Cultural Geographies*, 22(2): 369–374.

Jenkins, H. (2006). *Convergence Culture: Where Old and New Media Collide*. New York, NY: New York University Press.

Last, A. (2012). Experimental geographies. *Geography Compass*, 6(12): 706–724.

Latour, B. (1996). On actor-network theory: A few clarifications. *Soziale welt*, 47: 369–381.

Latour, B. (2005). *Reassembling the Social: An Introduction to Actor-Network-Theory*. Clarendon Lectures in Management Studies. Oxford, New York: Oxford University Press.

Law, J. (2004). *After Method: Mess in Social Science Research*. London: Routledge.

Law, J. (2009). Actor network theory and material semiotics. In B. Turner (Ed.) *The New Blackwell Companion to Social Theory* (3rd ed., pp 141–158). Oxford: Blackwell.

Licoppe, C., Rivière, C. A. and Morel, J. (2016). Proximity awareness and the privatization of sexual encounters with strangers: the case of Grindr. In C. Marvin, S. Hong and B. Zelizer (Eds.) *Context Collapse: Re-assembling the Spatial*. London: Routledge.

Marres, N. (2015). Why map issues? On controversy analysis as a digital method. *Science, Technology, & Human Values*, 40(5): 655–686.

Marres, N. and Weltevrede, E. (2013). Scraping the social? Issues in live social research. *Journal of Cultural Economy*, 6(3): 313–335.

McCormack, D. P. (2008). Geographies for moving bodies: thinking, dancing, spaces. *Geography Compass*, 2(6): 1822–1836.

McCormack, D. P. (2013). *Refrains for Moving Bodies: Experience and Experiment in Affective Spaces*. Durham, NC: Duke University Press.

McLean, K. (2012). Emotion, location and the senses: a virtual dérive smell map of Paris. *Proceedings of the 8th International Conference on Design and Emotion: Out of Control*, September 11–14. London: Central Saint Martins College of Art & Design.

Paglen, T. (2008). Experimental geography: from cultural production to the production of space. *Experimental Geography*, 28: n.p.

Pink, S. (2008). An urban tour: the sensory sociality of ethnographic place-making. *Ethnography*, 9(2): 175–196.

Pink, S., Horst, H., Postill, J., Hjorth, L., Lewis, T. and Tacchi, J. (2015). *Digital Ethnography: Principles and Practice*: Los Angeles, CA: Sage.

Quercia, D., Schifanella, R., Aiello, L. M. and McLean, K. (2015). Smelly maps: the digital life of urban smellscapes. *arXiv preprint arXiv:1505.06851*.

Rogers, R. (2009). *The End of the Virtual: Digital Methods*. Amsterdam: Amsterdam University Press.

Sicart, M. (2014). *Play Matters*. Cambridge, MA: MIT Press.

Taylor, T. L. (2009). *Play between Worlds: Exploring Online Game Culture*. Cambridge, MA: MIT Press.

Thompson, N. (2015). *Experimental Geography: Radical Approaches to Landscape, Cartography, and Urbanism*. Brooklyn, NY: Melville House.

Venturini, T. (2012). Building on faults: how to represent controversies with digital methods. *Public Understanding of Science*, 21(7): 796–812.

Weltevrede, E., Helmond, A. and Gerlitz, C. (2014). The politics of real-time: a device perspective on social media platforms and search engines. *Theory, Culture & Society*, 31(6): 125–150.

Wilmott, C. (2016). In-between mobile maps and media movement. *Television & New Media*. doi: https://doi.org/10.1177/1527476416663637.

Witkowski, E. (2015). Running with zombies: Capturing new worlds through movement and visibility practices with zombies, run! *Games and Culture*, 13(2): 153–173.

Wittel, A. (2000). Ethnography on the move: from field to net to internet. *Forum Qualitative Sozialforschung/Forum: Qualitative Social Research*, 1(1): n.p.

Woolgar, S. and Ashmore, M. (1988). The next step: an introduction to the reflexive project. In *Knowledge and Reflexivity: New Frontiers in the Sociology of Knowledge* (pp. 1–11). London, Thousand Oaks, New Delhi: Sage Publications.

# 2

# Affective analysis

*Laura U. Marks*

Over the years I have developed a simple method for analysing movies, artworks and other phenomena by working through affective, perceptual and conceptual responses. Affective analysis is a kind of aesthetic analysis that begins by analysing affective and embodied responses. Often critical analysis begins with formal analysis of the perceptible qualities of a work; or it jumps straight to the level of discourse. Affective analysis draws thought back to the body, forcing us to generate new thoughts, or face the fact that we do not yet have thoughts. It works as a 'reality check' to slow intellectual responses and to guarantee that, when we arrive at them, they will be well grounded and relatively free of ideology. It may generate what Spinoza terms adequate ideas, or ideas that align the powers of the body with the capacities of the mind in a given situation. I use it in encounters with a film or artwork, in studio visits to artists, when reading and in everyday situations. Over the years I have taught it to many students in classes and workshops, and it works well in itself, or as the basis for further research.

I first realized the need for this method some years ago when I was watching *Charlie's Angels* (2000). There is a scene where the brave Dylan, Drew Barrymore's character, is betrayed by her erstwhile lover moments after they have had sex. After he and his sidekick explain the conspiracy, he shoots at Dylan. She throws up her arms and falls dramatically backward through the plate-glass window of the high hotel room: presumably she falls to her death. The film cuts to another scene and then returns to explain, in slow motion, what happened: The bullet somehow strikes not Dylan but the window behind her. She falls back, in a cascade of shattered glass. The bedsheet catches on the window ledge, saving her life; and there she hangs, grasping the sheet, now completely naked, as the conspirators leave the hotel room. During these scenes I noticed that I got goosebumps and felt aroused! Even though the film was about 'empowered', sexy, fighting women, my joyful affective response arose not from these representations but from an image of a woman menaced and vulnerable (though managing to survive). This startling response showed me that if I analysed only the conceptual or narrative content of the film, I would entirely miss how the film worked.

Often artworks and other cultural phenomena operate differently at the molar and molecular levels, and our responses at these levels differ as well. As Elena del Río (2008) explains, the molar level deals with bodies as a whole; it supports identity politics, struggles against constraints and struggles for representation. The molecular level deals with energies that are not yet captured by these discourses of identity. It provides a source of energy for molar-scale struggles. Doing affective analysis we are working to identify our responses along continua from the molecular level to the molar level, from the non-discursive to the discursive; from those parts of experience that seem free of culture and ideology to those that are clearly cultural and ideological.

Affective analysis is grounded in the philosophy of Spinoza, Deleuze and Guattari, as well as existential phenomenology; its triadic method derives from the logic of C. S. Peirce. Affective analysis takes place in the body of a specific beholder, but it is objective, because it identifies empirical, sometimes physical data that arise in the aesthetic encounter. We are using our bodies to do philosophy. Affective analysis isolates the three analytical categories of affect, percept and concept (Deleuze and Guattari 1994). In Peirce's logic, Firstness is a possibility, 'a mere may-be', as redness is a possibility before it is embodied in something red (Peirce 1955: 81). This is the level of affect. Peirce's Secondness is the realm of actuality, in which one thing is constrained by another or two things struggle with one another: I identify this with perceptibles.[1] In Thirdness, a third element enters to carry out a relation between two things, as in comparison, judgement, prediction and interpretation: this is the level of concept. My method consists simply of comparing the affect and percept that arise together at a specific moment in order to create a concept that adequately explains how what we perceived gave rise to, or occurred simultaneously with, that affective response (e.g. why did the image of naked suffering Drew Barrymore thrill and arouse me?). That should generate a useful concept that can direct further research. When there is no noticeable affective response, we can carry out the analysis by accounting for other kinds of embodied response, as I will explain.

Affective analysis accounts for the experience within individual sensation of forces that come from without. Guattari describes aesthetic encounters as 'blocks of mutant percepts and affects, half-object half-subject, already there in sensation and outside themselves in fields of the possible'. Paradoxically, as he points out, affects that come from beyond are catalysed by representations (Guattari 1992: 92–93). Thus we usually experience affect, perception and concept all at once, balled up, as it were. Affective analysis draws this ball of responses into a line. Doing this might feel rather artificial, but it helps to slow the path from affect to percept to concept, which makes it possible to produce well-grounded concepts.

Before concept, perception. In the encounter with a work of art, critics are often under pressure to quickly come up with concepts. However, conceptual analysis tends to respond to an artwork as a representation. The first analysis might sound smart, but it is likely glib, reactive and unable to account for what the artwork *does* to the perceiver. That representational reflex, David Raskin argues, can be countered by the 'natural, realist position that our conjunctive conceptions and perceptions are enmeshed in an emerging and material world' (Raskin 2009: 69). Accepting for the moment Raskin's realist position, a first step in postponing the conceptual reaction is to focus trustingly on perception: to describe without judging, in the method of phenomenology.

Another reason to focus on perception is that contemporary media technologies treat perception as merely an interface to information. Thus phenomenological methods that enlarge our sensory capacities and skills constitute a strong defence against the cultural-economic tendency to make people information processors.

Before perception, affect. Unfortunately, even our sincerest acts of perception are menaced by habit. As I note elsewhere, 'Habit (Peirce), conventional perception (Bergson), and cliché (Deleuze) form the skin that holds an individual together in a predictable attitude' (Marks 2010: 17). Perception is, of course, shaped by history and culture. It does not give complete access to the world; in fact, as Bergson pointed out, perception *protects* us from the world by focusing on survival. Perception is colonized. The reactiveness of perception is exacerbated by technologies that inform how it is possible to perceive. Moreover, while the close bodily senses of touch, taste and smell may create a temporary private *Umwelt*, even these senses may deliver our body to capital.[2]

Thus an adequate analysis needs to begin with affect, noncognitive thought, which Deleuze defines as 'every mode of thought insofar as it is non-representational' (Deleuze 1988). In Spinoza's two terms often both translated as 'affect', *affectus* denotes the encounter between bodies, *affectio* the resulting modification 'by which the body's power of acting is increased or diminished, aided or restrained' (Spinoza 1901: Part III, prop. I). Affective analysis focuses on the modification, *affectio*, in order to identify the encounter, *affectus*. Affective analysis treats the encounter as capable of opening in two directions, both potentially infinite: 'outward' to thought and 'inward' to matter (Marks 2008). Affect indicates the fold between thought and matter, which Spinoza (1901: Part II, prop. VII) argued are the same thing, considered according to different attributes.

In Deleuze's adaptation of Spinoza, he emphasized that it need not be the whole human body that responds to an affective encounter, but rather that parts of our bodies may enter into combinations with the other entities we encounter. He and Guattari suspected that the body as a whole was overcoded by ideology. Hence they privileged the molecular nature of these encounters over the larger, molar scale at which meaning, narrative, thought and even emotion take place. This shift of emphasis to the molecular informs the influential argument of Brian Massumi (2002) that the activities of affect are best detected at the level of the autonomic nervous system.

Before affect. Autonomic responses such as goosebumps, arousal, blushing yield valuable data in affective analysis. However, in years of practice I have found that these and other autonomic responses can encode cultural ideologies. Moreover, contemporary media increasingly bypass perception to mobilize affect with unprecedented skill. Many argue that social media, computer games and other surveillant entertainments instrumentalize humans' very synapses and contribute to the production of what Pasi Väliaho (2014) calls the 'neoliberal brain'. For these reasons, we cannot assume that our affective responses yield adequate ideas. Therefore, we need to use critical precision to identify the relations that occur between affect, percept and concept – as well as the extra categories I suggest below of embodied response and feeling – in a given situation. Affective analysis works case by case.

Here's how to do it:

Choose a particular moment in your experience of the artwork (all art is time-based, because we experience it in time) or event that seems especially dense, like that ball of affect–percept–concept that I mentioned, or that especially pleases, excites or troubles you. Note and set aside any initial concepts you have about it. You will be making a triadic analysis, following Peirce's logic: in which affect is First, percept is Second and concept is Third:

Affect – Percept – Concept

However, you might need to interpolate a couple of half-steps: embodied response and feeling:

Affect – embodied response – feeling – Percept – Concept

1 *Affect.* Identify the affective responses or non-cognitive thoughts that you experience at that moment. First, you might have the good luck to experience autonomic nervous system responses. Shivering, goosebumps or hardened nipples; arousal; blushing; a rush of adrenaline; twitching of the forehead or upper lip; and other responses over which you have no control all constitute precious data. This response comes from something like the animal in you. However, as I noted above, even at the autonomic level our bodies are informed by culture.

It may be that you experience none of these. Thus the next step is to identify embodied responses that are likely learned and culturally grounded. Are there tears in your eyes? Is your throat constricted? Notice what else your face – that surface that gathers micro-movements but is unable to act, to move away, protect itself, or fight – is doing (Deleuze 1986: 87–88). For example, there are many kinds of smiles – a grin, a smirk, a rictus: which one is happening on your face? Similarly, there are many kinds of laughter, such as a belly laugh, snort, giggle or embarrassed laugh. (Embarrassment is very useful data!) Turn your attention to the Spinozan definition of affect as a movement to a greater or lesser power of action as you notice your bodily state. Cringing, grimacing, agitation; elation, a 'bursting' feeling; calm; feeling yourself open up or close down: these embodied responses are examples of Spinozan affects. Do you feel tickled? Slapped around? Such responses also call up Vivian Sobchack's point, drawing on the argument of George Lakoff and Mark Johnson, that metaphors are not arbitrary but based in embodied experience (Sobchack 2004: 53–84).

In this method I try to avoid the category of emotion, since it so often results from manipulation. However, my students' sensitive accounts of their feelings in response to a work of art taught me that feeling is a useful category to include in the expanded notion of affective response. I use this term in an underdetermined way to indicate responses that fall somewhere between embodied response and emotion. Feelings such as wistfulness, elation and longing correspond closely to Spinoza's terms and can still fall short of the more coded emotions telegraphed by the work under study.

2 *Percept*. Describe impartially all that you perceive with all your senses. Work to be as precise as possible, for it is likely the singularity of a colour, a rhythm, a shape, a scent, or another perceptible that gave rise to our affective response. At this point, a sophisticated phenomenology kicks in: one that attends to what the body becomes in the act of perception. Here we have to acknowledge that perception requires us to 'make assumptions about the world according to the systems that have already been given, according to a world that [precede us], that is given by others' (Fielding 2009). Perception is blurred by convention, but it is rich with singular data nonetheless. The longer you postpone recognition of what is before your senses, the richer and more precise your description will be.

3 *Concept*. Compare the affect and percept that arose at a given moment to move toward a concept tailored to that encounter. The well-formed concept might prove to be a Spinozan adequate idea, in that it matches the powers of the body and the capacities of the mind in a given situation (Deleuze 1988: 74, 85). In this case, affective response will give rise to an adequate idea that increases understanding. This Spinozist turn in the theory of affect draws on the thought of Deleuzian feminists Rosi Braidotti (2011), Elizabeth Grosz (2008) and Mai Al-Nakib (2013), who seek to identify practices that can increase joyful affects and develop adequate ideas.

Here's how affective analysis works on my *Charlie's Angels* example. My affective response occurred at the autonomic level: arousal and goosebumps. I described what happened narratively in the scene, but what I perceived that gave rise to those responses were the loud crash as the young woman's body smashes backward through the window, the whole window shattering into sparkling shards, and, later, her smooth naked body as she clings to the sheet hanging from the window ledge. These were the moments that displayed her greatest vulnerability. Comparing my affective response with what I perceived, I conclude that I was aroused by a spectacularly sadistic image of a woman in peril. This response dismays me, because it suggests that I share my society's general misogyny at a fundamental level – one that, in a Spinozan sense, inhibits my capacity to live and increase my powers. Now I can analyse *Charlie's Angels* as a movie that propounds a 'positive' image of women in its representations but draws its power from affects of gleeful misogyny. An irritatingly large number of movies work this way. Thus my affective analysis draws to a disappointing close.

However, sometimes what we arrive at in comparing affect and percept may be, if we are honest with ourselves, nothing. This result echoes Deleuze's observation, 'Not that the body thinks, but, obstinate and stubborn, it forces us to think, and forces us to think what is concealed from thought, life' (Deleuze 1989: 189). At the conclusion of the careful process of affective analysis, an incapacity to think, to bring together what we perceived and what we felt, can function as a painful marker for a thought yet to come.

At this stage we need to carefully distinguish our conclusion (or lack of one) from any initial concepts we had before beginning the exercise. Our initial concepts might be supported by the affective analysis, in which case, bravo! But if the affective analysis does not support the initial concept, most likely it was not our own concept but a habit of thought. For example, we may get affective responses not to the perceptible image but to an idea that it stimulates. Similarly, we might have responded affectively not to the perceptual event itself but to personal memories and associations to which it gave rise. Both of these are noteworthy responses, but on their own they will not give rise to a strong concept. It helps to take note of these responses, set them aside for later research and begin the process anew.

4 Finally, employ the resulting triadic concept, or the painful triadic marker in lieu of a concept, as the leaping-off point for research that sheds light on the affect–percept relationship you have discovered. Let the research revise your understanding and, in Peircean fashion, produce a new object for a next round of affective analysis.

## Notes

1 Here my categorization diverges from Deleuze's in *Cinema 1*, which identifies perception as a degree zero, affection as First, action as Second and reflection as Third.
2 As I argue in Chapter 4 of *The Skin of the Film: Intercultural Cinema, Embodiment, and the Senses* (Durham, NC: Duke University Press, 2000).

## References

Al-Nakib, M. (2013). Disjunctive synthesis: Deleuze and Arab feminism. *Signs*, 38(2): 459–482.

Braidotti, R. (2011). *Nomadic Theory: The Portable Rosi Braidotti*. New York, NY: Columbia University Press.

del Rio, E. (2008). *Deleuze and the Cinemas of Performance: Powers of Affection*. Edinburgh: Edinburgh University Press.

Deleuze, G. (1986). *Cinema 1: The Movement-image* (Trans. H. Tomlinson and B. Habberjam). Minneapolis, MN: University of Minnesota Press.

Deleuze, G. (1988). *Spinoza: Practical Philosophy* (Trans. R. Hurley). San Francisco, CA: City Lights.

Deleuze, G. (1989). *Cinema 2: The Time Image* (Trans. H. Tomlinson and R. Galeta). Minneapolis, MN: University of Minnesota Press.

Deleuze, G. and Guattari, F. (1994). *What Is Philosophy?* (Trans. H. Tomlinson and G. Burchell). New York, NY: Columbia University Press.

Fielding, H. A. (2009). Maurice Merleau-Ponty. In F. Colman (Ed.) *Film, Theory, and Philosophy: The Key Thinkers* (pp. 81–90). Durham, UK: Acumen.

Grosz, E. (2008). *Chaos, Territory, Art: Deleuze and the Framing of the Earth*. New York, NY: Columbia University Press.

Guattari, F. (1992). *Chaosmosis: An Ethico-aesthetic Paradigm* (Trans. P. Bains and J. Pefanis). Bloomington, IN: Indiana University Press.

Marks, L. U. (2000). *The Skin of the Film: Intercultural Cinema, Embodiment, and the Senses*. Durham, NC: Duke University Press.

Marks, L. U. (2008). Thinking multisensory culture. *Paragraph*, 31(2): 123–137.

Marks, L. U. (2010). *Enfoldment and Infinity: An Islamic Genealogy of New Media Art*. Cambridge, MA: MIT Press.

Massumi, B. (2002). *Parables for the Virtual: Movement, Affect, Sensation*. Durham, NC: Duke University Press.

Peirce, C. S. (1955). The principles of phenomenology. In J. Buchler (Ed.) *Philosophical Writings of Charles Sanders Peirce*. New York: Dover.

Raskin, D. (2009). The dogma of conviction. In F. Halsall, J. Jansen and T. O'Connor (Eds.) *Rediscovering Aesthetics: Transdisciplinary Voices from Art History, Philosophy, and Art Practice* (pp. 66–74). Palo Alto, CA: Stanford University Press.

Sobchack, V. (2004). *Carnal Thoughts: Embodiment and Moving Image Culture*. Berkeley, CA: University of California Press.

Spinoza, B. (1901). *The Ethics* (Trans. R. H. M. Elwes). Project Gutenberg.

Väliaho, P. (2014). *Biopolitical Screens: Image, Power, and the Neoliberal Brain*. Cambridge, MA: MIT Press.

# Data-sprinting

## A public approach to digital research

*Tommaso Venturini, Anders Munk and Axel Meunier*

It is controversies of this kind, the hardest controversies to disentangle, that the public is called in to judge. Where the facts are most obscure, where precedents are lacking, where novelty and confusion pervade everything, the public in all its unfitness is compelled to make its most important decisions.

*Lippmann 1927: 121*

## What's in a data-sprint?

Data-sprints are intensive research and coding workshops where participants coming from different academic and non-academic backgrounds convene physically to work together on a set of data and research questions.

Data-sprints have their roots in a series of organizational innovations introduced in the field of open-source development at the turn of the century (as a reaction to the previous 'waterfall approach' inherited from the engineering management (Raymond 2001)). Faced with radical uncertainty about how their project will develop and who will join them, open-source developers invented a form of coding event called 'barcamps' or 'hackathons' (or hacking marathons). Such formats consist of short events in which a group of developers and designers meet to work intensively and expeditiously on some digital object.

Many features of hackathons and barcamps fit the needs of interdisciplinary research extremely well. We appreciate in particular:

1   The heterogeneity of the actors involved. Hackathons and barcamps are generally organized to be open to many different types of actors. In part, this comes from the need to achieve deliverable results at the end of the event, which requires all the necessary competences to be brought together through all the phases of the project. In developing marathons, this translates into having experts from the entire programming stack: from setting up the server infrastructure, to designing the wireframes, from scraping the data to implementing the front-end. The push for heterogeneity also derives from the necessity to exchange with the potential end users of the projects, who should be at hand during the developing dash.

2   The effort to convene participants physically. The unity of time and place that characterizes hackathons and barcamps is an appropriate counterbalance to the dispersion of research

efforts often observed in international and interdisciplinary projects. One of the problems of working across disciplines is that experts in one field have a blurred appreciation of what experts in other fields might need as an input for their work. Such misunderstandings are normal in interdisciplinary projects and can become disastrous if discovered too late – a risk particularly salient for international projects. Yes, technologies for distant cooperative work can ease some of these difficulties, but nothing facilitates mutual supervision or speeds up collaboration more than direct presence. One more time, 'digital' turns out to be opposed to 'virtual'. Exploiting digital inscriptions demands the coordination of the efforts of many different disciplines and this in turn demands that they be brought together in the same space and time.

3   The 'quick and dirty' (or 'design to cost') approach. Though thriving on the increase in the availability of digital inscriptions, hackathons and barcamps are somewhat opposed to 'big data' approaches. The short and intensive nature of these events shields them from the dream of exhaustivity often associated with 'big data'. Participants know that they will only be able to treat a limited amount of digital traces and that they will achieve imperfect results, but they accept such constraints more as a challenge than as a weakness. Making the most out of light infrastructures, simple logistics and agile organization methods, participants are well aware that their work should hack code and information gathered in earlier projects and that their outcomes will become the basis for further ventures. It is not only hackathons and barcamps that foster iteration, but they are explicitly conceived as intermediary steps of a larger developing cycle.

With the format of the data-sprint, we tried to adapt hackathons and barcamps to the practice of academic research by adding the larger efforts of 'contextualization' before, during and after the event:

1   Data-sprints are always preceded by a long and intense period of preparation. When participants meet up, most of the research infrastructure should have already been collected and prepared for treatment. Time-consuming operations such as data cleansing or infrastructural setting-up should be accomplished beforehand, so that the days of the sprint can be dedicated entirely to the operations that require a more direct collaboration. Also, participation in data-sprints is not open: sprinting lineup and team formation need to be taken care of in advance to make sure that the working groups contain all the competences needed to achieve significant results.
2   Data-sprints are also generally longer and more structured than their antecedents. While hackathons and barcamps are usually organized to last two or three days, sprints work better when they extend over a full working week.
3   Finally, data-sprints require a greater follow-up than hackathons and barcamps. The 'quick and dirty' approach that characterizes the five days of a sprint should be complemented by an extensive work of refinement and documentation, in order to endow the results with the precision and robustness demanded by scientific research.

For the sake of clarity, it is possible to pull out six different phases of a data-sprint that, though mingled in the practice of data-sprints (because of their flexible and iterative nature), correspond to distinctive organizational concerns:

1   *Posing research questions.* Research questions are posed on the first day of the sprint by the invited issue experts. Besides suggesting research questions, issue experts are also invited to

help the other participants (most of whom have little previous knowledge of the issue at stake) to get to grips with the topic of the meeting. This can be done through Q&A sessions or panel discussions, but also (and often more fruitfully) through informal consultations as part of the running feedback on data visualizations.

2   *Operationalizing research questions into feasible digital methods projects.* In a sense, this process begins already before the sprint when the organizers try to anticipate what type of projects the sprint might lead to. We found that an excellent way of doing this initial vetting is to ask issue experts to suggest interesting datasets. This provides a chance to get back to the experts, explaining why the proposed dataset might be unsuitable for certain research questions, thus getting them attuned to what a digital methods project can and cannot achieve.

3   *Procuring and preparing datasets.* As mentioned above, while it is desirable to have datasets available in advance, this is sometimes at odds with the agility of the sprint and it is not uncommon that complementary data have to be searched for and collected in the first days of the sprint.

4   *Writing and adapting code.* Sprints are issue-specific (that is, they are meant to address the needs of the controversy actors) and their aim is less to develop generic tools than to adapt existing code to the research questions raised by the issue experts. This does not mean, however, that effort should not be invested in making datasets, scripts and visualizations re-usable beyond the original project. Sprints should remain faithful to their communitarian roots and ensure that all the data, code and contents produced are liberated through open-source, copy-left and open-publishing licences.

5   *Designing data visualizations and interfaces.* One of the driving forces of sprints is that they deliver tangible outcomes. These outcomes might have different forms, but they always share the characteristic of being directly usable by actors of the controversy. In many cases, this translates to issue experts leaving the sprints with tangible results that they can immediately mobilize in their debates.

6   *Eliciting engagement and the co-production of knowledge.* Data-sprints abide by the 'co-production of knowledge model' of social sciences advocated for by Callon, Lascoumes and Barthe (1999). This approach assumes that scientific activities should be pursued in a constant and genuine dialogue with their publics. If data-sprints take shape in the five phases described above, it is this final phase that is most significant for if they fail to create a common space for social scientists and social actors, they will have failed in all other respects as well.

## EMAPS and the example of climate adaptation

To illustrate a research situation in which data-sprinting can be useful, we draw here on a concrete experience of a three-year EU-funded collaborative project called EMAPS (Electronic Maps to Assist Public Science). EMAPS was a project in controversy mapping (Venturini 2010, 2012) with the specific objective of analysing public debate about climate change adaptation. Discussions about how to cope with the impacts of climate change have become particularly salient in the last few years after the recurrent failures to reduce greenhouse gas (GHG) emissions (Aykut and Dahan 2015).

Adaptation constitutes one of the most intricate controversies of collective existence: actors enter and exit the discussion as recklessly as the rise and fall of issues; coalitions form and dissolve hectically; and conflicts cross-cut each other making it difficult to identify opposing factions. In such overflowing complexity, existing institutions are so completely over-run by the shifting of alliances and oppositions that functionalist and critical approaches lose much of their value. In the debate on mitigation, investigating which international organizations are most suitable to regulate GHG emissions or which companies are most liable (Heede 2014) makes perfect sense. Not in the debate on adaptation. When it comes to imagining how to live through the radical changeover of global warming, distributing blame and praise is less important than working with actors to make new collective arrangements possible.

Yes, but what actors? Willing as we were, at the outset of EMAPS, to engage with the widest possible variety of actors, we soon had to recognize that we had little clue as to who these actors were or what they were concerned about. Not because of lack of candidates, to be sure, but because of their proliferation. International negotiators seemed an obvious target, but what about NGOs, local administrators, companies, climate scientists, activists, indigenous communities? What about the non-human actors involved: forests, rivers, shores, hurricanes, species threatened by extinction? To make things worse, none of these groups have clear-cut borders or evident spokespersons. Which of their members should we elect as representatives?

Had we had a clear view of how the adaptation debate was structured, we could have sampled its actors or contacted the most relevant ones. But the fluidity of the adaptation debate offered no clear landmarks for navigation. We were trapped in a vicious circle: since we had no informants, we could not improve our understanding of the controversy and, since we had only a vague appreciation of the debate, we did not know with whom to engage. We were lost because isolated, and isolated because lost.

As in all bootstrapping dilemmas, the solution comes from iteration. We cannot design good maps from scratch or summon large publics out of thin air, but we can design bad maps and then improve them, engage with small audiences and then extend them. And this is precisely what we did. We started by getting in touch with other research projects on climate adaptation (in particular, weadapt.org) and asking them how we could help. At first, they could not really tell because they had no clue what our methods could deliver. So they asked imprecise questions and we gave them back bad results. Slowly, mistake by mistake, the collaboration improved: they started to understand us and we started to understand them. More importantly, they put us in touch with other actors of the debate (negotiators, activists, climate scientists . . .) helping us start a new and larger cycle of consultation. By the end of the project, we had produced a decent set of diagrams of the adaptation debate (www.climaps.eu and Venturini *et al.* 2014) and compiled an address book spanning a variety of disciplines and societal sectors.

Turning a vicious circle into a virtuous spiral, however, required a fundamental change in our research practices. It made little sense to organize the research according to established protocol in which research questions, data collection, analysis, visualization and dissemination follow neatly after one another. This type of organization was just too linear and time-consuming. Had we followed it, we would have discovered at the moment of dissemination that our research questions were irrelevant for the controversy's actors and that our informants represented only a tiny minority of the debate's protagonists. What we needed instead was an approach allowing us to iteratively try, fail and improve our research intervention. And this is where, learning from the experience of the Summer and Winter School of the Digital Methods Initiative in Amsterdam (Rogers 2013), we turned to the iterative and intensive format of the data-sprint.

## The politics of interdisciplinarity

The EMAPS example illustrates how data-sprints entail a very specific approach to scientific research and its political contribution. Traditionally, social sciences have taken two opposing but equally valuable political stances. On the one hand, since Auguste Comte at least, researchers have supported the work of economic and administrative institutions, providing them with information to uphold the organization of collective life. On the other hand, since Karl Marx at least, other researchers have exposed the functioning of institutions, providing their opponents information to contest them. Though in opposing direction, both traditions assume that the structures of collective life are given and that the aim of social sciences is to strengthen or weaken them.

This assumption is reasonable in times of social stability, but it is unworkable in situations where collective institutions are 'under construction'. Public controversies, such as the one on climate change adaptation, are a classic example of such situations (Callon *et al.* 2009). In these situations, the problem is not to support or denounce previous equilibria, but to deal with their evaporation. In controversies, it is idle to argue about the fairness of earlier conventions, since it is precisely their breakdown that creates the dispute. What matters instead is to help social actors to work out a new cohabitation. If possible, one that is more durable and inclusive.

This is precisely the objective of 'controversy mapping' (Venturini 2010, 2012), an original research method developed within the tradition of Actor-Network-Theory (Latour 2005). Controversy mapping (CM) is interdisciplinary by construction. Any researcher aiming for political relevance ought to reach beyond her disciplinary boundaries, but in CM this obligation becomes extremely important. For scholars practising functionalist or critical research, it is not hard to identify the actors to engage with: they coincide either with the formal members of the investigated institutions or with their self-appointed opponents. Such leisure is not available for controversy mappers, as public debates arise precisely when the official actors (the experts, if you wish) fail to contain their disagreements. In the words of Walter Lippmann:

> Government consists in a body of officials, some elected, some appointed, who handle professionally, and in the first instance, problems which come to the public opinion spasmodically and on appeal. Where the parties directly responsible do not work out an adjustment, public officials intervene. When officials fail, public opinion is brought to bear on the issue.
>
> *Lippmann 1927: 63*

But if anyone who is concerned by the consequences of a controversial situation (as in the famous definition of John Dewey (1946)) should be considered a legitimate actor of that situation, then aren't controversy mappers forced to engage with a monstrous multitude and variety of actors? Yes, they are – and it is precisely to handle such extreme indeterminacy that the interdisciplinary format of the data-sprint has been introduced.

From our perspective, interdisciplinarity is not a value in itself. When things are stable enough, when uncertainty is limited and disagreement confined, disciplinary boundaries can have great virtues. They allow us to rely on previous paradigms, to advance faster and more surely. Yet, social researchers cannot limit their intervention to such convenient circumstances. Political responsibility does not stop at the frontiers of existing institutions, but extends crucially to moments of radical transformation. And these are also the situations where the contribution of social researchers is most needed, but also more difficult. Data-sprints are a modest but pragmatic suggestion to handle such moments.

## References

Aykut, S. C. and Dahan, A. (2015). *Gouverner le climat?* Paris, France: Presses de Sciences Po.

Callon, M., Lascoumes, P. and Barthe, Y. (2009). *Acting in an Uncertain World: An Essay on Technical Democracy.* Cambridge, MA: MIT Press.

Dewey, J. (1946). *The Public and its Problems: An Essay in Political Inquiry.* Chicago, IL: Gateway Books. Retrieved from http://books.google.com/books?id=IMkLAQAAIAAJ&pgis=1

Heede, R. (2014). Tracing anthropogenic carbon dioxide and methane emissions to fossil fuel and cement producers, 1854–2010. *Climatic Change*, 122(1–2): 229–241.

Latour, B. (2005). *Reassembling the Social*. Oxford: Oxford University Press.

Lippmann, W. (1927). *The Phantom Public*. New York, NY: The Macmillan Company.

Raymond, E. S. (2001). *The Cathedral and the Bazaar*. Sebastopol, CA: O'Reilly Media.

Rogers, R. (2013). *Digital Methods*. Cambridge, MA: MIT Press.

Venturini, T. (2010). Diving in magma: how to explore controversies with actor-network theory. *Public Understanding of Science*, 19(3): 258–273.

Venturini, T. (2012). Building on faults: how to represent controversies with digital methods. *Public Understanding of Science*, 21(7): 796–812.

Venturini, T., Meunier, A., Munk, A. K., Borra, E. K., Rieder, B., Mauri, M. and Laniado, D. (2014). Climaps by Emaps in 2 pages (A summary for policy makers and busy people). *Social Science Research Network*, ID 2532946.

# 4

# Digging

*Jussi Parikka*

## A situated practice

I would like to start with a hesitation about digging as a method: surely it is only a material activity with little epistemological value? It seems after all, an activity for hands and large machines and less for the cognitive production of knowledge. It leads to thinking about undergrounds instead of epistemologies, masculine connotations instead of methodological subtlety. What is the knowledge that comes out from digging, besides perhaps some sort of tacit understanding of the earth and its qualities? Despite the first reaction, the term brings to mind images of archaeology, excavation, material labour and depth. It starts to become even epistemologically interesting. Digging penetrates surfaces, opens up visibilities and distributes a new sense of the infrastructural underground that underpins the surface of what we take for granted as a subject of everyday experience. Exhumation and uncovering are closely related, but it is important to understand that considering digging as an active verb in the methodological sense of action both relates to situations of engaging with constructed material reality that can articulate new knowledge and can also function as a collective activity. This activity gathers participants around that shared object or material process, emphasizing how academic research is linked to creative-practice methods in art and design that can become collaborative design-workshops (Ratto 2011).

Digging opens up to what conditions experience. Digging can be understood as a cultural technique (Siegert 2015) that exposes distinctions often taken as fundamental (inside/outside, woman/man, up/down, sacred/profane) but which are also produced by a plethora of techniques. Digging exposes the (under)ground of the objects of academic and creative analysis. Digging becomes an operation that hovers between media theory, critical design and collective speculative work. But please do not be mistaken: it is not merely a word of metaphorical value, but refers to an activity increasingly central to media studies, critical studies of technology and science, art and design methods as well as an attitude that one finds in some of the humanities labs that are the new sites of not merely thinking but also of (experimental) making (see, for example, Drucker 2009; Ratto 2011).

As a method, digging opens up historically constructed material reality. It does not merely expose 'ruins' but the multiple historical realities where material infrastructures have been layered,

*Figure 3.4.1* DIY culture, electronic art and design focused Dorkbot group's visit to Norton Sales, an aerospace surplus supplier in North Hollywood, California. The aim was to find and buy something and 'make something interesting with it'. Photo: Garnet Hertz. Used with permission.

revealing different 'distinctive temporalities and evolutionary paths' (Mattern 2015: 14). In this sense, digging opens the different temporalities that are all the time layered in infrastructures of cities, in media technological objects and in everyday situations. It includes the literal sense of going under the surface to discover the infrastructures and material components. This sort of a focus stems, in part, from the mobilization of 'archaeology' as a field that runs through modern episteme of knowledge from Immanuel Kant to Walter Benjamin, Sigmund Freud to Friedrich Kittler (Ebeling 2012). It resonates with media archaeological methods that emphasize the productive nature of excavations both in the sense of using archives to uncover earlier forgotten paths of media history (Parikka 2012; Zielinski 2006) and as site-specific work on abandoned technologies and infrastructures, sharing a ground with methods of contemporary archaeology (Guins 2014; Piccini 2015; Reinhard 2015; Parikka 2015). Both senses of the word are effective and embedded in an understanding of material reality irreducible to textual interpretation.

Digging for the remains of media culture is not restricted to textual traces; besides a historical methodology, media archaeology is increasingly operated as a situated practice. For example, Berlin Humboldt University's Institute of Media Studies hosts the Media Archaeological Fundus which is a space for such pedagogy of learning epistemology hands-on. It is not merely a collection but an active meeting place for theoretical concepts and old media technologies that demonstrate media as material artefacts. Here epistemological knowledge is an activity: taking an oscilloscope from the 1950s, a radio from the 1940s, an optical toy from the nineteenth century or an educational computer from the 1970s to open it up, investigate, look at its insides, engage in media theory hands-on. It is digging but not merely with a shovel, nor only by way of solitary contemplation.

Similar operations related to hands-on humanities analysis are visible in other labs, such as the Media Archaeology Lab in Boulder, Colorado as well as the vast range of labs in the humanities and critical design that have emerged over the past years, demonstrating a new situated and collective way of engaging with knowledge creation. Indeed, digging becomes a term that mediates between media theory and critical design, archaeology and garbology (Rathje 2001; Gabrys 2013) as well as resonating with the non-academic institutional methodologies in hacklabs and hackspaces. *Opening up* becomes an exercise where technological skills meet up with socially concerned awareness of the implications of black-boxed technologies. The interest in critical making in the humanities has spurred practices that combine technical skills from software to hardware with an ethos of reverse engineering; the focus is more on the *process* of dismantling and remaking ('all making is remaking', as Natalie Jeremijenko argues) than producing specific objects (Hertz and Jeremijenko 2015). The hands-on approach is said to expose the object/material thing as something that ensures an awareness of the friction, the world pushing back (Hertz and Ratto 2015). This needs to be also related to the issue of wider infrastructures that remain unidentifiable in terms of solitary objects. Recent research on infrastructures (Mattern 2015; Parks and Starosielski 2015) has effectively demonstrated the relevancy of digging and how media studies can borrow methods from, for example, archaeology. And it has also been a context in which instead of merely production or consumption, issues of distribution are relevant, as Lisa Parks has noted (cited in Mattern 2013).

Digging is not merely a deconstruction of designed objects, but an awareness of the potentials of the subsurface layers as powerful realities in themselves. This mediation between invisible and visible is epistemologically significant and in situated practices becomes a shared matter of concern (Latour 2008) among the participants. Digging can be seen as a collective experience of sharing expertise and as mobilizing the speculative sense of 'what if?': what if the technological artefact could work otherwise, what if it would work this way, what if it could be made to work in alternative social assemblages?

## A collective dig

Hence, there is more to the colloquial version of 'to dig' referring to affective relationship: to intensively like something, have an affinity with, to be in close relationship with a thing or a person. Indeed, a lot of the digging that scholars and artists are engaged in is also embedded in an affective investment that often is prescribed as part of the maker and hacker culture (Ratto 2011, referring to Papert and Franz 1988). This sort of a stance relates to the pedagogies of constructionism; epistemic objects are always triggering and triggered by affective relations and environments, such as when co-working. This social bond is an important part of the material activity, and is one driver of how such methodological issues become part of the consideration of the social settings in which media research functions as process-focused. We also need to be aware of the gender implications in some of these contexts; so, for example, the masculine bias of maker culture and tinkering culture have been flagged as an issue (Chachra 2015).

In this context, digging becomes a hands-on approach to what seems the most ephemeral: the digital and technical media culture that is at the same time avoiding the sensory and yet can be approached by way of its physical location, characteristics and even technical features. The relation to such engineering practices as reverse engineering might at times be close, even if the idea of digging is also about more than just the reversal of the engineering process. By escaping such determinism, the artistic and hacktivist versions often include an idea of subversion as part of the activity of digging, hacking, opening up. The collective workshops that are starting to define a methodological – even if most often still outside academic settings – attitude to digital

*Figure 3.4.2*   Do-It-With-Others: Toy Hacking Workshop instructor and participant disassembling and modifying a battery-powered police car toy in Irvine, California. Organized by Garnet Hertz. Photo: Peter Huynh. Used with permission.

culture indicate an important trend: cryptoparties, hackathons, game jams and other sorts of activities that run over one night or multiple days (and sometimes nights) define a fan-styled enthusiasm that attaches curiosity, dedication and often a critical attitude to working with machines, whether in terms of coding, hardware hacking or the social and legal issues around digital culture from surveillance to economy (for example, copyright). Instead of mere Do-It-Yourself (DIY)-ethos, there are suggestions of more socially oriented hack and other activities of DIWO – Do-It-With-Others (Garrett and Catlow 2012).

Collective digging, making and remaking in technological culture can participate in a redefinition of social ties; of who is considered an amateur, who an expert 'allowed' to engage with the inner workings of machines; issues of credibility and actionable knowledge that are starting to define a field of civic technoscience (Wylie, Jalbert, Dosemagen and Ratto, 2014). The material activity of digging reveals an epistemological side that furthermore testifies to the possibilities of collective work that starts at a grassroots level and quickly becomes one part of a bundle of terms and activities: digging, deconstructing, (re)designing, challenging, activating, sharing, co-working, engaging and redefining.

## References

Chachra, D. (23 January, 2015). Why I am not a maker. *The Atlantic*. Retrieved from: www.theatlantic.com/technology/archive/2015/01/why-i-am-not-a-maker/384767/.

Drucker, J. (2009). *Speclab: Digital Aesthetics and Projects in Speculative Computing*. Chicago, IL: University of Chicago Press.

Ebeling, K. (2012). *Wilde Archäologien 1. Theorien der materiellen Kultur von Kant bis Kittler*. Berlin: Kadmos.

Gabrys, J. (2013). *Digital Rubbish: A Natural History of Electronics*. Ann Arbor, MI: University of Michigan Press.

Garrett, M. and Catlow, R. (2012). DIWO: Do it with others – no ecology without social ecology. In S. Biggs (Ed.) *Remediating the Social* (pp. 69–74). Bergen: ELMCIP.

Guins, R. (2014). *Game After: A Cultural Study of Video Game Afterlife*. Cambridge, MA: The MIT Press.

Hertz, G. and Jeremijenko, N. (2015). Engineering anti-techno-fetishism: Natalie Jeremijenko in conversation with Garnet Hertz. *Ctheory*, Conversations in Critical Making. Retrieved from: http://ctheory. net/articles.aspx?id=755.

Hertz, G. and Ratto, M. (2015). Defining critical making: Matt Ratto in conversation with Garnet Hertz. *Ctheory*, Conversations in Critical Making. Retrieved from: http://ctheory.net/articles.aspx?id=755

Latour, B. (2008). *What is the Style of Matters of Concern?* Amsterdam: Van Gorcum.

Mattern, S. (July, 2013). Infrastructural tourism. *Places*. Retrieved from: https://placesjournal.org/article/ infrastructural-tourism/.

Mattern, S. (2015). *Deep Mapping the Media City*. Minneapolis, MN: University of Minnesota Press.

Papert, S. and Franz, G. (1988). Computer as material: messing about with time. *Teachers College Record*, 89(3): 408–417.

Parikka, J. (2012). *What is Media Archaeology?* Cambridge, UK: Polity.

Parikka, J. (2015). *A Geology of Media*. Minneapolis, MN: University of Minnesota Press.

Parks, L. and Starosielski, N. (Eds.) (2015). *Signal Traffic: Critical Studies of Media Infrastructures*. Champaign, IL: University of Illinois Press.

Piccini, A. (Ed.) (2015). Media Archaeology-special section, *The Journal of Contemporary Archaeology*, 2(1).

Rathje, W. (2001). Integrated archaeology: a garbage paradigm. In V. Buchli and G. Lucas (Eds.) *Archaeologies of the Contemporary Past* (pp. 63–76). London: Routledge.

Ratto, M. (2011). Critical making: conceptual and material studies in technology and social life. *The Information Society*, 27(4): 252–260.

Reinhard, A. (2015). Excavating Atari: where the media was the archaeology. *The Journal of Contemporary Archaeology*, 2(1): 86–93.

Siegert, B. (2015). *Cultural Techniques: Grids, Filters, Doors and Other Articulations of the Real*. Trans. Geoffrey Winthrop-Young. New York: Fordham University Press.

Wylie, S. A., Jalbert, K., Dosemagen, S. and Ratto, M. (2014). Institutions for civic technoscience: how critical making is transforming environmental research. *The Information Society*, 30(2): 116–126.

Zielinski, S. (2006). *A Deep Time of the Media* (Trans. G. Custance). Cambridge, MA: The MIT Press.

# 5

# Issuecrawling
## Building lists of URLs and mapping website networks

*Richard Rogers*

## Introduction: making URL lists of right-wing populist and extremist groupings

According to claims made by the popular press and the think tank Demos we are witnessing the rise of a new kind of populist politics defined by opposition 'to immigration and concern for protecting national and European culture, sometimes using the language of human rights and freedom' (Bartlett, Birdwell and Littler 2011). This 'new right' movement is said to be supplanting the (fascist or neo-Nazi) old guard in a series of European countries, with an orientation distinctive from the 'blood and soil' pathos of old (Van Gilder Cooke 2011). This chapter describes how we might examine these claims empirically through an online, interdisciplinary approach that combines crawling techniques from web science and close reading of websites from media studies.

The 'how to' research protocol that follows describes how to build lists of URLs to seed link crawling software and ultimately make link maps of right-wing extremism and 'new right' populism in particular European countries. The maps show links between websites, or online networks of websites that can be analysed according to a series of technical characteristics, but here a substantive analysis is also undertaken to examine the claims made. These methods may be situated alongside reading party manifestos and favoured literature, going native by embedding oneself in the groups, interviewing imprisoned or former group members, and other qualitative techniques to distil significant content. The online mapping method of issuecrawling can thus be considered either as an exploratory step that provides leads for further in-depth analysis, or as a means to create country reports with a broad stroke, as is the intention of the longer analysis behind this piece (Rogers 2013).

The exercise commences with the collection of the URLs of populist right-wing and right-wing extremist websites in a series of countries named in popular press articles as well as the Demos study: Austria, Belgium, Bulgaria, Denmark, France, Germany, Greece, Hungary, Italy, the Netherlands, Norway, Portugal, Romania, Serbia and Spain. Lists of websites are made by following a heuristic known as the 'associative query-snowballing technique'. (For a step-by-step elaboration, see the research protocol below.) Queries are formulated, and made in the local domain Googles associated with the countries in question (such as google.at, google.be, google.bg and google.rs), in the respective local languages, largely in these styles: [populist right parties] as well as [right-wing extremist groups]. When the names of parties, groups or other related entities (e.g. a webshop selling right-wing t-shirts, music and literature) are found, they are entered as lists (each in quotation marks) into the search boxes of the respective local domain Googles, and the results are read. This process is repeated, until no new names are found. That is, lists of populist right and extremist groups are slowly built up from query results. Once the lists gathered from the web search engines are finished, they are compared to expert lists. To find expert lists, queries are made in Google Scholar, first in the home language, and subsequently in English. The queries made are similar to those entered in the local domain Googles, in the first round of list-building from the web. Any new groups found on the expert lists in the scholarly literature are searched for online, and if they have a web presence, they are added. Thus, the expert lists add to the web lists. For each group, actor or entity on the list there should be an accompanying URL or multiple URLs.

The work of locating URLs might be arduous for the new right's web presence could be 'on the move', dodging authorities, as is the case in many countries such as Germany where website owners regularly change URLs (and hosts to outside the country) and also move to social media such as Facebook, so as to attract a larger following and make it more burdensome for the authorities to take down what it construes, nationally, as 'illegal content' (Prodhan and Lauer 2016).

## Quanti-quali analysis of European right-wing formations online

The URLs of the populist right, the extreme right and the populist and extreme right together are crawled, per country, in three, separate analytical procedures, using the Issuecrawler (issuecrawler.net). Of interest are the comparative sizes of the populist and extreme right as well as other indicators of activity such as responsiveness and freshness. By responsiveness is meant whether the sites are online, and return a response code (or http status code) of 200, when loaded in a browser. Freshness concerns its last update, and its recent consistency in updating.

The two seed sets are crawled together, as well, to compare them and gauge their interconnectedness. Do they form one cluster, or are they (largely) separate? Doing this enables one to begin to examine the claims that the populist right is distinctive (clustered separately) and overtaking the old guard, at least according to online network analysis, including responsiveness and freshness. For the analysis, one asks, does the new right have larger, denser clusters and more active and fresher websites than those of the old guard? In most countries under study the answers are in the affirmative, thus largely confirming the popular press and think tank claims.

In terms of the method, for each set of populist, extreme right and combination of URLs, automated co-link analysis is performed, with 'privileged starting points' (a special setting), keeping the seeds on the map, if linked to, whereby those websites receiving at least one link from the seeds are retained in the network. 'Newly discovered' sites are required to receive two links to be included in the network (standard 'co-links'). The 'privilege starting points' feature gives the seeds an increased chance of remaining in the network.

Each of the networks is visualized as a cluster graph (according to measures of inlink centrality), and the findings are described. First, are there other (heretofore) undiscovered groups found through the link analysis? Co-link mapping is a procedure that discovers related URLs through interlinking. In the event, we found Facebook to be a large node in many countries, which not only is in keeping with the impression of groups 'on the move' to social media but also prompts the question of its (separate) analysis, for Facebook cannot be crawled as above. (Only links to Facebook are on the map, not outlinks from Facebook.) Second, which sites are responsive and fresh? Are they mainly the populist ones? Indeed, the old guard's web in a variety of European countries is often stale. It also might be of interest to inquire into where the websites are registered and by whom. Are they registered under aliases and hosted outside the country? Or are they registered in country, under one's own names? In certain countries, these are signs that groups are in hiding or operating in plain sight, so to speak. In Germany, the groups often mask themselves, while in Austria they tend to operate out in the open.

Apart from the 'technical' characteristics of the websites in the networks (network size and density as well as site responsiveness, freshness, geo-registration and use of alias) the qualitative analysis we conducted concerns the groups' orientation as well as activities, especially in their outreach, forms of communication as well as youth recruitment. Is there an active music scene? Where does one go to participate in person in populist and extreme right-wing culture? Generally, the substantive characteristics of the right-wing formations online specific to the country may be understood by spending significant analytical time reading the websites on the (clickable) Issuecrawler map of each of the national right-wing scenes in question, picking out significant themes, which vary from country to country. In Hungary, for example, the supposed Mongolian language roots have been appropriated by the right (old and new), and the question might be asked: how to take back the yurt. Unlike in Bulgaria (and Spain), where the old guard still thrives (online), in Serbia there is a new, right-wing civil society, with think tanks, which seek to shape the discussion on the future of Serbia around the questions of land and Kosovo. France, witnessing the rise of identitarian (youth) groups and ethno-differentialism, is a dividing line between northern and southern Europe in the sense that counter-jihadism (also referred to as anti-Islam and Islamophobia) is present but not a dominant theme in the new populist right. In Denmark, Norway and to an extent the Netherlands, counter-jihadism increasingly organizes the new right, and indeed here we find especially some of the language of the new right the London think tank described. The claim that the new right employs a vocabulary of immigration opposition borrowed from 'rights talk' is difficult to pinpoint, but the broader claim can be nuanced through the observation that the new right in question is geographically distinctive, and located in northern Europe. In Austria, contrariwise, the populist right's is an anti-capitalist critique (against lavish Austrian balls, and the storage of Austrian gold abroad). In Germany, there is (still) a preponderance of 'brown culture'.

In the following the list-building technique is elaborated in more detail, prior to a reflection on the types of lists that may be authored with the aid of search engines these days, now that the editorial practice of creating web directories has waned.

## Research protocol: URL list-making with the associative query-snowballing technique

The objective is to assemble three URL lists per country under study: extreme right, populist right and a combined list. 'Extreme right' and 'populist right' are broad terms not categorized in advance, but instead the authors of online lists classify them as such.

Below are the step-by-step instructions on how to make a list through what is termed associative query-snowballing. The example of list-building is for the 'extreme right' in Spain, however, the process is much the same for any country. The third list is made eventually by merging the first two.

### Part I: Making a URL list using the technique

1   Load the local domain Google search engine for the country in question in the browser, e.g. google.es. Design a broad query that will output extreme right groups in Spain. For example, we used: 'Grupos de Extrema derecha en España' (translation: 'Extreme right groups in Spain').

*Figure 3.5.1*    Google.es results of a query for right-wing extremist groups in Spain.
Screenshot, 4 September 2012.

*Figure 3.5.2*    Simple spreadsheet with names of groups and URLs per group. Screenshot,
4 September 2012.

2    After performing the query, the user is returned a set of results, some of which are lists. List
is meant in a broad sense. For example, a news article that reviews the most influential
extreme right-wing groups usually will name a number of them. One might find that the
article refers to parties or groups not only from the country in question but also to other
international groupings. From the pages and articles, the researcher needs to extract the
names of the groups that correspond to the country in question, and also find the URLs
and include them in a spreadsheet. Let us say in this first step two main groups have been
found: España 2000 and Plataforma per Catalyuña (see Figure 3.5.2).

3    Return to Google.es. Enter the names of the groups found in the previous search results as
a query using quotation marks: ['España 2000' 'Plataforma Catalunya']. The fresh set of
results returned contain ideally not only the two groups used in the query but also new ones
that will be associated with them (associative snowballing). Comb through the results, select
the names of the new groups and add them to the spreadsheet. For example, the first result
contains the new name, 'Democracia Nacional' (see Figure 3.5.3).

4    Enter the two initial groups ('España 2000'and 'Plataforma per Catalunya') together with
the new group ('Democracia Nacional') in the search box. Again, one will receive results
in which the three groups may be associated with other groups. Add the new ones, including
their URLs, to the spreadsheet.

5    Repeat until no new groups are found. For the purposes of robustness one might wish to
make queries that contain new combinations of fewer groups.

6    As a note, the last groups to make the lists could be thought of as marginal or historical.
It is advisable, as a last step, to query the marginal groups separately, which ideally will return
a new set of even more marginal groups, though these also could be from other countries.
Repeat until no new country-specific results are found.

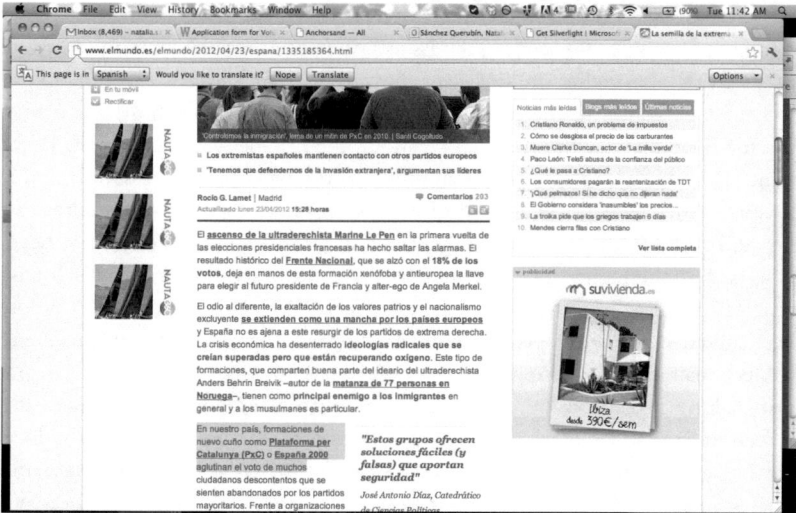

*Figure 3.5.3* Associative query-snowballing technique, second iteration. Results of the Google.es query for Plataforma Catalunya and España 2000 yield a third group, Democracia Nacional, which is then added to the spreadsheet, with its URL. Screenshots, 4 September 2012.

## Part II: Finding expert lists, compiling them, adding them to the web list, and making the final list (the web + expert list)

1   Search for academic literature that mentions the extreme right in Spain. Academic articles and grey literature case studies usually have their own collections of names. One may use Google Scholar to query in the original language or in English, again employing the broad search terms: [extreme right-wing Spain]. From the results explore and choose approximately three or more articles that you have detected containing lists. Recall that lists do not always look like lists.

2   Extract the names of the groups, and search for the groups' URLs, if (as is often the case) they are not included. Make a list of all groups and URLs. This is the expert list.

3   Compare the web list (from the associative query-snowballing technique) with the expert list. There is a list comparison tool, 'triangulation' at https://tools.digitalmethods.net/beta/triangulate/. It shows the URLs unique to each list as well as those that are common.

4   Take note of the groups or other entities that are unique to the expert list or to the web
    list. Query the unique groups' names in the search engine, and ascertain whether it
    has one or more URLs. Retain those groups on the expert lists that have a web presence,
    i.e. one or more associated URLs claiming to represent or give significant voice to
    the group.
5   Concatenate the URLs from the web list and the expert list.

Finally, one may take note of what the web yields in comparison to the experts. One may
compare epistemologies (how lists are made) as well as ontologies (types of lists). Expert lists
(including Wikipedia's) are often exhaustive and alphabetical, and include historical actors, while
web lists outputted by search engines are, in the main, hierarchical and fresh.

## Conclusion: web and expert URL lists

List-building in preparation for seeding the Issuecrawler or other link crawling software such as
Hyphe or VOSON often relies on 'link lists' (Jacomy, Girard, Ooghe-Tabanou and Venturini
2016; Ackland *et al.* 2006). In the past preferred starting points were those lists maintained by
Dmoz.org, the open directory project, and Yahoo!, the original web 'directory'. Both projects
are dormant. To a degree, directories of all kinds on the web have been supplanted by search
engines, which also author lists, albeit of query results rather than list of websites categorized
by human editors. Inter-governmental organizations as well as NGOs also have been keepers
of expert lists, but their curation practices (such as Amnesty International's list of human
rights organizations) have been in abeyance for years. Wikipedia continues to be one of the
few human-edited list-makers; given their encyclopaedic quality (and exhaustiveness) they
require subject-matter expert paring.

The list-making and query-building technique introduced above is designed for a post-
directory web. It strives to build lists anew, with the aid of search engines, first by locating lists
of mentions of groups, actors or entities (in this case of the right wing), and subsequently by
sourcing their URLs, again via search. It is a digital method dubbed 'associative query-
snowballing' because each of the actors found has been acquired by association to other actors
through iterations of query results.

## Acknowledgements

The project initiated at the 2012 Digital Methods Summer School, University of Amsterdam,
carried out by Andrei Mogoutov, Anton Sokolov, David Moats, Elena Morenkova Perrier, Ellen
Rutten, Johan Söderberg, Luis F. Alvarez-Leon, Saskia Kok, Simeona Petkova and Stefania
Bercu. A subsequent new right populism mapping workshop (September 2012) saw contribu-
tions by Jan Bajec, Federica Bardelli, Lisa Bergenfelz, Sharon Brehm, Alessandro Brunetti,
Gabriele Colombo, Giulia De Amicis, Carlo De Gaetano, Orsolya Gulyas, Eelke Hermens,
Catalina Iorga, Juliana Paiva, Olga Paraskevopoulou, Simeona Petkova, Tommaso Ranzana,
Radmila Radojevic, Ea Ryberg Due, Catherine Somzé and Lonneke van der Velden. Natalia
Sanchez Querubin, co-organizer of the workshop, assisted on the construction of the querying
technique. The analysis is written up in more detail as a report, supported by the Open Society
Foundations (Rogers 2013). This chapter is adapted from the study.

## References

Ackland, R., O'Neil, M., Standish, R. and Buchhorn, M. (2006). VOSON: A web services approach for facilitating research into online networks. Paper presented at the *Second International e-Social Science Conference*, 28–30 June, Manchester, UK.

Bartlett, J., Birdwell, J. and Littler, M. (2011). The rise of populism in Europe can be traced through online behaviour . . .: The New Face of Digital Populism: Lega Nord., London: Demos. Retrieved from: http://www.demos.co.uk/files/Demos_OSIPOP_Book-web_03.pdf.

Jacomy, M., Girard, P., Ooghe-Tabanou, B. and Venturini, T. (2016). Hyphe, a curation-oriented approach to web crawling for the social sciences. *Proceedings of the International AAAI Conference on Web and Social Media* (ICWSM-16), Cologne, Germany.

Prodhan, G. and Lauer, K. (25 February, 2016). Germans talk tough, fete Facebook's Zuckerberg, *Reuters*. Retrieved from: www.reuters.com/article/us-facebook-germany-zuckerberg-idUSKCN0VY2DD

Rogers, R. (2013). Right-wing formations in Europe and their counter-measures: an online mapping. Amsterdam: Govcom.org Foundation and the Digital Methods Initiative. Retrieved from: https://wiki.digitalmethods.net/Dmi/RightWingPopulismStudy.

Van Gilder Cooke, S. (29 July, 2011). Europe's right wing: a nation-by-nation guide to political parties and extremist groups. *Time Magazine*. Retrieved from: http://content.time.com/time/specials/packages/article/0,28804,2085728_2085727_2085712,00.html

# 6

# Moving methods

*Monika Büscher*

---

reality is movement

*Bergson 1960[1919]: 319*

Mobile methods have been developed to gain deeper insight into critical features of contemporary life, such as the growing global dependence on fossil-fuelled transport, manufacturing and energy systems; the secret flows of offshored finance; environmentally and conflict-related mass displacement and migration; the cultural lure and political economies of tourism; and the present absences and absent presences of digital communications. They often leverage the potential of new mobile technologies, from digital notepads to body-worn cameras, and perhaps because of this are often understood as mobile in a literal sense, requiring researchers to physically 'move along with, be with, sense with' their research subjects (Merriman 2013). Yet, as Merriman argues, the 'newness' of these methods is questionable, as is their emphasis on literally mobilizing researchers to study mobile subjects and objects. Mol and Mesman in their study of orderings in a neonatal ward provide an example of how mobile methods in this sense can get 'stuck': '[W]hat about the pieces of paper that travel from the ward to the dispensary? J couldn't enter the hospital's postal system with them, for its plastic tubes were . . . far too small for human bodies' (Mol and Mesman 1996: 422–423).

Literally moving with the subjects and objects of research can be difficult when they are too small or too vast, too slow or too fast, too complex or immaterial to follow. Research subjects might not be human or material objects on the move, but animals, diseases, ideas, atmospheres or whole mobility systems. Think of data in wireless networks, global flows of people, goods, finance, or resources, or the movement of emotions within a crowd or a dispersed media public.

But the contemporary motivation to make methods mobile for research arises in the context of the new mobilities paradigm, a transdisciplinary analytical orientation that sees the world, the social, the material, their rhythms, histories and futures as they are made in and through movement, blocked movement, stillness (Sheller and Urry 2006). It thus springs from a deep appreciation of 'how reality is movement' (Bergson 1960[1919]), as mobilities research, along with feminist, actor-network, and non-representational theories, ethnomethodology and process philosophies recognizes the emergent nature of reality, the way in which social and material phenomena are made and made durable in and through the inter- or intra-actions of many human and non-human agencies. Yet, mobile, inventive or 'live' methods (Sheller 2015) still struggle with the fact that 'the mobile flies for ever before the pursuit of science' (Bergson 1960[1919]: 317). This is, in part, because narrow interpretations of what 'mobile methods' might, or even should be, have constricted the analytical leverage sought and found (Merriman 2013). But many researchers have been moving methods in ways that make moving methods and methodologies open up new avenues for multiple mobile transdisciplinary forms of science. The examples below provide an impression of how moving as method and methodology can slow science to attune to the mobile and generate new analytical momentum for methodological assemblage.

## Moving and becoming

Moving researchers might 'prioritise "being there" . . . to understand phenomena' (Fincham, McGuinness and Murray 2009: 171), but this does not have to be conceived as an exercise in finding more 'authentic' ways of 'bringing back the data'. Instead, it can be developed as a way of creating deeper understanding of 'how places, spaces and subjectivities are constituted in and through motion', where moving as method might entail sensory ethnographies (Figure 3.6.1), experiencing and reflectively practising movement alongside others, and finding new ways of articulating the mobile in a collaborative research encounter (Brown and Spinney, in Fincham *et al.* 2009: 130). This does not necessarily even require physically going along with participants.

Laurier (in Fincham *et al.* 2009) shows how 'being there' vicariously, by inviting research subjects to document their journeys with a camera installed in their cars, can be a technique for moving alongside the mobile to study (im)mobile becoming. Laurier's concern is with the practical achievement of place, space, subjectivity, especially the doings, obligations, and responsibilities of 'passengering'. His approach seeks to increase the richness and agility of insight, by *not* being there in person (as another passenger). To understand naturally occurring conduct in a way that respects its situated performativity, the 'absent presence' of recording devices operated by research participants allowed lived practices of passengering to be performed and captured *in vivo*. But again, the aim of moving with the mobile is not to 'bring back the data' for positivist expert analysis, but to foster the patience needed to analytically exhibit social and material becoming and to show how, in the mundane everyday practices of movement, and the specificities of doing passengering and driving, socio-material realities and subjectivities are made.

Relatedly, albeit in a rather different way, studying archival records of past experiences of driving with an explicit analytical orientation to the mobile can be seen to leverage moving across different historical accounts as a method to mobilize sensitivities to the performativity of movement and enable a deeper understanding and richer description of emergent and

*Figure 3.6.1*   Sensory ethnography of the social order of traffic. A motorbike is audibly wishing to overtake in tight space. (Photo: Cosmin Popan, reproduced with permission.)

interconnected pasts, presents and futures (Merriman 2013). And moving with the rhythms of early twentieth-century postcard exchanges by combining multiple methods, ranging from historical ethnography of material culture to textual analysis can enable researchers to enter into past 'interweavings of personal mobilities and creativity in the uptake and shaping of a new communications technology' (Gillen and Hall, in Büscher, Urry and Witchger 2011: 34; see also Coleman, this volume).

## Moving and movement

The methodologies described above use literal, vicarious or analytical agility to apprehend mobile becoming, or the way in which places, spaces, subjectivities are made in lived (im)mobile practice. This moving as method is intrinsically multi-scalar, concerned with realities made of movement at different scales, exploring the micro-interactive orders of passengering as well as planetary effects of $CO_2$ emissions. In his study of planetary mobilities, Bron Szerszynski (2016) moves through a wide-ranging review of research in physics, biology, palaeontology and environmental sociology to trace how a 'sublunary far-from-equilibrium planetary becoming' makes realities in the multi-scalar movements of micro-particles, animals, technologically augmented anthropogenic mobilities such as planetary jet streams and ocean gyres, as well as interplanetary mobilities. He makes a move towards connecting the different scales of lived lives and geophysical processes and finds that through these mobilities Earth is 'self-organising over deep, geological time and thereby creating its own unique history and set of powers' (2016: 614). Together with a more empirical focus on the mundane practical achievement of socio-material orders, and of place, space, subjectivity in motion at different scales, rich reflections on the ethico-episteme-ontology of mobile becoming become both possible and necessary. Such personal and societal considerations resonate deeply with calls for more circumspect ethico-episteme-ontologies of research, such as that issued in feminist theorist Karen Barad's investigation of the agential realism of position, potential and momentum at the quantum physics heart of the universe (Barad 2007).

Researchers interested in how realities are constituted in and through the work of movement and its opposites of blocked movement or stasis at different scales have 'moved across' disciplines and 'moved in' to other disciplines as well as multiple lived contexts in which phenomena manifest, because 'getting' the multi-scalar mobile requires transdisciplinary, live, collaborative, experimental, critical and creative approaches. Artist Jen Southern's work on 'comobility', for example, was produced by moving her investigation with geographical positioning systems (GPS) into sociology, making new theoretical and methodological inroads, crossroads and vantage points for analytical and inventive endeavours that mix these disciplinary perspectives. Her investigations of an emerging 'new sense of comobility, of being mobile with others at a distance' (Southern 2012: 75) proceed reflexively through literary explorations (St Exupery), learning to fly, engaging with pilots, reindeer and their herders, developing a 'comobility app' and studying its uptake and shaping in everyday life, and working with mountain rescue dogs and handlers, and searcher kite fliers through participant observation (Figure 3.6.2). The co-mobile realities made by connecting GPS satellites with the feet, hooves and paws that bring position to life on the ground are not entirely new but deeply consequential transformations and transpositions of relational connectedness.

## Moving to be moved

In doing mobilities research, many researchers find themselves moved – by atmospheres, affects and injustices, and they react to this not only with rational analysis, but also emotional and

*Figure 3.6.2* Jen Southern 'Searcher' (2015), taking frames from rescue dog-mounted video up into the sky to explore connections between GPS satellites, grid coordinates, mountain rescue practices. (Photo: Jen Southern, reproduced with permission.)

creative modes of analysis; indeed, they may use moving as a method, placing their body into contexts as 'an affective vehicle through which we sense place and movement, and construct emotional geographies' (Sheller and Urry 2006: 216). David Bissell's work on quiescence, passivity, lethargy, tiredness, hunger and pain, for example, provides a much needed counterpoint to an emphasis on mobility in mobilities research (Bissell, in Fincham *et al.* 2009), for example. One of the most haunting examples of the moving potential of moving methods is Harry Ferguson's mobile ethnography of social work and his description of 'moving scenes from social work' (in Büscher *et al.* 2011: 78–80). Quoting from an inquiry report on 'Baby Peter', he reminds readers of the death of a 17-month-old boy. Peter died as a result of over 50 injuries, received at his home while he and his carers were under the supervision of social services in a borough of London in 2007. Ferguson contrasts the way in which social workers are blamed for their failure to exercise rational judgement at innumerable junctures in this case with a moving ethnographic analysis of the anxiety and fear that can undermine that very capacity of reasoning when visiting homes where children are abused. The social workers he accompanied were often met with deception, strategic obstructive placement of matter (dirty clothes, chocolate smears on children's faces), animals (large dogs) and aggressive human bodies, as well as a threatening atmosphere, which clouded their perception. Ferguson argues that by moving along with practitioners, mobilities research can trace the emotional geographies that so clearly have an impact on professional practice, and challenge institutional practices to provide more support for the mobile and deeply moving work of social work.

But emotions can also become the subject of more metaphorically moving methods, using textually based discursive modes of following phenomena of affect. Sara Ahmed (2004) moves through websites, government reports, political speeches and newspaper articles to 'feel her way' around the manifestation, naming, doing and doings of specific emotions: pain, hate, fear, disgust, shame, love, queer feelings and feminist attachments. She is moving with the circulation of emotions, because 'emotions after all are moving, even if they do not simply move between us' (p. 11). She traces the cultural politics of emotion to show how emotions circulate and powerfully transcend the personal: 'words for feeling, and objects for feeling circulate and generate effects: they move, stick, and slide' and, in the process, they 'create the very surfaces, boundaries and distinctions' that connect and divide societies. Ahmed shows, for example, how hate works by 'sticking "figures of hate" together, transforming them into a common threat, within discourses of on asylum and migration' (p. 15). Emotions such as pity, fear, disgust, shame, she shows, are deeply political.

Moving methods that enable emotional immersion, analysis, mapping, or a tracing of the political performativities of emotions can generate important insights. However, being moved, also has – and in my view should have – normative momentum.

## Moving for momentum

In her book *Staying with the Trouble*, Haraway (2016) discusses examples of troubles worth moving with. However, making normative moves while in the company of troubles can be a fraught undertaking and put people off, for very good reasons. For example, despite the prospect of 11 billion people on Earth in 2100,

> many feminists, including science studies and anthropological feminists, have not been willing seriously to address the Great Acceleration of human numbers, fearing that to do so would be to slide once again into the muck of racism, classism, nationalism, modernism, and imperialism.
>
> *2016: 6*

Moving responsibly, or response-ably within such quagmires, she argues, means to set aside these fears and 'think together anew across differences of historical position and of kinds of knowledge and expertise' (2016: 7). That is certainly a necessary part, but considering the abysses that such thinking needs to span, thinking is not enough.

Moving as method and methodology as practised by artists, designers and social scientists like Southern and others, including myself, suggests other modes of more hands-on and collective engagement that are useful, including literally 'moving into' troubles, such as those surrounding the surveillance potential of digital tracking technologies (Büscher, Hemment, Coulton and Mogensen, in Büscher *et al.* 2011). This involves becoming engaged in their design, and contested collective experimentations with prototypes, which is likely to reveal the emergence of unintended consequences in ways that can not only be felt directly, but that also make it possible to critically-creatively address these. Working with critique is often most productive when the grounds for change are contested vigorously and visions of how problems and challenges (as well as opportunities) should be addressed are translated into experimental but 'inhabitable' futures that allow collectives to 'move in' to explore unintended consequences (Figure 3.6.3). This makes it possible to reversibly move forward, sideways or back with innovations in – as Latour would put it – a simultaneously more radically careful, and carefully radical way. Other examples include Malene Freudendal-Pedersen and her colleagues' methods for sharing responsibilities for urban planning more widely (in Fincham *et al.* 2009).

Pivotal to such interdisciplinary engagements with practitioners, designers, policy-makers, planners and others beyond the academy are methods that can move people to be passionate, respectful and open to multiple forms of expertise, interests and motivations. They require new capabilities for negotiating the multiply emergent and unknown over the longue durée, contesting power, considering intergenerational justice and responsibility. Moving methods have close affinities to engaged, inventive, live methods and public sociology endeavours that share these ambitions (Sheller 2015). They add sensitivity to how reality is mobile, and could be otherwise. If reality is made in the work of movement, blocked movement, stasis, it can be made differently.

*Figure 3.6.3*   Moving with the troubles of digital tracking. Discovering unintended consequences of digital tracking technologies in networked risk governance and emergency response by experimenting with prototypes developed in the BRIDGE project www.bridgeproject.eu/en. (Photo: Monika Büscher 2014.)

## In conclusion

This (highly selective!) discussion traces the emergence of moving methods. These are not radically new methods constitutive of a new methodological paradigm. They seem to have taken shape in many different settings and seem to be coming together as a proposition to move methods onwards as well as sideways and into discoveries of new reversibilities, into deeper depths, and new collaborative engagements, because the troubles found in the world need more circumspect, critical, creative, collective, contested research. Even if still 'the mobile flies forever before the pursuit of science, which is concerned with mobility alone' (Bergson 1960[1919]: 317), moving as method and methodology can enrich an analytical orientation to movement, blocked movement and stillness and enable multi-scalar, critical-creative research.

## References

Ahmed, S. (2004). *The Cultural Politics of Emotion*. London: Routledge.
Barad, K. (2007). *Meeting the Universe Halfway*. Durham, NC: Duke University Press.
Bergson, H. (1960[1919]). *Creative Evolution*. London: Macmillan.
Büscher, M., Urry, J. and Witchger, K. (Eds.) (2011). *Mobile Methods*. London: Routledge.
Fincham, D., McGuinness, M. and Murray, L. (Eds.) (2009). *Mobile Methodologies*. Basingstoke: Palgrave Macmillan.
Haraway, D. J. (2016). *Staying with the Trouble: Making Kin in the Chthulucene*. Durham, NC: Duke University Press.
Merriman, P. (2013). Rethinking mobile methods. *Mobilities*, 9(2): 167–187.

Mol, A. and Mesman, J. (1996). Neonatal food and the politics of theory: some questions of method. *Social Studies of Science*, 26(2): 419–444.

Sheller, M. (2015). Vital methodologies: live methods, mobile art, and research-creation. In P. Vannini (Ed.) *Non-Representational Methodologies* (pp. 130–145). London: Routledge.

Sheller, M. and Urry, J. (2006). The new mobilities paradigm. *Environment and Planning A*, 38(2): 207–226.

Southern, J. (2012). Comobility: how proximity and distance travel together in locative media. *Canadian Journal for Communications*, 37(1): 75–91.

Szerszynski, B. (2016). Planetary mobilities: movement, memory and emergence in the body of the Earth. *Mobilities*, 11(4): 614–628.

# 7

# Playing with ethics

*Miguel Angel Sicart*

In the closing chapter of *Homo Ludens*, Johan Huizinga famously claims that play is 'in itself neither good nor bad', that it is 'outside morals' (Huizinga 1992). This is an argument that highlights the complicated relations between play and ethics. In these days of gamification and videogames, in which digital playful experiences are ubiquitous, the question of the relation between play and ethics is more urgent than ever.

In this short piece, I want to continue my own work on ethics and videogames (Sicart 2009, 2013), yet I want to go beyond the potential moral dangers of computer games culture to ask how play is valuable for our well-being (Deci and Ryan 2008). I will argue that play is a moral activity that can contribute to our flourishing as human beings; more specifically I will suggest that seen through the lens of virtue ethics, in which the emphasis is on the role of character rather than either doing one's duty or acting in order to bring about good consequences, playing can be a central element of a good life. The objects we play with, from games to toys, are part of a constellation of technologies that can be analysed as part of the way we develop our moral being by playing.

There have been some studies investigating the impact of ethical game design for players (Zagal 2011). These studies apply a conventional questionnaire-based qualitative methodology. These studies contribute important insights into the role of ethics in gameplay experiences, and the possible impact that ethical gameplay design has for our ethical discourses. However, to understand the role that playing has in our moral engagement with the world, it is necessary to combine these classic qualitative methods with research approaches that allow us to understand the ethics of play in context. A valid methodology would be contextual inquiry (Wixon, Flanders and Beabes 1996), a phenomenological research method that observes specific actions and then asks users to explain these actions in context. A more experimental methodology, such as cultural probes (Gaver, Dunne and Pacenti 1999), might also be useful to research on the attitudes that players have towards the ethics of the games they play, and the way in which they relate to play morality in their daily lives. In general, any methodology for the study of the relation between ethics and play needs to have access to the reflective work that players do when engaging with moral challenges. That is, the study of ethics in games needs to be qualitative and contextual, engaging users directly in the production of and reflection upon the research data.

Despite Huizinga's claim above, I suggest that play-ing is not a morally neutral activity. There are ethical risks when we play, in excessive play (Caillois 2001), in addiction (Schüll 2012), in deep play (Geertz 1972). However, playing is important for the moral fabric of society not only despite its potential risks, but also *because* of these risks. Playing is learning to navigate, playfully and deeply seriously, our own being in the world. Because playing is dangerous (Schechner 1988), and because it is also a creative, human form of expression, it has value for us; it has the potential to make us better human beings. To play, as an expressive, appropriative form of being in the world, is to assert ourselves in the world creatively, to explore it under rules we have accepted as valid, that we have agreed to submit to or that we have ourselves created.

In philosophical terms, I am taking a constructivist ethics approach (Bynum 2006). In this view, ethics is a practical science that helps us develop as human beings by practising virtues. To be a morally sound human being we must develop our potential, we must exercise, practice, test and expand our virtues, from empathy to love, to courage. We are ethical beings because we can develop those virtues through time and practice. That practice takes place in all instances of life: when we work, we love, when we are idle and when we exhaust ourselves (Burke 1971). A way of understanding this active, constructivist approach to ethics is to think about morality as another way of being in the world, including how we conduct research, one that determines how we engage with others and how we take decisions. It is therefore crucial for the study of the ethics of playing that we involve users qualitatively, so they are allowed to engage with and reflect upon their own morality when playing.

Playing is a way of being in the world that appropriates, and is sometimes mediated, by objects, things and circumstances. In this sense, the importance of playthings in our betterment is obvious: things and devices can help us play a good life (Waterman, Schwartz and Conti 2008). But here we find too the problem that the Huizingian theory of play poses: if play is considered to be outside the domain of ethics and morality, even though we acknowledge that it does foster some virtues, its lack of seriousness and lack of productivity condemn it as an empty leisurely act.

I claim that we need to leave behind the idea of play as something that happens separately from the world; as something that is not affected and does not affect the contexts and objects through which it is manifested. Playing is valuable *because* it is appropriative, expressive and disruptive – the values of play reside in the way play allows us to explore, train, investigate, study and develop our best potential as human beings. Given that ethics is a way of being in the world that underlies all of our actions, activities and ideas, its relation to play should be obvious. The ethics of play should be seen as *the value* of play, the way in which, through play, we live a good life.

This is not to say that all play is good, that there are no moral risks with play. Play can seduce us; through playthings we can forget that play is a mode of being in the world, and we can lose the relative distance between the action and the context that we need for playing to be ethically and culturally valuable (Henricks 2006). Play can become an addiction, the only mode of being in the world, not allowing us to develop relations that are not through play.

We need to play because we need occasional freedom and distance from our conventional understanding of the moral fabric of society. Play is important because we need to see values and practise them and challenge them so they become more than mindless habits. Games, toys, playgrounds (Seitninger 2006) are all instruments that allow us to explore, enact and develop our own different understandings of morality, not because they are separate from the real world, but because they are things we play with. When play is *about* ethics, it is so because it appropriates and explores values. But play should not be reduced to being ethically significant when it *explicitly* addresses morality. Like any other way of experiencing and expressing the world, play is *always* moral. Play is the expression of a moral being in a world.

So how do we study the ethics of playing? Play is an appropriative activity that helps us explore our values. The experience of play is mediated by technologies and social contexts, all of which influence how the activity of play configures our moral being. To understand the ethics of playable things, we need to look at the ways they enable, or constrain, the appropriative capacities of players, their occasional freedom (Danzico 2011). To develop our moral being, to flourish and live a good life, playable things should be open to appropriative play.

The two main vectors that we can adopt to understand the ethics of play as mediated by objects are those of submission and of resistance. Most games are playable things that want their users to submit to the world of rules and systems and mechanics that create a social encounter. Our understanding of the ethics of games, then, will have to do with the way that submission allows for the players' ethical being to reflect and develop towards its full potential. Games like *This War of Mine* or *Papers, Please*, show how morally sound playthings can contribute to our moral development by giving us worlds in which our act of play requires moral reflection.

On the other hand, the play of resistance can be a form of expression, an appropriation of the world through play that allows for the practice of values. For example, the playful appropriation of tracking technologies to subvert, challenge or ridicule their properties, from drawing penises with Nike+ to cheating fitness trackers, highlight the questionable ethics of everyday surveillance by making fun of them, in a carnivalesque reversal of the meaning of objects. Playing with the world to reveal, and rebel against its power structures is a form of asserting the moral power of play. This is precisely the kind of attitudes that methodologies such as cultural probes can help reveal and understand.

Play is important for our moral life because it can turn our own assumptions and ethical principles into props for play. Play gives us distance to, but also engagement with, our own moral fabric. To live a balanced life, to explore and become who we can become and flourish as ethical human beings, we need to understand our values and principles. And play, because of its appreciative nature, allows us to do precisely that: appropriate, estrange us from our own moral being, and allow us to explore what our values are.

We *need* to play to be better human beings. There is much talk and importance given to games and other playthings because they can address serious topics. But that is an unnecessary argument: play in itself is already important, necessary for living a good life. The values of play reside in how play is done, in playing. Play is necessary to be human not only because as humans we play, but also because through play we better express what it means to be a moral human being.

# References

Burke, R. (1971). 'Work' and 'play'. *Ethics*, 82(1): 33–47.

Bynum, T. W. (2006). Flourishing ethics. *Ethics and Information Technology*, 8(4): 157–173.

Caillois, R. (2001). *Man, Play and Games*. Champaign, IL: University of Illinois Press.

Danzico, L. (2011). BETWEEN THE LINES: what we talk about when we talk about happiness. *Interactions*, 18(1): 11–12.

Deci, E. L. and Ryan, R. M. (2008). Hedonia, eudaimonia, and well-being: an introduction. *Journal of Happiness Studies*, 9(1): 1–11.

Gaver, B., Dunne, T. and Pacenti, E. (1999). Design: cultural probes. *Interactions*, 6(1): 21–29.

Geertz, C. (1972). Deep play: notes on the Balinese cockfight. *Daedalus*, 101(1): 1–37.

Henricks, T. S. (2006). *Play Reconsidered: Sociological Perspectives on Human Expression*. Champaign, IL: University of Illinois Press.

Huizinga, J. (1992). *Homo Ludens: A Study of the Play-element in Culture*. Boston, MA: Beacon Press.

Schechner, R. (1988). Playing. *Play & Culture*, 1: 3–19.

Schüll, N. D. (2012). *Addiction by Design: Machine Gambling in Las Vegas*. Princeton, NJ: Princeton University Press.

Seitinger, S. (2006). An ecological approach to children's playground props. *Proceedings of the 2006 Conference on Interaction Design and Children*, pp. 117–120.

Sicart, M. (2009). *The Ethics of Computer Games*. Cambridge, MA: The MIT Press.

Sicart, M. (2013). *Beyond Choices: The Design of Ethical Gameplay*. Cambridge, MA: The MIT Press.

Waterman, A. S., Schwartz, S. J. and Conti, R. (2008). The implications of two conceptions of happiness (hedonic enjoyment and eudaimonia) for the understanding of intrinsic motivation. *Journal of Happiness Studies*, 9(1): 41–79.

Wixon, D., Flanders, A. and Beabes, M. A. (1996). Contextual inquiry: grounding your design in user's work. *Conference Companion on Human Factors in Computing Systems: Common Ground*, pp. 354–355.

Zagal, J. P. (2011). *The Videogame Ethics Reader*. San Diego, CA: Cognella Academic Publishing.

# 8

# Sensing atmospheres

*Sasha Engelmann and Derek McCormack*

Atmosphere has become one of the most alluring of concepts across the social sciences and humanities. Its appeal is manifold, but it is particularly important because it allows us to grasp the affective materiality of spacetimes that are diffuse and excessive of bodies yet also palpable through the sensory capacities of those bodies (Anderson 2009). The emergence of atmosphere as an alluring concept within the social sciences and humanities also raises important method-ological and empirical issues (Anderson and Ash 2015). Not least of these are the issues that revolve around the promise and limits of sensing.

A first issue concerns what is being sensed when we invoke either atmosphere or the atmospheric. When we speak of the force of atmosphere as something that registers in bodies of different kinds, are we referring to the quality of an entity, or to variations in a process that is never reducible to the category of entity? Is atmosphere an entity, the quality of an entity, or something excessive of entities? This issue is important not least because it has implications for any claim about the possibility of atmospheres as not only distributed but also shared.

A second issue concerns the problem of how to sense atmosphere. On one level, the appeal of atmosphere is that it suggests immersion in a spacetime that, while vague, is somehow directly and immediately sensed in and for human bodies. But we cannot take for granted this capacity to sense atmospheres: no less than other registers of sensing, such as seeing, it depends upon the complex relationship between experience, technique and technology.

In turn, this leads to a third issue: if we cannot take the capacity to sense atmosphere as a given, then how can we cultivate it as part of interdisciplinary methodological experiments? Here experiments with performance practices offer important possibilities. In many ways performance practices offer researchers in the social sciences and humanities a repertoire of techniques for generating atmospheres and for sensing variations in their intensity and distribution (see Böhme 1993; Thrift 2008; McCormack 2013). Performance practices foreground how atmospheres emerge in the relation between forms of skilful embodiment, techniques of stagecraft, and objects of different kinds. In some ways, of course, these practices amplify and intensify other styles of sensing atmospheres that have long been honed through ethnographic modes of attunement to the structures of feeling of the ordinary. These styles of sensing are not necessarily staged, even if they involve attending to scenes of life in which atmospheres become palpable with particular force (Stewart 2011).

A fourth issue concerns the problem of how to produce accounts of atmospheres that in some way evoke a sense of the atmospheric. The craft of writing remains an important domain of expertise through which to do this (Stewart 2011). At the same time, researchers in the social sciences and humanities have collaborated with a range of performance-based and creative researchers to generate atmospheric spacetimes (Engelmann 2015a; Hawkins 2016).

A fifth, and final issue concerns the political or ethical ends to which any interdisciplinary experiment with sensing the atmospheric might be put. The condition of being immersed within atmospheres is not necessarily benign (Philippopoulos-Mihalopoulos 2015). Given this, we might ask if the goal of any collaboration is to critically demystify the atmospheric qualities of spacetimes, to create new kinds of immersive atmospheric spacetimes, or to engage in some affirmative combination of both. In a world where atmospheres are becoming the 'object-target' (Anderson 2014) for various forms of intervention, should we be aiming to produce new forms of atmospheric politics, or be content to critique the operationalization of atmosphere across different domains of life?

These questions would be challenging enough if we could restrict atmospheres to the affective orbit of human sensory capacities. The atmospheric is not a domain circumscribed by phenomenological modes of conscious sensing, however. Indeed, much of the data and processes that can now be sensed operate below and before thresholds of human awareness. The domain of what Mark Hansen calls atmospheric media is becoming ever more infrastructurally ambient, and is fading into a background that nevertheless continues to shape what shows up as a foreground (Hansen 2012). But it is not only the expanding domain of technical sensing that should cause us to question the primacy of the human in relation to the question of sensing atmospheres. It is also the fact that atmosphere does not refer only to a field of affective experience. It refers also to an envelope of gases that surrounds the Earth, and other planets for that matter.

To speak of atmosphere is to invoke a gaseous medium in which different forms of life are immersed and to which they are exposed in a relation of respiration. Variations in gaseous atmospheres are meteorological: they show up as gradients in temperature, pressure, humidity, etc. These variations are part of the affective turbulence of the world. Sometimes these variations can be, and are, sensed in human bodies, or in the experiential texture of what Tim Ingold calls 'weather worlds' (2015). These variations do not need to be sensed in human bodies to make a difference, or to be considered affective, however: they can be sensed in non-human bodies and devices of various kinds through the capacities of those bodies to be affected or perturbed (Bryant 2014; Ash 2013).

The question of how to sense atmosphere involves exploring possibilities for sensing the elemental materiality of spacetimes that are both affective and meteorological, and variations in which can be sensed across particular arrangements of bodies and devices. Methodologically this never just involves a kind of static immersion within a milieu that reveals itself fully to a body. It involves, instead, the problem of how to sense variations in an expanded elemental milieu, while also finding ways of moving with these variations to enhance our capacities to act. This can be understood as the elaboration of an expanded sensory ethology, in which the properties and qualities of elemental atmospheres are sensed through different assemblages of bodies and devices.

Elements of this kind of approach characterize ongoing experiments with practices and politics of sensing across a range of empirical, conceptual and political domains. Here we could point to the work, for instance, of Jennifer Gabrys (2016), who highlights how the scope of sensing is expanding in all kinds of ways that complicate the primacy of human agency or experience. Equally, we could point to the kinds of practices of participatory sensing undertaken by organizations such as Public Lab (see https://publiclab.org). Central to this work is the employment of a range of relatively simple devices (including cameras, technical instruments and kites) with which a range of processes, including different kinds of air pollution, can be sensed. These are processes that might not be ordinarily available for monitoring and scrutiny. Equally, as far as possible, the devices are not black-boxed: they remain open for hacking and modification by a growing community of users.

Some of these devices used by groups like Public Lab are particularly useful for sensing atmosphere. Simple things such as kites, sails and balloons have all long been used for feeling and moving with the variations in elemental atmospheres (Ingold 2009; Serres 2012; Flusser 1999). These devices might appear anachronistic in a world in which capacities to sense have become ever more diffuse, algorithmic and pre-phenomenological. Nevertheless, both alone and in combination, they have the capacity to generate opportunities for sensing atmospheres in important ways.

Consider one of these devices: the balloon. As a device for atmospheric sensing, the balloon interests us in a variety of ways. It has long been implicated in the emergence of new forms of sensing, both as a platform for human journeys into the atmosphere, and as a vehicle for various forms of remote sensing and atmospheric sounding. At the same time, the balloon has been used to generate aesthetic works responsive to the elemental conditions in which they are immersed. The multiple possibilities of the balloon in this regard make it a particularly interesting object for exploring interdisciplinary experiments in sensing atmospheres. In making this claim we draw upon involvement in ongoing collaborative work with Berlin-based artist and architect Tomás Saraceno (see Saraceno, Engelmann and Szerszynski 2015; Engelmann, McCormack and Szerszynski 2015).[1] Saraceno's work is multiple and manifold, but central to this work are experiments with different ways of being and becoming airborne, organized around the speculative promise of collective forms of life in the air. There are many devices through which Saraceno experiments with this promise. One of these is the solar balloon, which, unlike other balloons, uses no helium, hydrogen, solar panels or burners of any kind: it relies instead only upon energy from the Sun during the day, and infrared radiation during the night (Figure 3.8.1).

The range of Saraceno's experiments lies far beyond the scope of this short piece, but there are a number of important points worth making about them insofar as they reveal what it involves to develop an interdisciplinary approach to sensing atmospheres. The first is the way in which these experiments foreground how the elementality of atmosphere is sensed in ways that do not obey any strict division between the natural or the social, or the affective and meteorological. Instead, in the shape of what Saraceno calls a solar sculpture, the balloon envelope becomes a device that senses variations both in the meteorological conditions in which it is immersed and in the elemental force of the Sun.

Then, and second, these works reveal how such sensing is a collective assemblage of both human and non-human participants, energies and forces (Serres 2008). Clearly, once aloft, a solar sculpture can remain in the air both day and night by absorbing short-wave energy from the Sun during the day and infrared radiation from the Earth during the night. However, the process of its taking to the air involves the enactment of a form of distributed expertise in which knowledge of the prevailing meteorological conditions, including wind-speed, cloud-cover, etc. is crucial. In turn, and third, such experiments reveal how sensing atmospheres is both technical and aesthetic. Sensing atmospheres is not about experiencing elemental natures in the raw: rather, it is about how our capacities to sense can be enhanced through as simple a technical operation as the folding and stitching of a fabric envelope. Equally, this sensing is not only something that takes place in the air: the process of fabricating and inflating the envelope on the ground generates atmospheres of involvement and participation.

Our own experience of participating in experiments with these devices reminds us of how interdisciplinary methods have the potential to be political insofar as they generate novel

*Figure 3.8.1*    Tomás Saraceno *Aerocene*, launches at White Sands (NM, United States), 2015. The launches at White Sands and the symposium 'Space without Rockets', initiated by Tomás Saraceno, were organized together with the curators Rob La Frenais and Kerry Doyle for the exhibition 'Territory of the Imagination' at the Rubin Center for the Visual Arts. The sculpture D-OAEC is made possible due to the generous support of Christian Just Linde. The artistic experiment achieved two world records of the first and the longest solely solar flight by a lighter-than-air vehicle. (Courtesy the artist; Pinksummer contemporary art, Genoa; Tanya Bonakdar, New York; Andersen's Contemporary, Copenhagen; Esther Schipper, Berlin. © Photography by Studio Tomás Saraceno, 2015.)

distributions of sensing atmospheres. On one level, this is evident through the kinds of atmospheres that gather around the prospect of a launch, atmospheres with the potential to generate new orientations in bodies towards the elemental conditions in which those bodies are immersed. It is also evident in the way in which, even when inflated on the ground, these envelopes provide spaces in which to gather. We can see this in one of the projects in which Saraceno is involved, *Museo Aero Solar* (Figure 3.8.2). This participatory project assembles around a solar balloon fashioned and fabricated from reused plastic bags. Individuals who donate plastic bags can participate in the solar balloon's fabrication and its eventual launch; both of these processes draw bodies in affectively through the shaping and sensing of something becoming airborne.

But the distribution of atmospheric sensing is also evident in other, less immediate ways. This is perhaps more evident in Saraceno's recent open-source artistic project, *Aerocene* (Figure 3.8.3). The larger aim of the *Aerocene* project is to generate the conditions through which new forms of life in the air might become possible. On the way to this, Saraceno has undertaken various tethered flights with solar sculptures, and some free flights without passengers. These latter flights rely upon technologies for tracking the movement of the solar sculpture. These technologies and their diagrammatic renderings of movement can also provide lures, in the shape of alluring abstractions that allow individuals and groups to track and trace the movement of solar sculptures.

*Figure 3.8.2*   *Museo Aero Solar* in Prato, Italy. *Museo Aero Solar* is a collaborative art project initiated by Tomás Saraceno, Alberto Pessavento and many other friends and collaborators around the world. (Photo: Janis Elko *Museo Aero Solar*, 2009 www.museoaerosolar.wordpress.com.)

*Figure 3.8.3*   Tomás Saraceno. *Aerocene Gemini*, Free Flight, 2016. (Courtesy the artist; Pinksummer contemporary art, Genoa; Tanya Bonakdar Gallery, New York; Andersen's Contemporary, Copenhagen; Esther Schipper, Berlin. © Photography by Tomás Saraceno, 2016.)

Experiments with solar balloons or solar sculptures are not the only interdisciplinary method for sensing atmospheres. But they remind us how the affective-meteorological materiality of atmospheres can be sensed and generated as part of the methodological repertoire of the social sciences. This sensing can be understood as a form of sounding (see Dyson 2014; Engelmann

2015b), where sounding is both the assaying of a milieu and its collective enunciation. Developing this, and drawing upon the kinds of experiments undertaken by Saraceno, it might be possible to imagine and devise new methods for sensing atmospheres in which the meteorological atmosphere itself becomes part of the infrastructure of sensing. These experiments would necessarily employ different devices and practices for learning to be affected by the force of the atmospheric as an elemental variation in both meteorological and affective spacetimes. To be sure, these experiments would afford opportunities for generating envelopes of experience that we might understand as particular kinds of atmospheres of immersion. But they would also stretch the envelope of atmospheric sensing far beyond the limits of the human body. In doing so, these experiments would provide opportunities for what, following Felix Guattari, we might call the 're-singularizing' (1995) of capacities to sense the elemental conditions in which diverse forms of life take shape. They would generate situations of co-fabricated assembly in which atmospheric conditions are made explicit for sensing by bodies whose very forms of life are dependent upon those conditions.

## Note

1 For Sasha, this involves ongoing ethnographic immersion and collaboration in the creative life of Saraceno's studio as part of the first doctoral project undertaken specifically on his work. For Derek, this has involved a series of invited pieces of writing and speaking (some with Sasha and others) that contribute to the elaboration of the ecology of ideas around these projects.

## References

Anderson, B. (2009). Affective atmospheres. *Emotion, Space and Society*, 2(2): 77–81.

Anderson, B. (2014). *Encountering Affect: Capacities, Apparatuses, Conditions*. Farnham, UK: Ashgate.

Anderson, B. and Ash, J. (2015). Atmospheric methods. In P. Vannini (Ed.) *Nonrepresentational Methods: Re-envisioning Research* (pp. 34–51). London: Routledge.

Ash, J. (2013). Rethinking affective atmospheres: technology, perturbation and space-times of the non-human. *Geoforum*, 49(1): 20–28.

Böhme, G. (1993). Atmosphere as the fundamental concept of a new aesthetics. *Thesis Eleven*, 36: 113–126.

Bryant, L. (2014). *Onto-cartography: An Ontology of Machines and Media*. Edinburgh: Edinburgh University Press.

Dyson, F. (2014). *The Tone of Our Times: Sound, Sense, Economy, and Ecology*. Cambridge, MA: MIT Press.

Engelmann, S. (2015a). Toward a poetics of air: sequencing and surfacing breath. *Transactions of the Institute of British Geographers*, 40(3): 430–444.

Engelmann, S. (2015b). More-than-human affinitive listening. *Dialogues in Human Geography*, 5(1): 76–79.

Engelmann, S., McCormack, D. and Szerszynski, B. (2015). Becoming aerosolar and the politics of elemental association. In T. Saraceno, *Becoming Aerosolar*, Exhibition Catalogue, 67–101. Vienna: 21er Haus.

Flusser, V. (1999). *Shape of Things: A Philosophy of Design* (Trans. A. Mathews). London: Reaktion Books.

Gabrys, J. (2016). *Program Earth: Environmental Sensing Technology and the Making of a Computational Planet*. Minneapolis, MN: University of Minnesota Press.

Guattari, F. (1995). *Chaosmosis: An Ethico-aesthetic Paradigm* (Trans. P. Bains and J. Perfanis). Sydney: Power Publications.

Hansen, M. (2012). Ubiquitous sensation or the autonomy of the peripheral: towards an atmospheric, impersonal and microtemporal media. In U. Ekman (Ed.) *Throughout: Art and Culture Emerging with Ubiquitous Computing* (pp. 63–88). Cambridge, MA: MIT Press.

Hawkins, H. (2016). *Creativity*. London: Routledge.

Ingold, T. (2009). The textility of making. *Cambridge Journal of Economics*, 34(1): 91–102.

Ingold, T. (2015). *The Life of Lines*. London: Routledge.

McCormack, D. (2013). *Refrains for Moving Bodies: Experience and Experiment in Affective Spaces*. Durham, NC: Duke University Press.

Philippopoulos-Mihalopoulos, A. (2015). *Spatial Justice: Body, Lawscape, Atmosphere*. London: Routledge.

Saraceno, T., Engelmann, S. and Szerszynski, B. (2015). Becoming aerosolar: from solar sculptures to cloud cities. In H. Davis and E. Turpin (Eds.) *Art in the Anthropocene: Encounters among Aesthetics, Politics, Environments and Epistemologies* (pp. 57–62). London: Open Humanities Press.

Serres, M. (2008). *The Five Senses: A Philosophy of Mingled Bodies* (Trans. M. Sankey and P. Cowley). London: Athlone.

Serres, M. (2012). *Biogea* (Trans. R. Burks). Minneapolis: Univocal.

Stewart, K. (2011). Atmospheric Attunements. *Environment and Planning D: Society and Space*, 29(3): 445–453.

Thrift, N. (2008). *Non-representational Theory: Space, Politics, Affect*. London: Routledge.

# Section 4
# Of interdisciplinarity

# 1

# Of interdisciplinarity

*Angela Last*

As a geographer, one of the first things you learn about your discipline is that it is not necessarily considered to be one. In fact, the subject was nearly eradicated in North American universities in the 1940s for lack of academic integrity. The geography department at Harvard, for instance, was closed for being 'hopelessly amorphous' (Neil Smith, cited in Paglen 2009). Getting rid of geography meant getting rid of unscientific descriptions of places, more fit for the school classroom or the fanciful travelogues of aristocratic explorers. Instead of physical and human geography, the subject could simply be split into geology and sociology. This struggle for legitimacy continues and now extends to geographical information systems (GIS), which are, again, seen as 'unscientific' and lacking disciplinary integrity (Drummond, cited in Last 2015).

There is still much debate over what constitutes the essence of geography, despite claims that disciplinary debates have been 'railroaded' in favour of debates around interdisciplinarity (Griffin, Medhurst and Green 2006: 7). For decades, human geographers have been indoctrinated to answer 'space' to any enquiries about their central concept or identity (Johnston, Gregory, Pratt and Watts 2000: 353; Massey 2005). More recently, some geographers have controversially proposed the Earth and its processes – the 'geo' of geography – as a more appropriate, or at least additional characteristic that takes the material relations of Earth systems more strongly into account (Clark 2011), drawing predictably hostile reactions from the 'space defenders'. Such boundary making becomes even more pronounced when you enter geography as an outsider and are constantly scrutinized for the correct performance of 'geography'. Despite such stabilizing impulses, geography continues to be a discipline that is amorphous enough to allow for a great amount of interdisciplinary work.

Another piece of geographical history that is helpful in understanding interdisciplinarity, and its link to innovation and internationalization, is the discipline's association with geopolitics and colonialism. Geographical methods are not innocent and often hide violent histories around resource extraction, land appropriation and the construction of unequal infrastructures and divisions, whether we are talking about cartography or the use of statistics (Warren and Katz 2015). What becomes apparent is that methods can never be divorced from their context, and if they are, the context will keep surfacing. Academics and other commentators from colonized countries have repeatedly noted how the objectification of indigenous and other colonized peoples continues, as well as the imposition of European concepts and categorizations (Deloria 1991; Kukutai and Taylor 2016; Smith 2012). Linda Tuhiwai Smith (2012), for instance, argues that not only research questions and methods need to be revised, but also questions relating to the power structures in which these are embedded. For her, such questions should be extended to include:

> Who defined the research problem?
> For whom is this study worthy and relevant? Who says so?
> What knowledge will the community gain from this study?
> What knowledge will the researcher gain from this study?
> What are some likely positive outcomes from this study?
> What are some possible negative outcomes?
> How can the negative outcomes be eliminated?
> To whom is the researcher accountable?
> What processes are in place to support the research, the researched and the researcher?
> *2012: 175–176*

While researchers already think they address these questions, often in front of ethics committees through which their proposals need to pass, dynamics between researcher and researched/ other researchers can work out very differently in practice (see Newell *et al.* in this section). Smith herself uses her book *Decolonizing Methodologies* to document and propose 25 example methods – from claiming to sharing – that have been used as alternatives by indigenous communities (2012: 143). In such circumstances, methodological innovation can literally become a lifesaver.

Both disciplinary identity and problematic histories point to a fundamental question, implicit in Linda Tuhiwai Smith's set of questions: how does a method come into being? This question is important, because it is easy to lose sight of our research processes and what has shaped and continues to shape them. Take, for example, interviews. How come we are using interviews? How did the methodological guidelines for interviews come into being? How can interviews be done differently?

When students learn about methods, they are told that these are needed to produce 'new knowledge'. What counts as knowledge is, again, determined by historical, geographical and social context, and this, in turn, affects the conception of methods. There is no stable ground or yard stick. This is a problem that has occupied many researchers and writers. One of them is Georges Perec, a French writer whose parents 'legally' died violent deaths in the Second World War – on the battlefield and in a concentration camp. This personal history played a key role in his questioning of the processes that go on in the world and count as 'normality' and acceptable practices. One of his questions could be described as: if we take methods to be the outcome of a decision of what we want to measure and how we want to measure it, the issue that remains is: what tells us that our method is appropriate? As he emphasized in his striking methodological experiments, such as those found in the collection *Species of Spaces* (Perec 2008), there is nothing we can rely upon in terms of external constraints that can function as a marker for appropriate conduct. Of course, there are temporary markers: at the present time, for instance, we cannot go out and ask random people about a specific topic and call it 'research'. We also cannot undertake research with particular populations (such as children or patients) without complicated processes of consent. However, at a different point in time, this could have been – or could be – entirely appropriate. As Perec proceeds to undertake apparently absurd projects – from experimenting with alternative address systems to minute descriptions of spaces and events that do not seem to matter – he communicates an urgency: to 'question the habitual' (2008: 210) and to always ask what circumstances led to something becoming normalized.

Across the sciences and social sciences, we now look upon a significant chunk of past research with horror, in terms of the methods and ethics applied. Books from anthropology to the history of science are full of examples of what we now recognize as systematic violence, especially against marginalized populations (for example, Brandt 1978; Nelson 2011). If past methods are not violent, they often appear naive or silly. They did *what* to measure psychological disorders? They did *what* to demonstrate the inferiority of women and people of colour? How could researchers adopt such appalling methodological standards? Although we like to think that methodological insanity is a thing of the past, headlines about problematic studies are not likely to disappear. For all we know, the methods that we are presently employing might be ridiculed only 20 years later. On the other hand, methods that now seem utterly ridiculous to people who have undergone standard methods training – some performative methods spring to mind – might become perfectly acceptable in the future.

At present, for instance, there is considerable debate around the benefits and hyperbole of interdisciplinarity. To adherents to disciplines, the investment in interdisciplinarity as a superior approach to problem-solving often seems overstated – and overfunded (Barry, Born and Weszkalnys 2008; Thrift 2006). Many such researchers would even agree that there cannot be such a thing as interdisciplinary methods – perhaps a negotiation of methods or a common practice such as a survey or an experiment. The good news is that instances of 'methodological insanity' are not always a bad thing. In some fortunate moments, some of us might get thrown into such a state right in the present. Often, the occasion is indeed cross-disciplinary and international collaboration, where colleagues from different methodological traditions – most likely including yourself – make seemingly unspeakable proposals. In the moment when everyone is taking turns to justify their method, you might suddenly realize not only why a particular method is weird, but also why it could make sense in a particular context.

Further, one of the key obstacles to any method is what is commonly described as 'the field'. 'The field' – understood as the site of study, whether this be a lab, an archive or a geographical region – is where all good methodological intentions seem to fail. Whether you look at field notes, published government reports, articles, theses, monographs – there are numerous documentations of how methods did not work, or had to be 'tweaked' to produce any kind of useful research. Some of the resulting 'rogue methods' are more controversial than others. Here, the work of Nancy Scheper-Hughes on organ trading (2004) and Irish bachelors (2000) is often cited as a key reflection on how a researcher moved into a field with noble intentions and ended up in an ethical and methodological car crash – albeit not without educational and 'rogue ethical' side effects. Such examples show not only how methods are challenged by the environments in which they are placed, but also how they are potentially reshaped into something unexpected.

In this context, I would like to come back to the earlier question of 'how does a method come into being?' If method is not a neutral practice, but a practice that is shaped by context – by the field as well as wider systems in which research is embedded – one should really ask the question: what gives rise to interdisciplinarity and interdisciplinary methods? Is it the research problem, the collaboration, the specific institutional context and availability of resources, the field, economic drivers or even geopolitical dynamics? Although wider drivers for interdisciplinarity are in place – such as the demand for interdisciplinary research from research councils and other funding bodies – the answers can be very different from project to project. This section provides space for reflection on the role of the contexts – the varying geographies, institutional and personal negotiations, field provocations and economic considerations – *of* interdisciplinarity.

## Collaborations

Of the things that give rise to interdisciplinarity, collaborations form an important point of methodological emergence. While interdisciplinarity does not have to take place between people – you can be an interdisciplinary researcher by drawing on resources from different disciplines – exchange between researchers remains the most visible basis. It also remains the most institutionally supported mode, as innovation is imagined to take place through cross(disciplinary)-fertilization. Since collaborations tend to be complex negotiations, many pages have been filled with studies of cross-disciplinary and interdisciplinary projects, often in grinding detail. How do you make yourself understood in a collaboration? How do you manage relationships between many, often speedily thrown-together project partners? How do you co-write a paper with a multitude of other authors? How do you innovate?

When I taught on the MA in Art and Science at Central Saint Martins, London, for instance, the art students frequently struggled to keep the science students interested in their work. Since there were already a growing number of collaborative project documentations online I pointed them to a variety of 'behind the scenes' reflections on the success or failure of joint research. One of these reflections was by the collaborators behind *Micespace*, a project on rodent welfare in experimental science between the artist Helen Scalway and the geographer of science Gail Davies. The project began as a spontaneous exploration and has very much retained the spirit of its origins throughout its 'mutations', including its online presence (www.micespace.org/). As Scalway and Davies describe it:

> This website is our own mutated mouse repository: it is offered as an 'anti-archive'. Micespace is a provocative, messy and speculative collection, inviting reflection and hesitation around what the conventional curation of scientific data may elide, miss out, or skip over: incommensurable logics, unanswerable conundrums, and aporias in meaning.

So far, this 'anti-archive' has turned out to be valuable for both collaborators – as a means of reflection and communication – as well as for outsiders such as my students who appreciate it as an example of how a collaboration can not only be documented but grow through processes of private and public reflection. Davies' and Scalway's entry also beautifully illustrates how methods that at first appear very different, such as scientific and artistic experimentation, can emerge from very similar objectives and be brought together with a shared vocabulary – in this case, that of diagramming.

The scale of the project – an organically evolving experiment between two people that is not so much confined by the start and end date of a set research project – also impacts on the way the research is conducted. Despite the serious topic, the project has a playful feel, with which Scalway and Davies explore the different connections between human and nonhuman research subjects.

> We were looking together at a drawing I'd made of an alert mouse and Gail said something like, the problem is, this is just a picture of a mouse. Then she had to leave the studio to make a phone call and in a trice my mind somehow somersaulted me into banging a large piece of paper up on the wall and frantically scribbling, as if from a mouse-eye view, the inside of a lab cage with bars and nesting materials and scent-trails, so that when she came back we were both in a charcoaly, smudgy, very crudely suggested mouse world.
>
> *Scalway 2015*

This does not have to mean that large-scale and more time-constrained projects cannot be playful. In fact, a report by Blackwell, Wilson, Street, Boulton and Knell (2009) recommends that all interdisciplinary projects should 'embrace chaos' and play, no matter at what scale, to overcome disciplinary habits (2009: 121). The authors suggest, for instance, 'coin[ing] novel, playful terminology' as a means of overcoming disciplinary jargon: 'be playful in the early stages, engage in experiments and avoid theory' (op. cit.). A good example of this approach is the *Hearing the Voice* project, an interdisciplinary medical humanities project that explored the phenomenon of voice hearing, based at Durham University. Some of the project conditions were described by Angela Woods, co-investigator of the project, in an article on interdisciplinary authoring:

> The 17 members of the 'Interdisciplinary approaches' working group were based in 7 different time zones and brought expertise from 19 different fields, ranging from medieval history to clinical psychology to neuroimaging. While a majority of authors were able to meet in person at the ICHR, most of our discussions, and particularly the planning and revising of the article, had to be conducted via email. It's not difficult to imagine some of the challenges that teamwork at this scale presents, and some of these are clearly specific to, or heightened by, working across disciplines.
>
> *Woods 2015: 5*

As a starting point to manage such challenging conditions, the project members created 'Voice Club', fortnightly meetings in which a variety of media, from games to talks, were employed to get to know one another's perspectives (Robson, Woods and Fernyhough 2015).

At the same time, there are other challenges in collaborations that are less easy to negotiate and often extend beyond the scale of the team. An added pressure, for example, was the desire to undertake methodological development. In the project: the researchers neither wanted to 'unnecessarily "tweak" existing methods' nor necessarily stick to established methods, but instead used the multi-disciplinary environment as a gauge as to where established methods might be insufficient and could benefit from mutual exchange (Wilkinson and Smailes 2015: 3). A downside of this approach to developing interdisciplinary methods was the required work and time intensity, which, the participants observed, resulted in fewer, albeit potentially more innovative outputs (op. cit.). As Wilkinson and Smailes (2015) argue, such differences need to be taken into consideration not just by researchers, but also by funders.

As indicated at the beginning of this essay, the shape of collaborations is dependent on many things, including what normally become labelled as 'external factors'. I realize that I follow a reverse movement to recent propositions that interdisciplinarity might perhaps be better understood if the focus moved away from institutional structures. As Felicity Callard and Des Fitzgerald (2015b) ask: 'How would our understanding of – and capacity to improve – interdisciplinary research change if we focused less on funders and journals and universities, and more on the mundane, day-to-day lives of collaborative researchers?' From my own experience, however, these 'mundane, day-to-day lives' and activities are so much infused by wider dynamics that I would like to zoom out again to highlight issues such as geopolitical divisions, and what is generally referred to as the 'knowledge economy' (how knowledge can be measured, valued and made to count in economic terms).

## Institutions

When it comes to its relationship with institutions, interdisciplinarity is often placed within a narrative that can be framed in terms of 'global challenges' and 'global competitiveness' (Ledford

2015; Science Europe 2012). According to this narrative, interdisciplinarity is attributed with 'a heightened significance due to a more general association of innovation with processes of boundary crossing, collaboration, and the integration of different kinds of knowledge' (Blackwell *et al.* 2009). Innovation and interdisciplinarity can end up, as Blackwell *et al.* remark, being used more interchangeably than can be justified (2009: 10–11). Not only might interdisciplinary outputs take longer to emerge, as stated in the preceding part on collaborations, but they are apparently also slower to manifest in the form of citations. As a recent report by the publisher Elsevier to the UK Research Councils found, 'it takes longer for interdisciplinary research impact and value to be recognized' (Elsevier 2015: 3). Moreover, while potentially advantageous in economic terms and globally recognized as a research priority (Global Research Council 2016), interdisciplinarity is facing institutional obstacles. UK research councils, for instance, have noted issues within their own structures, as well as other academic infrastructures. Similar issues persist elsewhere, too (Bozhkova 2016). At present, attempts are being made to remedy this situation, through cross-institutional funding programmes and also through transnational programmes, since international interdisciplinary research has been identified as having a greater impact (Elsevier 2015: 3).

The reality is that the uptake of interdisciplinarity, while seemingly universally desirable, is uneven. This not only has to do with geographical differences in university systems, but with resources and purpose-orientation: an institution or system has to decide what inter-disciplinarity is and what it is for. Achille Mbembe, in his talk about the future shape of South African universities and knowledge production, argues that some fundamental decisions about what counts as knowledge and discipline need to be made first, both for local and global contexts (2015). Similarly, academics have decried the emphasis on interdisciplinarity that is driven by economic outcomes, rather than problem orientation (Thrift 2006). Some problems are less profitable than others, a fact that has already been an issue in disciplinary research, most notoriously medical research, where diseases affecting poorer people or women remain under-researched (Morel 2003).

In this section, Nina Lykke stresses the importance of the history of interdisciplinarity in the social sciences and charts the concept's recent trajectory from radical intervention to institutional convention. Over the last 50 years or so, interdisciplinarity has not only been perceived differ-ently, but served very different purposes. As Lykke emphasizes, one history of interdisciplinarity is closely associated with projects such as critical gender and race studies that emerged bottom-up – from activist and student movements. This is a different situation from today, where inter-disciplinarity is often a top-down demand, despite lack of adequate infrastructures (British Academy 2016: 5). In the interview, Lykke argues that, on the one hand, institutional uptake can be regarded as a success – interdisciplinary research is now widely supported, from local to global research councils issuing statements on their valuing and handling of interdisciplinarity (e.g. Global Research Council 2016). On the other hand, she argues that institutionalization and economic imperatives can lead to uncritical and unethical practices that undermine activist goals, cross-cutting inequalities and institutional elitism.

Such institutionalization of radical values is often hitting the very people it was initially intended to support and protect, as their demands increasingly grate against the new norms of interdisciplinarity. In their book on interdisciplinary research with social scientists and neuro-scientists, Callard and Fitzgerald (2015a) flag how some counter-productive dynamics can be amplified through interdisciplinarity, for instance around status and value. When the emphasis is on negotiating disciplines, they suggest, other factors such as race, gender or class can end up becoming side-lined. At the same time, when it comes to the distribution of roles within a project, these factors tend to creep back in, especially in the distribution of low- and high-status work. In response, Callard and Fitzgerald offer a checklist of questions that are aimed at making valuation transparent. Among the topics are administration or 'housework', conceptualization versus data collection, public engagement work, publishing outlets (high ranking/low ranking; open access/standard publishing model; online/print) and author hierarchies (2015: 110–111). Such lists can be a useful component of project design, especially when hierarchies are not clearly pre-established as in most interdisciplinary projects. Again, additional administrative clout such as equality charters can be brought into the discussion as a means of holding people accountable. In practice, success depends very much on individuals as well as on structures, but even if such attempts at questioning value prove unreliable, they at least constitute one available tool.

Another issue is metrics and their use in evaluating interdisciplinary work. This can be witnessed, for instance, in the UK's Research Excellence Framework (REF), a university audit that in part determines the distribution of research funding. An issue of the last REF was the inability of the system to adequately deal with interdisciplinary work (Bhandar 2016; Department for Business, Energy and Industrial Strategy 2016). For many interdisciplinary researchers, it was considerably more difficult for their work to be included in the REF (British Academy 2016: 32), or to find a job during the period of strategic hires that preceded the exercise. Anything from publishing in interdisciplinary journals to publishing with project partners in journals outside of the Global North became a problem in the face of the metrics employed (Bhandar 2016; RACE 2016). While the criteria are currently being reviewed to take this kind of work into consideration, this process is not likely to be straightforward, especially when an international dimension is added. Whose evaluation criteria are going to dominate the project? This also leads to a bigger issue in knowledge production that could be described as the 'geopolitics of method'.

## Geopolitics

How should the 'geopolitics of methods' be understood, and why should we care about it? Interdisciplinary scholar Richa Nagar offers a reconsideration of the fashion for seemingly bounded 'border crossings':

> Crossing borders is on the agenda. Foundations wish to fund projects connecting the global with the local, academics with practitioners. Editors are excited by cutting-edge scholarship that blurs disciplinary boundaries, and emphasises hybridity. This is energising, but the popularity of these border crossings in the present political climate should give us pause.
>
> *Nagar 2014: 86*

What Nagar talks about is the impact of global inequalities on the way we conduct research. In her book *Muddying the Waters*, she calls for a feminist scholarship that takes such issues into account – issues that not only surface in international collaborations.

When thinking about the 'geopolitics of method', 'geopolitics' can be thought of as a method as well as a mode of power – the two cannot be separated. As a disciplinary field, geopolitics imagines territorial relations, and often claims to predict political developments across the world through an analysis of geography, including demography. It has been the subject of controversy, because of its association with environmental determinism, imperialism and fascism. A few decades ago, critical geopolitics developed as a way of critiquing the concepts and worldviews of classical geopolitics while re-emphasizing the importance of geography (Smith 2000: 370). As a mode of power, geopolitics impacts through its portrayal of power connected to geography. The starkest geopolitical division today, perhaps, is the imagination of the Global North and South, with the Global North presented as an aspirational model for the Global South.

I would like to argue that methods, too, are tied to geopolitical imaginaries in ways that impact on how they are practised and taught. In *Science and an African Logic*, Helen Verran (2001) argues for the cultural relativity of methods and measurement. Reflecting on her employment in teacher training in Nigeria, she shows how methodological hegemonies, in this case brought about through colonialism, can negatively impact on those whose methods continue to be devalued:

> Mrs Babatunde had an interest in the discipline sought by the official mathematics curriculum, which says that measuring volume should be achieved through the prior notion of a singular uniform extension. The curriculum insists on a very specific sequence of small bodily gestures as the proper way to measure volume. Those who are authorities in the matter of quantifying, insist that *that* is the way to proceed. Mrs Babatunde might suffer if she fails to instil in the children the specific, required routines, developing in them the proper bodily habits of enumeration. If she failed my subject because she taught incorrectly, she might lose her job. At the same time, Yoruba children embodied an important set of interests for Mrs Babatunde. She knew that they might not get it if she presented volume in the prescribed way, the little routines, gestures with hands and eyes, words and chart, might not congeal as number, and then she will have confused and rowdy children on her hands.
>
> *Verran 2001: 8–9*

This is but one example of how geopolitical power impacts on methods. Another one is the teaching that many of us have received in the social sciences where methods are routinely described as having emerged from a particular genealogy of thought and practice, mainly populated by white European men who brought about paradigm shifts that, in turn, brought about new methodological requirements.

While this critique is by now well-known through the work of de-colonial and feminist scholars, it also helps put matters into perspective (Harding 1993; *The Black Scholar* 1974). For many white Western scholars who are interested in a different perspective, this has meant trying to look at other methodological systems, vocabularies and practices. In their paper 'Provincialising STS', John Law and Wen-yuan Lin (2015) struggle with exactly this issue: how to *not* take Western concepts as universals, in their case in the field of science and technology studies (STS). What would a postcolonial STS look like? Like the experimenters who have developed 'Asia as method' discussed by Celia Lury, their suggestion implies applying concepts from other knowledge traditions to Euro-American contexts, as they may allow for a different story to be told. As, to some scholars, this might look like another form of cultural appropriation, it becomes clear that the effort to provincialize European knowledge is not one that can easily distance itself from the usual dynamics. Helen Verran, too, has experimented with responses to the issue of knowledge inequalities. Through her experience of different knowledge systems and her dialogue with STS scholars, she has come to see everything in the world as emerging out of practices. Thinking in terms of 'bundles of practices' (Verran 2014: 530) gives rise to a 'flatland' where even basic (Western) knowledge boundaries between the sciences and humanities disappear. In her writing, she pursues the strategy of 'attempting to provoke readers to think "ontologically" – to question and experiment with the very foundations of knowledge that we take for granted, and also our place within them' (2014: 531).

A further layer to the debate around knowledge hierarchies is added through the top-down demand for internationalization by university managements and publishing companies. These, too, are embedded in geopolitical dynamics and hierarchies that directly shape research practices. Not only are British universities increasingly reliant on recruiting international students, they have over 200 satellite campuses in other countries. International students constitute a growing market both in the UK and abroad – in terms of tuition fees and consumers of publications. The same is true for research grants. It is not only interdisciplinary research that is increasingly desired, but international research as well. Due to unequal geopolitical and economic dynamics between the so-called Global South and Global North, there is often a greater support for knowledge from the Global North and a devaluation of knowledge from the Global South. This, for instance, translates into funders from the Global North having a greater say in project decisions. The critique 'Participation: the new tyranny?' (Cooke and Kothari 2001) gives examples of the problems of such imposed paradigms. Documented in this edited collection is the role of 'participatory methods' for development – fashionable in the Global North – and its unintended reinforcement of power structures through group dynamics and access issues.

In this section, paradigm failure is represented by the reflections of the DIRTPOL team. The project, on cultural perceptions of dirt, put together by researchers from different disciplines at Sussex, Lagos and Kenyatta universities, was accompanied by a series of blog reflections on what could be called methodological failure, re-evaluation and adaptation. This documentation showed not just the cultural conditioning of methods, but also the counter-productivity of insistence on methodological universality. In their chapter, the DIRTPOL team takes this argument further and argues that non-conforming, or 'dirty' methods can be not just a more appropriate, but a more ethical practice.

Ethics remain a serious issue in international interdisciplinary research. What a consideration of geopolitical dynamics demands of us is to ask some very basic questions such as: why am I collaborating with a particular researcher and what are the conditions of this exchange? In the case of this handbook, for example, the relatively standard production conditions became problematic for me as an editor when it came to soliciting international contributions. Apart from asking for many hours of sometimes unpaid labour for a book that most people, let alone libraries, will not be able to afford, I had to be sensitive to differences in authors' working conditions across the world. I was also aware of the conflict that many academics from the Global South experience around assimilation into the Global North, which further seems to feed its cultural dominance (Joseph 2015).

Tahani Nadim's entry on seed archives describes the lingering colonial histories and paradigms that resurface in present-day museum narratives. Her text makes apparent the constant supply of energy that is required to maintain and renew these narratives and ensure the future of convenient exclusions. In her context of archives, collections and museums, interdisciplinarity and the global dimension open a connection to other epistemologies and ontologies around the world that perform different sorts of 'disciplining' of materiality (see Todd 2016). An explicit concern in her entry is the role of the nonhuman, something that rarely seems to figure in discussions of either interdisciplinarity (Latour 2004) or globalization (see Clark 2002; Mbembe 2015). What provocations can different considerations of the nonhuman issue towards not only method but geopolitical considerations of method?

## Methodological futures

From the contributions in this section, a variety of questions and provocations arise. A theme that runs across all the contributions for me and which, I feel, is central to the design and practice of method in general, is negotiating exclusion. This is also reflected in the work of many philosophers who have written on method, such as Isabelle Stengers (2010), Michel Serres (1995) and Edouard Glissant (2010). Stengers' 'cosmopolitics' specifically attempts to cultivate a sensitivity to what is excluded from politics, which, for her, means expanding our notion of the political and to where it extends. For me, this includes our daily practices as researchers, our 'doings'. A method excludes by necessity, but there are always different choices that can be made, whether these choices simply concern techniques or seemingly prescribed delineations between life-forms or geopolitical territories.

Michel Serres cautions against the rigidity and exclusionary violence of the academy, as enacted in method. He describes his own method as 'anti method' (Serres, Harari and Bell 1982: xxxvi), as being 'fertilely inventive in the middle of chaos' (Serres and Latour 1995: 117). Deliberately unsystematic, Serres creates excessive, intertwined texts, which he hopes will enable readers to make their own connections. The 'noise' of Serres' texts further serves the purpose of building immunity to being stripped to reveal something that is manageable as a tool. Rather, a method, especially an interdisciplinary one, is supposed to be 'less a juncture to control than an adventure to be had' (Serres, cited in Brown 2003: 189). As discussed earlier, an element of adventure and play is always desired as a condition for innovation – Goldsmiths College in London, for instance, has a 'Unit of Play'. The question here is: how shall we, and can we, play under the current institutional-economic conditions? Or perhaps: why do we play?

The biggest challenge is perhaps issued by Edouard Glissant who challenges us to perform a different kind of globalization which he calls 'globality' or 'worldmentality' (Diawara 2015).

> What we call globalisation, which is the standardisation to the bottom, the reign of multi-nationals, the standardisation, the ultra-liberalism in global markets, for me, this is the negative side of a wonderful reality that I call globality. This globality is an extraordinary adventure given to all of us who today live in a world that, for the first time, in real and immediate, sudden ways, without wait, is simultaneously multiple and unique, just as much as it presents a necessity for everyone to change their ways of perceiving, of living, of reacting in this world.
>
> *Glissant 2005: 15; my translation*

While globalization represents increasing restriction, the drive to universalize in ways that create further inequalities, 'globality' represents the openness that globalization supposedly promises. This openness emerges from a willingness to being changed by the other (Glissant 2009: 66). Throughout his work, Glissant reminds us that this openness is not some idealistic grand gesture that needs to be performed to garner attention. Rather, it is something that should permeate our most mundane, everyday practices.

Against this background, interdisciplinarity, especially one with the aspiration to enable global exchange, can be a place to practise this openness to be changed by the other. This might seem ridiculous, as the conditions for exchange in the academy often feel instrumental, contrived or, due to their rootedness in ideas of globalization, even counter to 'globality'. However, in their very banality and commonplaceness, they can still be a provocation to our practices, if we allow ourselves to be open. Whether we want it or not, our seemingly geographically confined practices are entangled in much wider processes. This technically means that any 'ridiculous' experiments could reach some unexpected places. Says the geographer.

## References

Barry, A., Born, G. and Weszkalnys. G. (2008). Logics of interdisciplinarity. *Economy and Society*, 37(1): 20–49.

Bhandar, B. (2016). The Stern Review. LRB Blog. Retrieved 27 March 2018 from: www.lrb.co.uk/blog/2016/08/02/brenna-bhandar/the-stern-review/

Blackwell, A. F., Wilson, L., Street, A., Boulton, C. and Knell, J. (2009). Radical innovation: crossing knowledge boundaries with interdisciplinary teams. University of Cambridge, Technical Report No. 760. Retrieved 27 March 2018 from: www.cl.cam.ac.uk/techreports/UCAM-CL-TR-760.pdf

Bozhkova, E. (2016). Interdisciplinary proposals struggle to get funded. *Nature News* 29 June 2016. Retrieved 27 March 2018 from: www.nature.com/news/interdisciplinary-proposals-struggle-to-get-funded-1.20189

Brandt, A. M. (1978). *Racism and Research: The Case of the Tuskegee Syphilis Study*. The *Hastings Center Report, 8*(6): 21–29.

British Academy (2016). Crossing paths: interdisciplinary institutions, careers, education and applications. July 2016. Retrieved 27 March 2018 from: www.britac.ac.uk/sites/default/files/Crossing%20Paths%20-%20Full%20Report.pdf

Brown, S. C. (2003). Natural writing: the case of Serres. *Interdisciplinary Science Reviews*, 28(3): 184–192.

Callard, F. and Fitzgerald, D. (2015a). *Rethinking Interdisciplinarity across the Social Sciences and Neurosciences*. Basingstoke: Palgrave.

Callard, F. and Fitzgerald, D. (2015b). Why it's time to get real about interdisciplinary research. *The Guardian*. 14 October 2015. Retrieved 27 March 2018 from: www.theguardian.com/science/political-science/2015/oct/14/why-its-time-to-get-real-about-interdisciplinary-research

Clark, N. (2002). The demon-seed: bioinvasion as the unsettling of environmental cosmopolitanism. *Theory, Culture & Society*, 19(1–2): 101–125.

Clark, N. (2011). *Inhuman Nature*. London: Sage.

Cooke, B. and Kothari, U. (2001). *Participation: The New Tyranny?* London: Zed Books.

Deloria, V. (1991). Research, redskins, and reality. *American Indian Quarterly*, 15(4): 457–468.

Department for Business, Energy and Industrial Strategy (2016). *Building on success and learning from experience: an independent review of the Research Excellence Framework: July 2016*.

Diawara, M. (2015). *Edouard Glissant's Worldmentality: An Introduction to One World in Relation*. South as A State Of Mind Journal #6 (documenta 14 #1). Kassel: documenta und Museum Fridericianum. Online Version. Retrieved 27 March 2018 from: www.documenta14.de/en/south/34_douard_glissant_s_worldmentality_an_introduction_to_one_world_in_relation

Elsevier (2015). *A Review of the UK's Interdisciplinary Research Using a Citation-based Approach: Report to the UK HE Funding Bodies and MRC by Elsevier*. Retrieved 27 March 2018 from: www.hefce.ac.uk/media/HEFCE,2014/Content/Pubs/Independentresearch/2015/Review,of,the,UKs,interdisciplinary,research/2015_interdisc.pdf

Glissant, E (2005). *La cohée du Lamentin*. Poétique V, Paris: Gallimard.

Glissant, E (2009). *Philosophie de la relation: poésie en entendue*. Paris: Gallimard.

Glissant, E. (2010). *Poetics of Relation*. Ann Arbor: University of Michigan Press.

Global Research Council (2016). *Statement of Principles on Interdisciplinarity*. Retrieved 27 March 2018 from: www.rcuk.ac.uk/documents/documents/GRC2016Interdisciplinarity-pdf/

Griffin, G., Medhurst, P. and Green, T. (2006). Interdisciplinarity in interdisciplinary research programmes in the UK. University of Hull. Retrieved 27 March 2018 from: www.york.ac.uk/res/researchintegration/Interdisciplinarity_UK.pdf

Harding, S. (1993). *The 'Racial' Economy of Science: Toward a Democratic Future*. Bloomington and Indianapolis: Indiana University Press.

Johnston, R. J., Gregory, D., Pratt, G. and Watts, M. (2000). *The Dictionary of Human Geography* (4th ed.). Malden, MA: Blackwell.

Joseph, A. (2015). Scholarly publishing in South Africa: the global south on the periphery. *Insights*, 28(3): 62–68.

Kukutai, T. and Taylor, J. (2016). *Indigenous Data Sovereignty: Toward an Agenda*. Canberra: Australian National University Press.

Last, A. (2015). Notes on GIS & methods. *Mutable Matter*. Retrieved 27 March 2018 from: https://mutablematter.wordpress.com/2015/06/15/notes-on-gis-methods/

Latour, B. (2004). *The Politics of Nature*. Cambridge, MA: Harvard University Press.

Law, J. and Lin, W.-Y. (2015). Provincialising STS: postcoloniality, symmetry and method. 2015 Denver Bernal Prize plenary.

Ledford, H. (2015). How to solve the world's biggest problems. *Nature*, 525(7569): 308–311.

Massey, D. (2005). *For Space*. London: Sage.

Mbembe, A. (2015). Decolonizing knowledge and the question of the archive. Retrieved 27 March 2018 from: https://wiser.wits.ac.za/system/files/Achille%20Mbembe%20-%20Decolonizing%20Knowledge%20and%20the%20Question%20of%20the%20Archive.pdf

Morel, C. M. (2003). Neglected diseases: under-funded research and inadequate health interventions. *EMBO Reports*, 4(Suppl. 1): S35–S38.

Nagar, R. (2014). *Muddying the Waters: Co-authoring Feminisms across Scholarship and Activism*. Urbana, Chicago & Springfield: University of Illinois Press.

Nelson, A. (2011). *Body and Soul: The Black Panther Party and the Fight against Medical Discrimination*. Minneapolis: University of Minnesota Press.

Paglen, T. (2009). *Experimental Geography: From Cultural Production to the Production of Space*. The Brooklyn Rail. Retrieved 27 March 2018 from: www.brooklynrail.org/2009/03/express/experimental-geography-from-cultural-production-to-the-production-of-space

Perec, G. (2008). *Species of Spaces*. London: Penguin.

RACE (Race, Culture & Equality Working Group of the Royal Geographical Society with Institute of British Geographers) (2016). Comments on the Research Excellence Framework (REF). Retrieved 27 March 2018 from: https://raceingeographydotorg.files.wordpress.com/2016/03/race_ref_comments_final.pdf

Robson, M., Woods, A. and Fernyhough, C. (2015). 'Voice Club' in Working Knowledge: Transferable Methodology for Interdisciplinary Research. Retrieved 27 March 2018 from: www.workingknowledgeps.com/

Scalway, H. (2015). 'About the comic strip' MiceSpace. Retrieved 27 March 2017 from: www.micespace.org/about-the-comic-strip/

Scheper-Hughes, N. (2000). Ire in Ireland. *Ethnography*, 1(1): 117–140.

Scheper-Hughes, N. (2004). Parts unknown: undercover ethnography of the organs-trafficking underworld. *Ethnography*, 5(1): 29–73.

Science Europe (2012). Position Statement Horizon 2020: Excellence Counts. Retrieved 27 March 2018 from: www.scienceeurope.org/wp-content/uploads/2014/05/SE_H2020_Excellence_Counts_FIN.pdf

Serres, M. (1995). *Genesis*. Ann Arbor, MI: University of Michigan Press.

Serres, M., Harari, J. V. and Bell, D. F. (1982). *Hermes-literature, Science, Philosophy*. Baltimore, MD: Johns Hopkins University Press.

Serres, M. and Latour, B. (1995). *Conversations on Science, Culture and Time*. Ann Arbor: University of Michigan Press.

Smith, L. T. (2012). *Decolonising Methodologies*. London: Zed Books.

Smith, N. (2000). Is a critical geopolitics possible? Foucault, class and the vision thing. *Political Geography*, 19(X): 365–371.

Stengers, I. (2010). *Cosmopolitics I*. Minneapolis: University of Minnesota Press.

*The Black Scholar* (1974) Science and Black People (Editorial Statement Black Science). *The Black Scholar*, 5(6).

Thrift, N. (2006). Re-inventing invention: new tendencies in capitalist commodification. *Economy and Society*, 35(2): 279–306.

Todd, Z. (2016). An indigenous feminist's take on the ontological turn: 'ontology' is just another word for colonialism. *Journal of Historical Sociology*, 29(1): 4–22.

Verran, H. (2001). *Science and an African Logic*. Chicago, IL: University of Chicago Press.

Verran, H. (2014). Working with those who think otherwise. *Common Knowledge*, 20(3): 527–539.

Warren, G. and Katz, C. (2015). Gwendolyn Warren and Cindi Katz in Conversation. Retrieved 27 March 2018 from: https://vimeo.com/111159306

Wilkinson, S. and Smailes, D. (2015). 'An interdisciplinary dialogue' in working knowledge. Transferable Methodology for Interdisciplinary Research. Retrieved 27 March 2018 from: www.workingknowledgeps.com/

Woods, A. (2015). 'Interdisciplinary Authorship' in Working Knowledge. Transferable Methodology for Interdisciplinary Research. Retrieved 27 March 2018 from: www.workingknowledgeps.com

# 2

# Diagramming

*Gail Davies and Helen Scalway*

In this exchange, artist Helen Scalway and geographer Gail Davies reflect on collaborative work, which has been running in different intensities from 2010 to today.[1] They discuss diagramming as a form of spatial ethnographic practice, located at the interstices of geographical enquiry, ethnographic methods and visual analysis. This practice emerged in Helen's work on the 'Fashioning Diaspora Space' project, as a way of creatively mapping worlds mediated through the materiality and meaning of translational clothing, patterns and ornament. Practices of diagramming acted as a creative and conceptual provocation throughout the analysis and dissemination of Gail's subsequent work on the spatialized aspects of knowledge production in a very different realm: changing experimental practices in animal research and the coordination and commodification of international research resources in biology (Davies 2012a, 2012b, 2013a, 2013b). Diagramming practices became essential to the work of conceptualization from Gail's ethnographic research. Sharing written accounts of ethnography and working collaboratively on the production of diagrams enabled the complex and often closed spaces of biological research to become legible for wider dissemination and discussion. These diagrams also offered a critical space for opening up the topologies of scientific translation, in both biology and social theory. These alternative diagrams hold flows of materials and meanings together, productively placing the tension between simplification and the complex interactions in both contemporary biology and the social analysis of biology in view.

According to Sunder Rajan (2006: 20), the analysis of biocapital

> cannot simply be a network analysis that traces the various types of technoscientific or capital flows that occur in order to produce and sustain this system. Such an account also needs to understand how these flows are constantly animated by multiple, layered and complex interactions between material objects and structural relations of production, on the one hand, and abstractions, whether they are forms of discourse, ideology, fetishism, ethics, or salvationary or nationalist belief systems and desires, on the other. These abstractions may be hard to pin down and map in the same diagrammatic fashion as networks and flows, but it is essential to acknowledge them.

The term 'diagramming' implies schematising; while places, in their density and richness of meaning, seem to defy schematisation. Yet in relation to the visual investigation of place, is there some expansionary potential in the 'diagram' to become a visual medium through which meaning pours out everywhere, rather than being pared too much away in simplification?[2]

## Diagramming ethnography

*Gail*:  Can I start by asking you about how you use practices of diagramming in your work?

*Helen*:  I think of myself as a visual worker. I was an art student in a very interesting art college, and had tutors who said to me 'politicize, don't aestheticize'. But the aesthetics come out if one thinks with integrity, then the work will have a strength which is of a different kind. I think the word 'artist' is just too vague, a word like a screen to project anything on to, whereas if I think of myself as a visual worker it directs me towards thinking and thinking through doing, thinking through my practices.

The word 'diagram', I think carries a load which can be misunderstood in that it can refer to quite technical things, work that's always digital, for example. It might be taken to mean digital visualizations of data, datascapes and things like that and I have looked with interest at datascapes. But of course, what they deal with always is quantification, they metricize, they are looking at quantities, whereas I'm interested in the much more messy flows which turn space into complex place.

My diagrams tend to start as hand drawings because that's the way I think. I suppose the idea first came to me as a result of coming across Gilles Deleuze's formulation of the diagram as a map of the relationships between forces.[3]

He insists, and others have insisted and expanded after him, that such diagrams will always be very abstract, they're not pictures and they really do abstract away from anything figurative, to just focus on the actual forces. Whereas I rather transgress this.

I would more often use the word 'chart' for what I do because I often find it helpful to take, for example, an architectural floor plan and look at that. That can be the ground plan of an entire site, like an animal research facility where I'd be looking not just at the actual mouse house and the places where procedures are carried out and the quarantine rooms and so on, but also such things as the barriers and the café and the sports facilities and the larger context.

*Gail*:  We have in front of us a diagram of a shop [Figure 4.2.1], which was the first piece of your work I encountered. Can you tell us a bit more about this chart first, and how you put it together?

*Helen*:  This drawing is an attempt to chart a South Asian textile shop serving a diasporic community in Green Street E13 [London], and is based on numerous visits and conversations in such shops. The proprietors, their assistants and the clients became collaborators on this piece as their input was incorporated. The diagram sought to engage as an economical but intriguing way of presenting the multi-dimensionality of the place and is perhaps the single most innovative piece of work I made during the Fashioning Diasporas project (Breward, Crang and Crill 2010).[4]

I became aware of this place as an incredibly shifting, rich, diverse, multiple space that was full of different histories. London, like every other big cosmopolitan city, is in a state of becoming, it's in a state of flux and change and you can't still it. It's very likely that in a couple of decades, these communities will have moved away just as other communities have done, like the Huguenots who came in the seventeenth century and the Jewish communities who came in through Spitalfields in the nineteenth. So this was something to try and seize hold of, the space at a moment of shifting time.

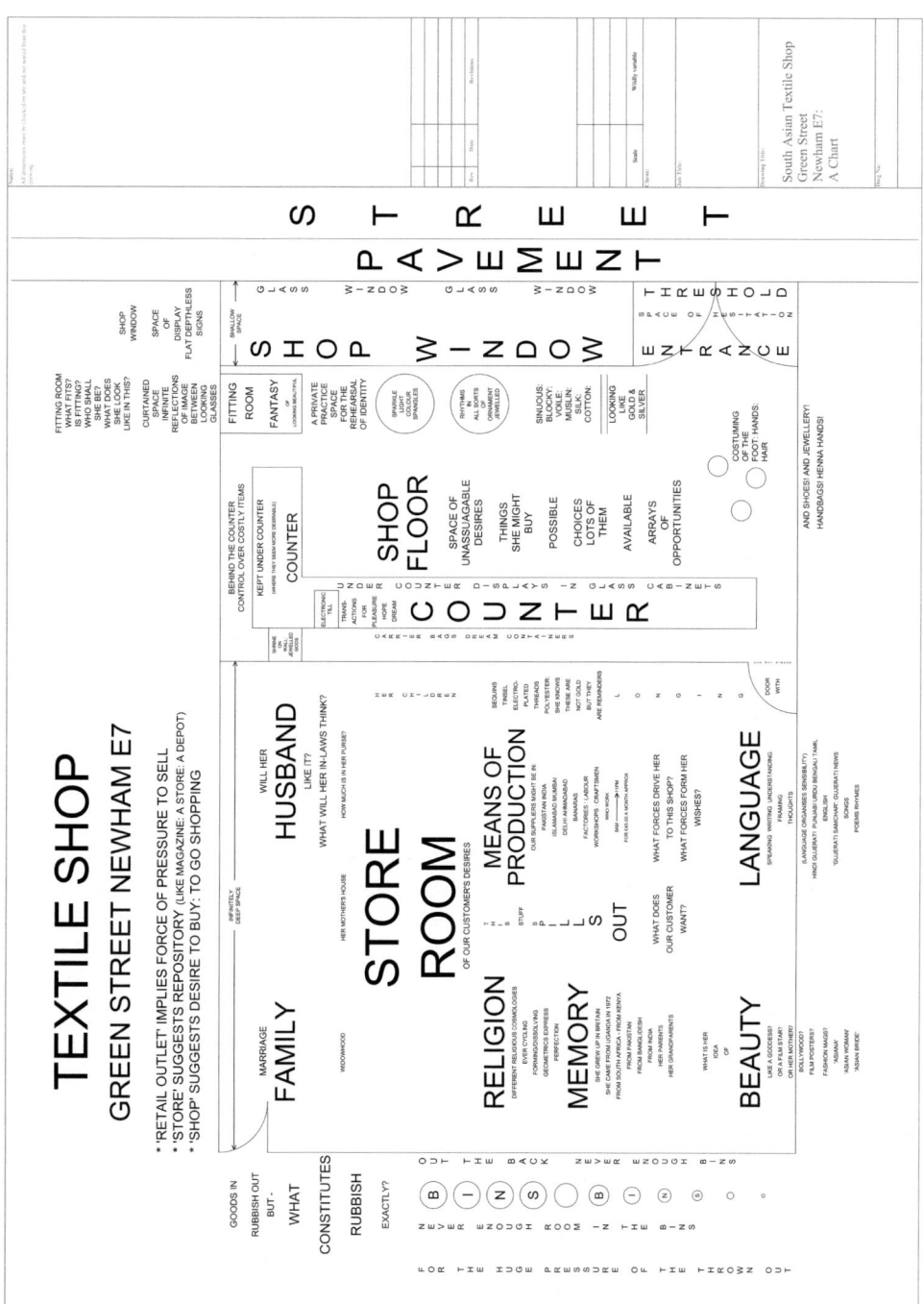

*Figure 4.2.1* Charting a London South Asian textile shop (2009). Copyright Helen Scalway. There is a zoomable version of this image available at www.flickr.com/photos/22894783@N08/3398749790/

I used to look at things like weather maps repeatedly, and those digital models which are animated, so you can see the storm coming and elements feeding in. Maps of course are often beautiful but they provoke me by eliminating all the traces of the journey and I wanted to make a diagram that somehow could contain all that shifting life.

The work is not literal though it riffs on the conventions of the architectural drawing of actual floor plans. Where it departs from the literally descriptive is in its labelling. So, for example, the street has got these plate glass windows with shallow display space behind. You look in from the street and there's a shallow space of display. Then the threshold, the door, is a space of hesitation. The fitting room is the space where the question is asked, 'What do I look like in this?', 'Who am I when I wear this?' – that is the question which makes that little cubbyhole space into a place, a fitting room, that and the infinite reflections between looking glasses; the shop floor is a space of unassuageable desires.

Then there's a one-way mirror – the shops I used to go into, had one-way mirrors between the front of shop and the store room. Go through them into the store room, and what is actually stored in there? Here the way the diagram works changes, the register changes. It's not just the boxes and the extra stock, though of course it's that as well. The diagram suggests the store room is a hidden reservoir of customer's desires, of the forces which have formed her taste. The influence of her family, is she a widow, will her husband like it, what will her in-laws think, what is her religion, what are her memories, what is her idea of beauty? These are some of the unseen forces which have produced the front of the shop, a real place with its arrays of choices.

Then, every shop has a front door but also a back door leading to the dustbins. The diagram asks what constitutes rubbish, we can get into questions of abjection here, what is it that might get put out the back, thrown out, buried?

This shop diagram is mapped in black and white, apparently simple, on to a pared-down architectural floor plan. So what you've got is something that's almost like a document for discussion, it's not meant to be beautiful, it is meant to get people talking.

Then I took a very early draft version of this document back to Green Street, by that time I'd learned which were the three or four shops where the people were really helpful to me, and we spread it out on the counter. I was on one side and they were behind the counter and we pored over this, the clients and shop assistants and proprietors were all helpful, so they became my colleagues and my co-workers. I was really dependent on them.

*Gail*: Do you get any sense about what new insights this allowed you to include or how it allowed you to see how they were reading or understanding the diagrams?

*Helen*: They gave me a very positive response. I had been concerned at what they might say 'let alone' think but they seemed to recognize what I showed them with a warmth in their recognition. They said, 'Oh yes, we understand this'. They took on board the various registers and different dimensions in the diagram at once, as though they had a real inwardness with it or it made something visible which they knew all about from the inside. They also made suggestions for which I was grateful, for example when it got to ideas about beauty one of the assistants said, 'The magazines are really important to us, *Asian Bride*, you've got to put that in', and someone else said, 'And the film posters because we are passionate about films'. The Bollywood films are very important and so I was actually able to put more things in. Those conversations really made it possible for me to begin to populate the diagram.

The diagram really got them going. Of course, there were some language difficulties. Not all the older ladies in Green Street speak English, they would be speaking in Hindi or Bengali, that is the ones who might be in their 60s or older, their daughters might be in their 40s and they would be speaking accented English but their grand-daughters, these are the 14/15/16 year olds were speaking as East London as they come, 'innit'. So this was fascinating especially when they started arguing with each other over the diagram – that was great.

*Gail*: When I saw this diagram, I immediately thought there were similarities with what I was struggling with in my work. This challenge about how you think about relationships within a space where everyone's talking slightly different languages. In the world I was researching, you had researchers experienced in molecular biology; care staff concerned with animal welfare; patients hoping for clinical translation; institutions invested in economic outcomes. There are people with desires for careers, hopes for cures and concerns about animal experiences. All of that's in there. It has different inflections, intensities and visibilities in different spaces, but with potential relevance for every space. The ways they then relate – or don't translate – becomes important for how science is operating now, and so is interesting to social science too. But given these are often unfamiliar and inaccessible spaces – both literally and in terms of their technical languages – this can be hard to convey in writing. I was drawn to your diagramming as a way of thinking through and representing these complexities.

*Helen*: You spoke for five minutes at Exeter University and I thought, 'I need to speak to this lady'! Because you are approaching spatiality through lithe gymnastics in language, a kind of place-making in language itself, a complex set of connected and related places which are performed in writing, and this is a striking way of evoking this emergent space of becoming and of translation.

    I'm very interested in the performativity of your work, the way that in the length of a sentence or a paragraph, you will travel across different terrains of language, different registers of language, so that your writing is spatialized not only in what it writes about but in what it performs, in the gymnastics that it performs between one register and another.

*Gail*: Thank you! Well, we have carried on a conversation ever since, which has varied in its own intensities given our other complex commitments. But it seems to me it endures as it is based on shared interest in a productive, but ultimately irresolvable tension, that comes from trying to clearly communicate the provocations of complexity. The conversation has carried on over email, meetings when we have chatted through my publications and your images. And I am also so grateful for your reading of my work and have gained a huge amount from this, particularly in how diagramming opens up ways of thinking about ethnographic practice and conceptualization too.

## Diagramming experiments

*Gail*: Can you say something about how you start work on these kinds of diagrams?

*Helen*: By first visiting the place with the intention of observing in an extremely active way. Absolutely not in a foreclosing way: rather, in a wide open one, even a vulnerable one, prepared to be surprised, even taken far aback, out of one's comfort zone. And by watching people's behaviours, that is vitally important, wondering to myself what is it that is important to them? What does the place mean for them, in their terms? Then, thinking about the questions raised by all this observation, questions which one might never have foreseen. There are masses of notes and query marks and bracketty bits trailing about untidily at this stage.

The next step involves finding or making an initially sparse diagram, and making it burgeon and leak with all those layers of meaning. It's somehow about holding on to those untidinesses and irreconcilabilities. So, what gets called 'labelling' can be an intensely thought-provoking activity. But it all originates with the habit of letting the space decentre one's own vision: of letting oneself be thrown by the potentially many different ways there might be of seeing this space, or door, or barrier or whatever. You can bring this approach to any place, and so to any diagram of a place, asking what complexities are layered up in there and how you might convey these.

*Gail*: One reason the charts are so interesting to me is they're clean in their look, they're black and white and they're a mix of flow diagrams, arrows, boxes, text and spaces, which have of them something of the qualities of an architectural diagram. Or their system. The one that I am looking at now is taken from a process diagram produced by the NIH, which is an idealized version of how laboratory spaces should be organized and flow of materials circulated to ensure a separation between the dirty and clean spaces in an animal research facility. What you have then done is added into these the other flows that animate these spaces, making the multiple and complex layering of materials and meanings values within these spaces legible. So, when you first look at them, you think 'oh yes I know what I'm looking at here'. It seems familiar. But when you read into them, you find unexpected things. Instead of the apparent or idealized transparency of a scientific diagram, which you might aim for in science, you get a rendering of overlapping perspectives, material practices and amidst these social anxieties too.

Can I ask how your process changed in our case, when at least at first, you were working through my ethnographic writing and description, before you were able to visit some sites?

*Helen*: I recall that soon after we first met you sent me some of your writings. Amongst these was a set of extremely scrupulous and sensitive interviews with different kinds of scientists working in animal research labs, concerning, as I understood it, what their work meant to them. These interviews, with their different perspectives, suggested a kind of Babel. But the very fact that the interviews were like a Babel, and that this was exactly what you were grappling with, was what appealed to me, whose interest as an artist has often been in the conveying of multiple perspectives. I was intrigued. How on earth could there be a visual map of a work place when all these people understood their work in that place so differently? How could one make sense when the sense was so multiple?

Then you sent me your paper on 'mutant mice' (Davies 2013a). I sat greedily over it – why? Well, for this reason. I was a student and teacher of literature before I embarked on art studies, and I have always been completely entranced by early nineteenth-century English literature with its implications for re-envisioning and for embodying understandings of world's 'becoming'. I have always thought Mary Shelley's novel of that revolutionary time, *Frankenstein*, a profound work about responsibility, care, consequences. I was always teaching it and every time found more in it. Your engagement with contemporary ideas of the 'monstrous' in relation to the biologically unfamiliar context of lab animals, resonated at so many multiple points with my own imaginative understanding of the 'monstrous', that I was plunged willy-nilly head over heels into dialogue and drawing.

Subsequent exchanges between us served all the more to engage and intrigue. For me it has been the way thought and making are provoked in me by your writing, which

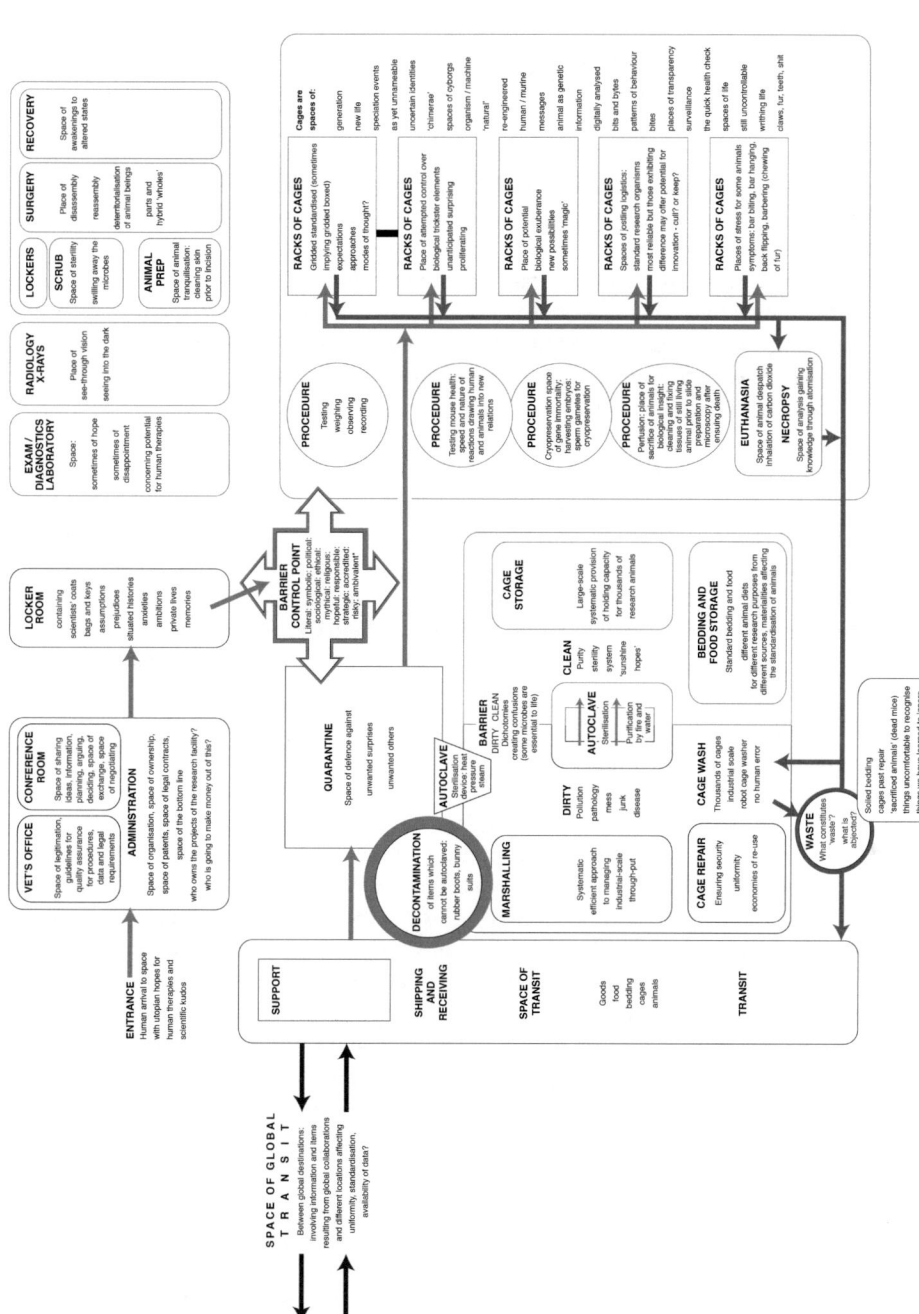

*Figure 4.2.2* Animal research facility: re-diagramming relations (2013). Copyright Helen Scalway. There is a zoomable version of this image available at www.micespace.org/ffcw.jpg

has kept the conversation glowing. I think in any co-working there has to be this kind of real interest in and by the work of the other, and enough *stuff* on both sides to feed it, for dialogue to deepen and become true collaboration.

Gail:    I think our collaboration is also sustained by this mutual interest in the uneasy and shifting relations between words and diagrams and meanings.

Helen:   In your interviews you present all these different kinds of scientists, some will tell their work as though it's a matter of bio informatics, just completely at one with the idea of DNA as encodable information, bits, zeros and ones, while a vet or an animal handler or an animal technician, each have different ways of telling the story and it's those slippages which the diagrams could be really interesting in helping to make clearer.

Gail:    We've got the locker room from an animal research facility represented here, where researchers leave items they can't take into the facility, prior to going through barrier controls. These may contain scientists' coats, bags and keys; it is also the receptacle for assumptions, prejudices, situated histories, anxieties, ambitions, private lives, memories and so on. These are essential aspects that animate the space of the facility itself too; they are never fully contained here.

If you go back to the idea of the diagram being maps of relations, these are not necessarily between abstract forces here. There may be those present, but these are also deeply social, material, humanized and non-human spaces in which the incongruous entities that inhabit these spaces interact. Later, in the 'non-human spaces' we get the jostling together of bites and byte, as animal bodies are made into bodies of data. Following the process on, there are similar questions around disposal as there are in the sari shop – what is it that might get put out the back, thrown out, buried? Here are things that are sacrificed, wasted, past repair, ignored or uncomfortable. I have to say, these are not necessarily languages I would use. When you write with these as anthropological terms, you bring certain commitments centre stage. But they fit here, as diagramming is able to capture divergent dimensions, which are hard to keep in play in the right way in a written narrative.

Helen:   Yes, I think the great thing that these diagrams do is that they make writing spatial. It would have taken perhaps 50 pages of prose or more to describe all that's going on in this animal research station. What the diagram does is to let you see at least some of it in one place, on a screen or sheet of paper. It lets the eye wander around. You can go around it in a directed way but also you can scan around it in your own way, in your own time. The other thing that it does, I think, is it breaks the sequential flow which written verbal language always has, and that sequentiality really does structure thought in a particular way, but it can also be a very containing and controlling way. I would suggest that the reason these diagrams can allow thought to behave differently is that sequentiality is broken, allowing things to be seen simultaneously or in a different order, the components have been dispersed in a way which lets the viewer or reader pick them up and rearrange them in their own way. And it's also quite surprising to see what goes side by side with what.

Gail:    What these diagrams do as well, I think, is they allow you to say something very important about the role of the social, the biographical if you like, in these spaces, but without reducing science to the social. They're actually deeply humanistic diagrams if you don't mind me saying.

Helen:   They're meant to be . . . I have to say that these diagrams became even more social after a conversation with Kaushik (Sunder Rajan) who said to me that the nature of scientific collaboration might also be much complicated by the social: for example, there might be two people, scientific colleagues who might have started as friends but have since quarrelled and they've still got another two years of the collaboration to run, and that tension gets into what they still have to produce.

Gail: That's something I have found difficult to write about in my work, because of commitments to preserving anonymity, the partiality of the knowledge that you always encounter, and the question of how to give recognition to the biographies, communities and socialities that matter without somehow reducing the scientific practices to a social explanation. You allow people to make their own connections between infrastructures and social processes, so they remain open in a way which is tricky to do in academic texts.

I wonder if I could ask you what your encounter with contemporary biomedical science practices has been like and how your work has changed in the shift from sari shops to scientific infrastructures?

Helen: I think the encounter that really struck me was at one of the facilities that I gained access to in the UK. I drove to it. You take a turn off a road which is discreetly marked, in such a way that the place is invisible from the main road. Having got through the initial security check from the car park, where the security men phoned ahead, I was then accompanied to a rather beautiful building with shallow, wide steps to a quite grand, simple, contemporary building, lots of glass and light. You go into this corporate entrance and it's a really quite splendid atrium, glass and light and transparency.

It connected to another aspect of fashioning diaspora, working in the archives at the V&A (Victoria and Albert Museum, London),[5] which had already sensitized me to the idea of how institutions produce 'knowledge', how important is the framing to what they are going to declare as 'knowledge', whether the framing is by a grand nineteenth-century ceramic staircase (as in the V&A) or a grand twentieth-century corporate atrium. Framing is utterly important and the value which is perceived in the knowledge, it relies on this grand context, this is one way in which things gain their authority.

At this facility, I was told that I certainly could not get within smelling distance of the mice which had been my big request. Instead I was taken to the press and education office and I was kindly treated but, certainly, we were not going to go anywhere near the mouse house. On the walls there were various lush decorative art works, the walls were high and white and light, there was a sense of corporate well-being, the library, the lecture theatre, the provision for refreshments, all of these things were very well appointed and I just had this feeling of having been on the receiving end of a presentation, perhaps a Utopian one.

With the lab diagrams, I think I'm conscious of wanting to put in all the things that might complicate what offers to be pure reason or to have the authority of 'scientific knowledge' or certainty or uniformity. The textile shop hadn't got any uniformity, it was not about science, it was all to do with desire, the customers' love of vivid orange or crimson with loads of silver and beads, so those issues were not there. But with the lab diagram, perhaps because I had been reading Bruno Latour (2005) I was really looking at the ways that these objectivities, so called, are really anything but. Here there are the break-out rooms and corridors and café spaces and the sports halls, squash courts and swimming pools and so on, the kinds of social spaces where love affairs might start or quarrels break out, where the collaborations and competitions and contests of the labs might be continued in other ways.

Gail: They are drawing our attention to interplay of hopes and desires, in personal biographies and the social imaginaries that animate science. They are also intimately connected to

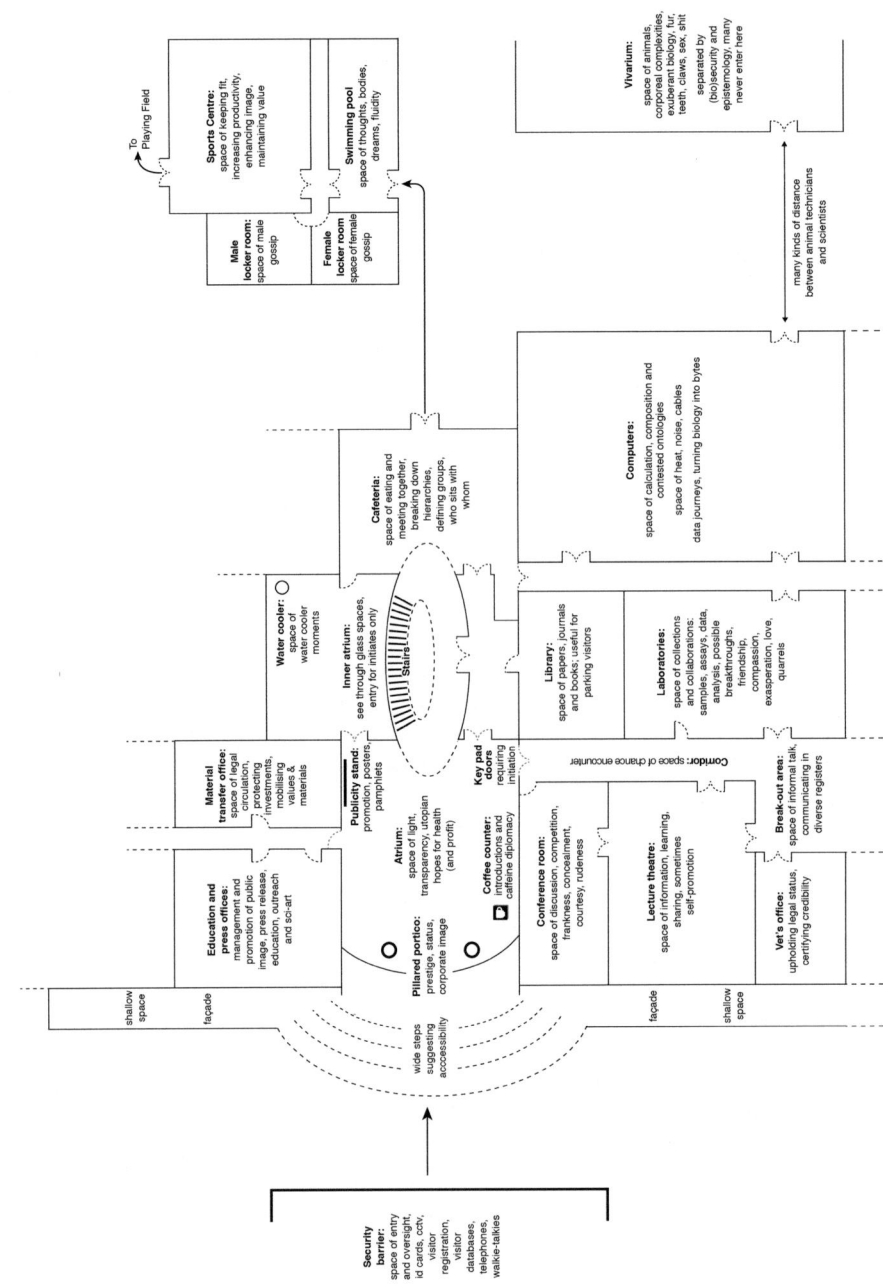

*Figure 4.2.3*  An international biology research facility (2013). Copyright Helen Scalway. There is a zoomable version of this image available at www.micespace.org/Utopian.jpg

their wider geographical setting. I think that's another comparison with the Victoria and Albert Museum. That institution acknowledges the fact it's also a geographical project, about collecting, which has had to engage in a post-colonial shift. This is one reason artists are being brought in to work with them in new ways, to trouble this colonial narrative, or encourage us to think differently about it.

However, the geography of science is, I think, both essential to, but often erased from these kinds of animal facilities and international infrastructures. Geography inflects the practices of science and scientists in important ways. A collaborative project may be underpinned by an idea of 'universality', but requires people in different countries to work together in ways defined through one place. These architectures also embody national and regional investments, whilst at the same time signalling openness and placeless-ness.

Animals and biological materials are also often transformed through their circulations across space. Mice change from place to place, due to microbial factors or genetic drift. So, if you want to think about translating science from these 'placeless' places, you have to think about how translations are made; how do linguistic, organizational, social and material aspects of experiments connect; and what do these institutions have to do to manage shape patterns of biological potential and emergence. There is thus also a temporality to these diagrams.

*Helen*:    I think there's a great deal to be said on the question of translation, to pick out just one strand from several in what you have just said. It invariably involves transformation. One of the most fascinating pieces you wrote was on the Rosetta Stone where you were actually really querying that metaphor, of the Rosetta Stone, for translation in this context of bioscience (Davies 2012b). The Rosetta Stone metaphor suggests there is something which can be translated bit for bit into some exact equivalent, as though it didn't change in translation and of course, it's not like this. In terms of language, for example, people who are bilingual seem to be able to move between different sensibilities, for example the English and French languages embody completely different sensibilities, it's almost like inhabiting two different worlds.

To some extent I've tried to address this, coming away from the architectural diagrams and using maps. There are a great many map artists and I think a lot of map art is not very interesting, but by bringing the atlas into play and working with the scalpel, cutting mouse outlines from maps say, of Japan, Singapore, North America, in the area where the Jackson lab is, Maine and the UK, Harwell and the Sanger labs, these

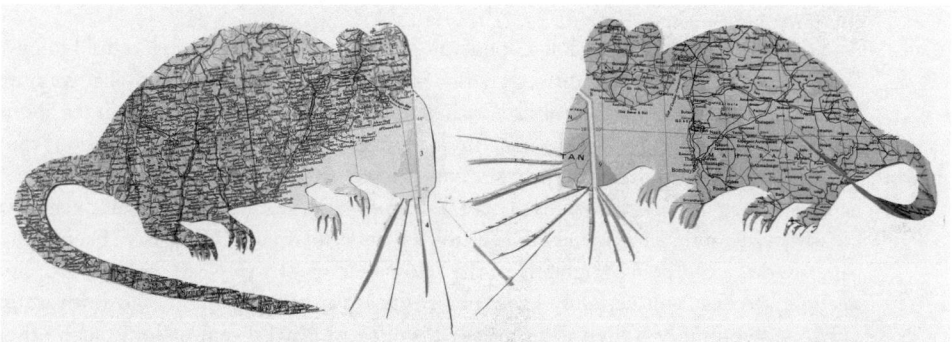

*Figure 4.2.4*   Map Mice (2012). Copyright Helen Scalway

different pages of the atlas, each speak. People will see the mouse outline, but they also read the different place names and feel the huge gust of other worlds. Whole other cultural hinterlands can be activated with the use of maps.

I certainly wouldn't claim originality in cutting up and rearranging maps, it's just that cutting mice out of an atlas is a bizarre thing that this project has provoked me into. Mice are very interested in each other, so when you put these mice together and their whiskers touch, they seem to be sensing each other. Geographically what does that mean? That distance and that intimacy? The UK mouse is maybe starting something with a Singaporean or Chinese or American mouse, these are issues I'm trying to refer to with a very light touch, the images are meant to suggest that the mice bring a whole lot of other stuff with them, how they, as animals have been fed, how they have been handled, the history of their genetic inheritance, their lives in those facilities and in those different parts of the world, all of that becomes animated, I hope, through these map images. A bit ambitious, perhaps.

*Gail:*    I think it does raise intriguing and critically important questions around what a model is. What does it mean to model human disease in a laboratory mouse? And what does it then mean to move the model from one lab to another, or towards clinical application?

That's another whole question we've engaged through your work, which is the way science often uses arrows in textbooks, policy guidance and papers, to indicate the steps to translational research as separate and sequential, and suggest materiality is not transformed through processes of translation.

## Diagramming translation

*Gail:*    When we look at the arrow, it's something that's given to us, which we don't tend to think about very much. Diagrammatic arrows are embedded in our thinking and in our computers, and in the ways we write and do analysis. I guess I'm interested in how the grammatology or the semiotics of the arrow forms such a strong way of thinking in all of these biological and social scientific representations of translations, which as you've indicated, are just so much more messy than the arrow can ever indicate.

*Helen:*    Latour (1998) seems to suggest that the way that scientific diagrams and illustrations come into being should be subject to as much scrutiny as is given to the process whereby old master art works have come into being. This is because scientific diagrams have all that mess and different trials and the passage of time elided so they can sometimes appear speciously clean and orderly and that gives them authority. But in fact, they are often embodiments of inflected narratives like other narratives, just as entangled in their place and time.

*Gail:*    And the arrow has become such a dominant icon of what we might call 'trans' biology. But it leaves so much out and is suggestive of linearity, when so much of what we now know about translational research is that it is dynamic, nonlinear, sometimes about expanding fields (Davies 2012b), and often about embodied skills and interactions too.

*Helen:*    I think that it's debated, but the arrow seems to have started with the image of a pointing hand as we see on old street signs. There is one in Hampstead in London left over from the early twentieth, late nineteenth century, a wonderful image. There have been entire papers written on the iconography of the arrow as it appears in places like airports, but in those places it will certainly have had its genesis in the graphic design software in someone's computer. Often for example it will be in PowerPoint or some such other digital program, where what we're given is a selection, but the choice we are offered is

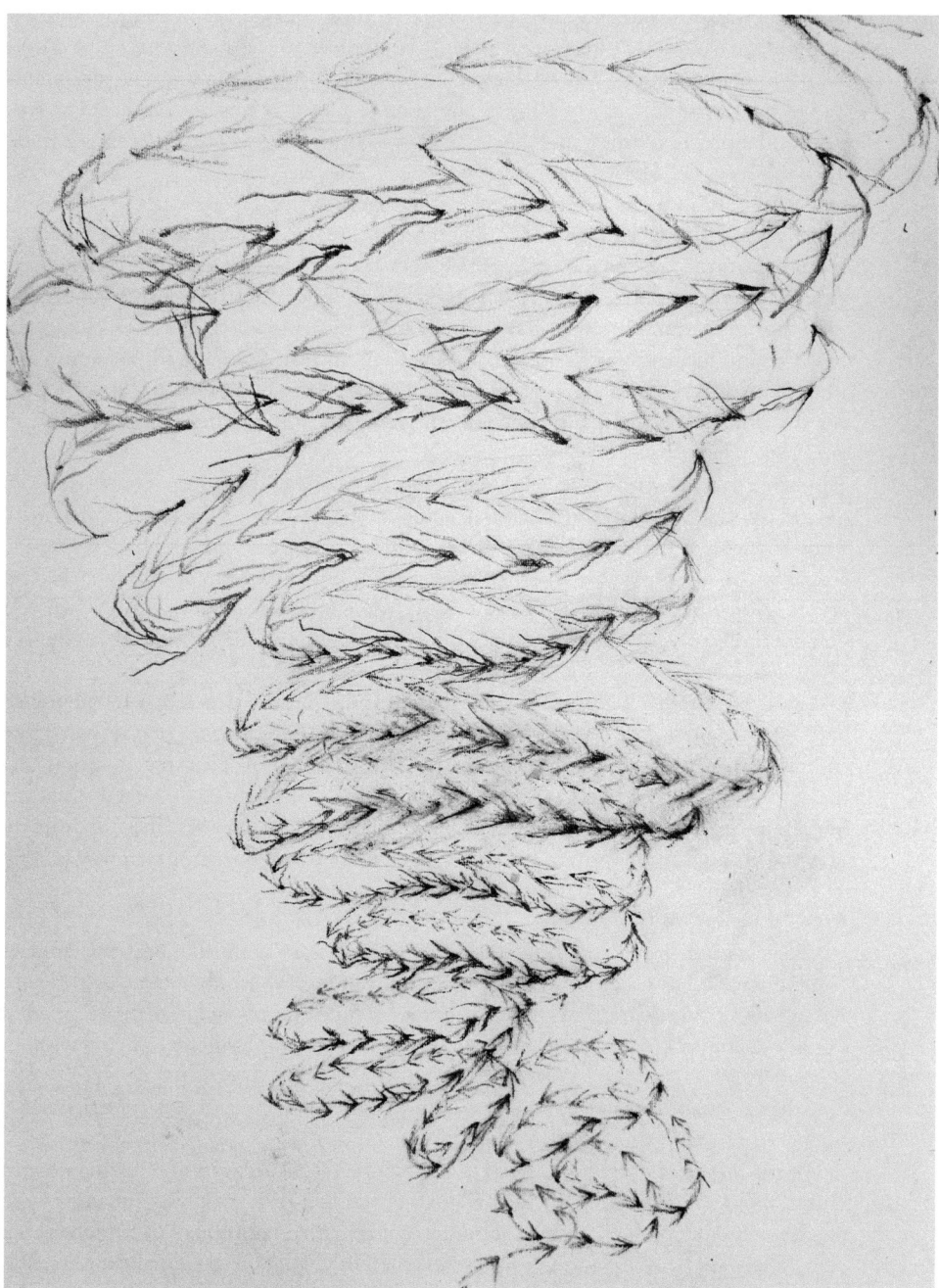

*Figure 4.2.5* Swarm (2014). Copyright Helen Scalway

somebody else's idea of what is adequate, somebody else's drawing and it only works through those pixels that they've already laid down.

I think that there's a lot to say about this, it goes back to the time when I idly and unthinkingly put an arrow linking two separate diagrams and you very rightly picked

221

me up on this icon and said, 'I've been thinking about the scope and possible meaning of this arrow' and of course, at that point, I realized there was more to it and I'd have to give it more thought.[6] The problem with the arrow icon comes with the unthinking way in which we idly use in a hurry any familiar tool that comes to our hands. And more and more, in business, in science, in architecture, everywhere, diagrams are made using this icon.

Very often also, the icon is used in a context where it's concealing a mess by simplifying it. But often we need to live more in the mess or at least to be aware of it because the arrow can be too often grossly reductive, it indicates certainty where there isn't certainty, it can suggest unity where there isn't unity, it can erase histories of mess and histories of unresolved and unresolvable mess.

Of course the arrow itself also suggests passage, in that it has a head, a column and a tail. The notion of passage comprises more than one dimension. An arrow might not always indicate simple direction; in other contexts it might indicate the passage of time also.

*Gail:*  The arrow emerging from the hand, seems like the ultimate reduction. You move from something that was touched, that was embodied, that had a subject identified with it and suddenly it becomes abstracted and abstracted. You lose the touch, you lose the embodiment, you lose the sense that the gesture emerged from relations, it had an origin, a someone who was gesturing towards something for someone else. When you think of it this way, the arrow becomes this way of figuring a movement that's not only about a direction, but always about a relation.

Until you started talking to me about them, I hadn't begun to look at how often such devices were used, not only in visual representations but also in the way you write and think. Just to give you one anecdote, I had some proofs back for an article and the copy editor suggested changing one of the words that I'd used. I'd talked about the importance of 'recognizing' something quite challenging, which was the impossibility of living without some form of suffering. The phrase was taken from Haraway (2008). The extract is:

> Recognising shared suffering is not simply to suggest symmetry between humans and animals, to raise the status of one versus the other in subsequent cost-benefit equations. It is more complicated than this, for both animals and humans are inserted into complex knots of relations, which recognise the impossibility of living without suffering.
>
> *Davies 2012a*

To me this term 'recognition' was really important. It had layers in it about encountering, about valuing, about acknowledging. It also had at its core this encounter with the other. So for me the point was about recognizing the impossibility of living without being implicated in some form of suffering, whether in relation to animal experimentation or in recognizing our interdependencies, of my health being dependent upon the corporeal experience of other people in other places participating in food production, biomedical, experimental or clinical processes. So this word 'recognition' implied to me a relationship to the other, a process of knowing, and the two together, of recognizing the implications of this knowledge in the context of the other.

However, the copy editor didn't like the repetition of the word 'recognize' and changed it to 'highlighting'! And I suddenly thought 'where does highlighting come from?' It is a visual term. It's the brightest part of a painting, or a device to highlight what already exists. It's a felt tip pen or it's the mark that you make on computer on a text. It suddenly seemed to remove everything I was trying to think about in terms of embodied relations and replaced it with an indication of emphasis. It was extraordinary. I got quite angry!

Helen:   The encounter had gone. It's the encounter with the other which had gone.

Gail:   That same manoeuvre goes on when you look at an arrow. That's lost the fact that it was a hand that had a relationship to a subject, or a potential relationship, to the other within it. So the arrow just becomes the way of highlighting either a sequence, we move from this to this. It's lost the encounters within it. And I think what you're doing in your work, by putting arrows alongside other sorts of intensities and depths and emotions, is you're replacing that. And I guess it comes back to what I started out by saying, is that these are quite humanistic forms, in the sense that they return multiple forms of subjectivity back into a system that you're trying to understand.

Helen:   The graphic software arrows can produce distance which should not be there. Often, there might be much to be said for returning to the expressivities of hand drawing, where you can suggest that something is less certain, cloudier, more dissolving, more resolved in one part than another.

What you say about the hand is absolutely beautiful and it's a thought that I hadn't had; a very potent thought. In the light of your comment I think the use of a digital hand pointing as a cursor, a connection between human user and digital interface, could be seen as a disturbing misappropriation, almost a kind of theft. There are sites discussing the hand icon on digital interfaces as connection between user and digital content, and a lot of concern that these are usually *white* hands with all the implications that carries.[7] But it also leads to something that's a problem, something that's unresolved for me which is that with icons like the arrow, we're using some kind of code. The 'clean' software arrow too simply encodes complexities but if I make a more nuanced cloudy or multiple arrow or an arrow which is drawn, the problem then is how is it going to be read? How is it going to be understood?

If I hand-draw then I've got to make sure that I'm still making a communication, unless I want to be deliberately laconic or mysterious, which I don't. In the hand drawing of code, looking to use the way of drawing a part of the code's communication, I'm slipping between several forms of . . . well if you like epistemologies, modes of communication. Allied with this problem is another, I'm looking at a diagram right now and, it's static. All over it is written and printed, things that are about the liveliness, the teeth and the claw and the furs and the hopes and the memories, etc., the place has obviously got a swarming life and yet the map itself looks very static.

The question of how you animate either a map or an arrow, make it richly communicative, give it whatever nuance is needed, is an unresolved one, probably unresolvable!

I've been returning to using materials, very much trusting the materials to try and create the idea of a living organism, as a sort of timeline. Like a river or a flow into which various things pour, but with the unpredictability of the sort of materials I use: inks, dyes, pigments, powders, muds, earths, gravels, sands. These of course get carried into the stream and they will have their odd inter-reactions which are unforeseeable and often quite exuberant.

Gail:   That's the tension we indicated at the start; it's the tension that's in people like Borges and the ultimate map being the map of all the territory that's fundamentally useless because it has so much in it, that it is the world, not a useable representation of it. This

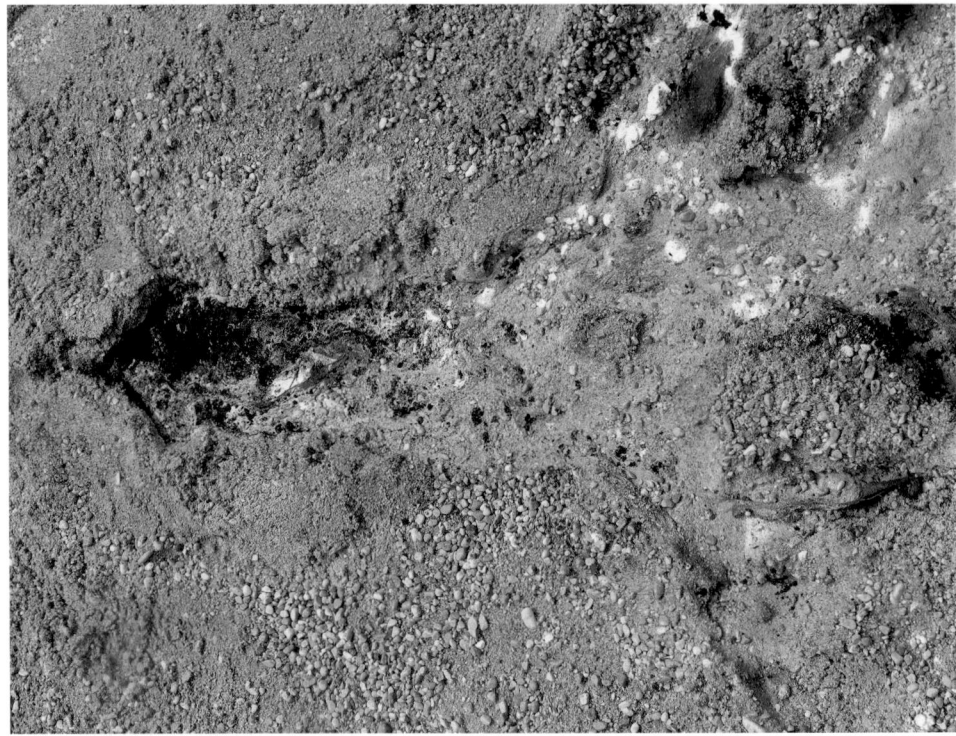

*Figure 4.2.6* Symbol and stuff (2014). Copyright Helen Scalway.

is the same tension in science – you need to abstract, to reduce, to understand and to render things in a way that makes them manipulable and intelligible. But also you need to hold onto those parts that have the potential for new understandings, for something unexpected, for emergence to come through, otherwise if either one of those aren't there, then you've either got a static system or you've got a totally excessive arrangement that's not understandable. And it is the same challenge for the work of ethnographic analysis.

So in some ways what you're doing is always moving between the two and I guess reminding us, at different points, that there are things that do have this potential for excess and maybe sometimes we're focusing on the wrong ones of those and we need to go back and open up certain points. But it is not that everything should be open at every possible point.

*Helen:* At every point Gail, you yourself are stitching back and forth between different systems, holding together multiple ways, things which have to be held in view together when we're looking at how something like biology travels. So at every point, yes I am querying the closedness. This visual work comes straight out of yours.

Some of my provocations about these diagrams and the word 'diagram', comes with this big history as well. I find that Bruno Latour uses diagrams and I get very . . . I think 'Just don't do this, put these back in the computer and shut the lid'. Why do those diagrams not help? I think it is because they are too reductive, they're too simple, they are so flat and elliptical as to be incomprehensible. I would like to . . . point to, these impasses of communication . . .

*Gail:* I've said this before but those diagrams from books like *Laboratory Life* and *We Have Never Been Modern* have been picked up quite widely in geography, and in some ways do produce generative outcomes, but then mostly they have travelled widely without their complexities. People using actor-network-theory started drawing these linear networked topologies in the same way as he renders the spaces of translation as simple networks of translation . . .

*Helen:* And they don't change. Translation involves such changes.

*Gail:* And they don't change and they lack all sorts of things that geographers talk about through a concern about flattened topologies. Latour (1999) himself then goes onto say 'actually it should be actant rhizome theory' but there's little attempt to follow that through with a new visual vocabulary. And it is difficult. A lot of actor-network-theory gets stuck with these reductive visual imaginaries.

*Helen:* One of the things I'm interested to produce is a mapping or series of mappings that are rhizomatic and full of arrows doing different things, but this could be an undertaking.

*Gail:* But in many ways you are already doing this I think. Your experimental working with materials, pushing the qualities of materials like graphite powder or flows of inks to the point where they tip over into something else that's potentially monstrous, and they begin to, if you like, point to the issues in holding onto openness and messiness in forms that are not just about narratives but about materiality too.

This is perhaps most evident where you have combined practices of diagramming with experiments with materials. I am thinking of the piece you did re-diagramming

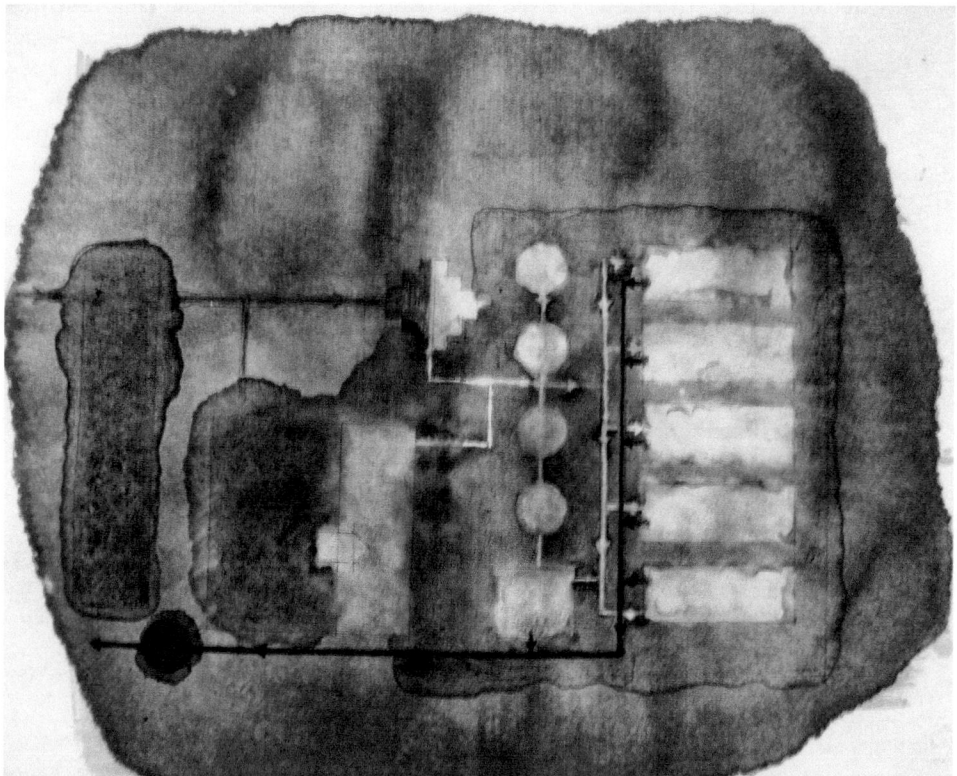

*Figure 4.2.7* Lab Cleanliness: ink and bleach (2013). Copyright Helen Scalway

225

the animal research facility. This involved drawing with sepia ink and then to quote from your description of this piece 'attempting with inevitable lack of success to wash out the designated "clean" areas of the floor plan with bleach. Sepia ink was used as it separates easily into its constituent pigments and reacts strongly with bleach in unforeseeable ways'. This was an extraordinary image for me, which must have emerged from some of my writing about the challenge of managing and understanding microbes in animal facilities. But I hadn't or couldn't have seen it in this way without your help.

*Helen*: I think we're both very much involved with the incommensurability of different people's meanings as they jostle in one space and that's where I find your writing particularly evocative, in that one can see that live trapeze artist swinging from one ring to another, over gaps. But if you have to write 10,000 words which have a beginning, a middle and an end, it's different from a diagram. The diagram is like a single freeze frame of a space at a particular moment, it hasn't got a beginning, it hasn't got an end but nonetheless the incommensurable meanings are there, written in, but it hasn't got to have that linear structure of time.

These practices of diagramming do have a specific relation with aesthetics – a subversive one. They work ironically and self-consciously to exploit the visual rhetoric, the aesthetics of modernity, the pared down, the uniform and replicable – in that they riff off the idea of the industrial flow-chart, in their use of digitally generated clean lines, boxes, circles and arrows. I wanted the viewer of these documents to think at first glance, 'I understand what this is, this is an image of some kind of familiar industrial space' based on received visualizations of flow-chart thinking. But then gradually, as the viewer goes beyond the purely visual elements and begins to engage with the text on the documents, I hope that that familiarity breaks down, as the text works against those over-familiar elements and we realize that the closed boxes and circles are unable to contain all that is swarming out. One set of meanings essentially from an earlier time, with simpler understandings, cannot possibly contain, despite its offer to do so, the much more exuberant and complex meanings we are groping towards today. This image points to both the overflow of materials you mention, but also of these meanings.

## Notes

1 This conversation was first recorded in 2012 prior to the joint presentation of work in the Knowledge/ Value Seminar series organized by Kaushik Sunder Rajan in Beijing in September 2012. It was reviewed and updated in 2017 for publication in this handbook. Further images from this collaboration are available via www.micespace.org/galleries/ (last accessed 2/5/2017).

2 Scalway, H. (2011). Concept note for Charting Place: Diagrams, charts and the visualisation of the making of places, 'Landscape Surgery' seminar at Royal Holloway University of London.

3 'The diagram . . . is the map of relations between forces, a map of destiny, or intensity, which proceeds by primarily non-localizable relations and at every moment passes through every point' (Deleuze 1988: 36).

4 For more information on the Fashioning Diaspora project, see www.vam.ac.uk/content/articles/f/ fashioning-diaspora-space/ (last accessed 2/5/2017).

5 For more information on Helen's work in the V&A archives, see www.vam.ac.uk/blog/section/ moving-patterns (last accessed 2/5/2017).

6 For more on Helen Scalway's residency around visual experiments with arrows, see www.c4rd.org.uk/ RESIDENCY_ARCHIVE/Pages/Helen_Scalway.html (last accessed 2/5/2017)

7 For more reflections on the iconography or arrows, see www.uxmatters.com/mt/archives/2009/10/ the-ever-evolving-arrow-universal-control-symbol.php (last accessed 2/5/2017).

# References

Breward, C., Crang, P. and Crill, R. (Eds.) (2014). *British Asian Style: Fashion and Textiles/Past and Present*. London: V and A Publishing.

Davies, G. (2012a). Caring for the multiple and the multitude: assembling animal welfare and enabling ethical critique. *Environment and Planning D: Society and Space*, 30(4): 623–638.

Davies, G. (2012b). What is a humanized mouse? Remaking the species and spaces of translational medicine. *Body & Society*, 18(3–4): 126–155.

Davies, G. (2013a). Mobilizing experimental life: spaces of becoming with mutant mice. *Theory, Culture and Society*, 30(7–8): 129–153.

Davies, G. (2013b). Writing biology with mutant mice: the monstrous potential of post genomic life. *Geoforum*, 48: 268–278.

Deleuze, G. (1988). *Foucault* (trans. S. Hand). Minneapolis: University of Minnesota Press.

Haraway, D. (2008). *When Species Meet*. Minneapolis: University of Minnesota Press.

Latour, B. (1998). How to be iconophilic in art, science and religion? In P. Galison and Caroline A. Jones (Eds.) *Picturing Science, Producing Art* (pp. 418–440). London and New York: Routledge.

Latour, B. (1999). On recalling ANT. *The Sociological Review*, 47(S1): 15–25.

Latour, B. (2005). *Reassembling the Social: An Introduction to Actor-Network-Theory*. Oxford University Press.

Sunder Rajan, K. (2006). *Biocapital: The Constitution of Postgenomic Life*. Durham, NC: Duke University Press.

# 3

# Conversation between Angela Last and Nina Lykke

*Nina Lykke and Angela Last*

---

AL: In your monograph, *Feminist Studies: A Guide to Intersectional Theory, Methodology and Writing* (Lykke 2010), you talk about the institutionalization of feminist studies and how the debate about how to engage with the disciplinary nature of the university system was key to feminist politics in Academia. I get the sense, also from examples from critical race studies, that disciplines were perceived as domains and divisions where white men traditionally held power and continued to do so. You discuss how feminist research initially mainly took place in interdisciplinary centres, partly, because of the struggle with the tension between the desires to integrate gender into the existing disciplinary system and to let gender remain an outside 'irritant' to the system. Can you talk more about this tension that influenced your experience of interdisciplinarity?

NL: I was part of a socialist feminist group that started the first Women's Studies Centre in Denmark in 1981. The format and organization of the centre was set up with US women's studies centres as the model. This meant, first of all, that we saw interdisciplinarity as an appropriate way to study gender perceived as an area of study that was embodying severe societal problems. Using the gender lens to focus these problems meant addressing them from below, from an activist point of view, and not from above.

We also saw gender as an issue that could not be addressed properly unless it was done in an interdisciplinary manner. For instance, if you take the problem of gender-based violence, you could look at it as a psychological issue and locate it in psychology, but what activist practices showed was that gender-based violence is not just about people's subjectivities; it is an issue that has links to wider social structures which need to be addressed as well. You could make the same argument around race, class or dis/ability issues. From activism, we learnt that societal problems are reduced if they are solely treated within the horizon of one discipline – they cannot really be contained and treated to the benefit of those who are hit by them from the perspective of a single discipline. If you had a traditional disciplinary outlook, you might think that you could solve these problems. With gender-based violence, for instance, you could say: I can come up with ideas about how to intervene here from a psychological point of view. But, activist practices, to which all these studies – gender studies, black studies, queer studies, etc. – had and have such strong links, generated the knowledge that these issues need to be analysed and addressed in a much broader way to benefit the people centrally concerned.

The activist connection was also a strong background for creating interdisciplinary gender studies units. In Denmark, they were created in the 1980s. The background for establishing gender studies units has, of course, been different in different contexts, but interdisciplinarity, generated through activist practices, was in many ways a shared feature. My activist practice came out of the socialist feminist women's and students' movements of the 1970s, and in both movements, interdisciplinarity was something that was promoted very strongly – precisely, in order to address societal problems which could not be contained by single disciplines. The 'view from below' was also very important: if you look at these problems from an activist point of view, from the point of view of the people who are experiencing the oppression or the problems, then it becomes clear that they cannot be properly understood against the background of existing knowledge divisions. These divisions prevent important interconnections from being grasped.

AL: At the moment, there are a few universities that do not just want to implement inter-disciplinary studies or projects, but that are actually attempting to restructure the entire institution in an experimental way. How was the situation at Linköping University – how did the restructuring take place?

NL: Tendencies towards restructuring in a more interdisciplinary direction have been outspoken at Linköping University, and this is the reason that I applied for a professorship in gender studies there in 1999. But, of course, mainstream structures and disciplinary identities have also been very strong. Interdisciplinarity did not just come about without political struggle. At Linköping University, the Department of Thematic Studies (TEMA), which now includes a unit of gender studies and three other interdisciplinary units (for environmental studies, science and technology studies, and child studies), was originally inspired by the 1970s' students' movement. A particular inspiration was a university in Denmark, located a bit outside of Copenhagen (Roskilde University), which back then was a new university. There, the students' movement was particularly strong, and had a strong impact on the organizing and running of the university.

Interdisciplinarity was considered very important there, and interdisciplinary Bachelor degrees in humanities and social sciences became the foundation of the university. The whole Bachelor level of study at this university was built on interdisciplinary grounds back in the 1970s. This happened just before the introduction of thematic studies (TEMA) at Linköping University. A group of radical teachers from Linköping University basically went to Roskilde University in Denmark to see how the interdisciplinary degrees were set up there. It was pretty impressive back then to have organized all of the humanities and social sciences on an interdisciplinary basis: no learning processes split up into disci-plines until the Master's level. However, after the Linköping group had visited Roskilde University, they, nevertheless, decided to take a different route to interdisciplinary organ-izing than the one chosen in Roskilde. Instead of making the Bachelor degree inter-disciplinary, the Linköping group decided to set up an interdisciplinary PhD programme. So, in Linköping, they started the other way around so to speak.

The interdisciplinary PhD degree, set up in Linköping, became very famous and successful. The PhD programme has always gotten very good evaluations and the hundreds of PhDs, who have graduated from the programme over the years, have got good jobs inside and outside of Academia. TEMA became part of the branding of Linköping University as an interdisciplinarity-friendly university. It's not that all of the university is organized in an interdisciplinary way – TEMA is just one, albeit big and influential

229

department – but the success of the TEMA department in terms of research, PhD degrees, research funding, new Master's programmes, etc., have made a strong impact on the university.

AL: So the success of Thematic Studies influenced the other parts of the university?

NL: Yes, it definitely did. Because in these neoliberal times, what speaks is very much money, research funding, and successful candidates, and the TEMA department has been very successful throughout the years of its existence. This led to the university becoming more interdisciplinarity-friendly overall. You do not have to fight so much for interdisciplinarity here. If you have an idea for an interdisciplinary project, it is approved and encouraged.

AL: Earlier on, you mentioned political struggle around interdisciplinarity. Can you expand a bit on that?

NL: Yes. I don't know if you are familiar with the reflections on knowledge production in terms of mode 1 and mode 2, that is, the conceptual framework put forward by Michael Gibbons, Helga Nowotny and others (Gibbons *et al.* 1994; Nowotny, Scott and Gibbons 2001). As you may know, mode 1 is the disciplinary model, organized against a background of knowledge-seeking for the sake of gaining new knowledge, that is, basic research. The mode 2 model of knowledge production is more issue-focused, inter- and transdisciplinary and also focused on applicability. When you have a university structured according to a mode 1 model, there are likely to be clashes when a mode 2 model is introduced. The mode 2 model is very much in tune with neoliberalism. But it is also worthwhile noticing that there are certain resonances between the politically radical interdisciplinarity that I'm arguing for and that goes with social movements and activism, and what, by the authors mentioned above, is defined as the mode 2 university, even though this latter university model is also very much part of neoliberalism. We may talk about strange bedfellows here.

There are certain features of neoliberally organized universities that actually provide a better context for interdisciplinary work, not only in a technocratic sense, but also when it comes to the deeply critical, activist inspired, radical interdisciplinarity that I am talking about in relation to gender studies, queer studies, black studies, etc. What I argue, in the context of my long academic feminist career, is that it is possible to do activist feminist, queer, anti-racist hit-and-run-interventions in the neoliberal university. By this I mean doing interdisciplinarity in different kinds of ways which hopefully can make a difference.

AL: This reminds me of Kwame Anthony Appiah's comparison of Utopia U and Utility U (2015) – the different ways that the university is imagined, either as a mind expansion for the greater good or as a means to simply get a better job and help improve the economy. Your phrasing of 'hit and run' also makes me think of the debate around the spectrum of positions towards the academy that one can occupy, especially as a politically engaged and/or marginalized scholar. Possible positions include rejection, decolonization and integration. In particular, I think about Stefano Harney and Fred Moten's conclusion in their book *The Undercommons* (2013) namely that the only possible relationship with the university nowadays is a criminal one. They talk about being in, but not *of* the academy and its system.

NL: That sounds like the things that I've been talking about. I also think a problem lies with what you talked about before we started the interview, about people paying lip service to interdisciplinarity in addressing their research problem. This is the dark side of the coin – that neoliberalism reduces interdisciplinarity to essentially a technocratic device.

AL: Yes, you cannot really think education and topics outside of the context that they are embedded in. You also speak about interdisciplinarity being an 'empty signifier'. Do you feel it can still have a radical potential, given what has been happening?

NL:   Yes, I think that if you frame problems in a political, activist, social movement fashion, and if you want to address these problems properly, you cannot do it from the point of view of one discipline, and you also cannot solve a problem if interdisciplinarity is treated merely like a technocratic device. I think lots of exciting things are happening in areas such as, for instance, gender/medicine/health, gender/climate change, anti-racism, trans-feminism, decolonial studies, critical disability studies, etc., that are highlighting currently vibrant political dimensions and alliances between academic research and activism. Again, the argument here is that you cannot just do research in these areas from the view of a single discipline – you have to involve other disciplines and their points of view. I often think of Bruno Latour's famous question (Latour 1993): What is the hole in the ozone layer about? Chemistry, meteorology, politics, economics? And the answer is, of course: all of them! They are entangled. If you then look at how climate change research is often specialized, and if you compare that with what happened around Hurricane Katrina in New Orleans, you really see how this 'weather phenomenon' is intertwined with race issues, class issues, gender issues, disability issues, etc. Nancy Tuana (2008) wrote a brilliant, critical analysis of this. You cannot do a proper analysis of these problems without interdisciplinarity, and I also believe that interdisciplinarity can work as a political tool in such contexts.

AL:   Could you tell me more about how this relates to what you call 'visibility raising'? It seems to relate to your political approach.

NL:   Well, visibility raising is about recognizing the problem. For me, there is a strong connection with Kimberlé Crenshaw's classic article, 'Mapping the Margins' (1995) that takes intersectionality as a point of departure. In this article, Crenshaw talks about 'stories that resist telling'. The stories that she referred to here were stories about violence against women of colour. In the article, Crenshaw highlights how, on the one hand, white feminist movements against gender-based violence are not taking the specific issues of women of colour into account, and how, on the other hand, anti-racist movements are not very keen on going into the issue of gender-based violence, arguing that this might take a bad racist turn. This way, the stories of gender-based violence that affect women of colour 'resist telling'. I think this is a very good example of the ways in which certain problems need to be made visible, before they can actually be addressed, theoretically as well as politically.

But of course, you need to do more than just make things visible. And I think this also links back to what we talked about earlier concerning the relationship between epistemology and the structure of the university. If you are embedded in a traditional disciplinary structure, and you become confronted with a problem like this, you may be forced into a mere 'add and stir' approach. But 'add women and stir', 'add race and stir' do not really do the job. You need to reorganize the university so that you can address problems without being hindered by traditional, disciplinary divisions of knowledge production.

I have been discussing these divisions also with my colleagues from the Medical Faculty at Linköping University with whom I collaborate, and we agree that there are such strong borders between, for example, the approaches used in medicine and those used in humanities/social science. To transgress the borders is really hard work. We do, indeed, have hybrid research areas such as medical humanities now, but, as I see it, we are still only scratching the surface in terms of actually understanding how comprehensively universities have to be restructured in terms of being able to respond adequately to the challenges posed by a complex world which cannot be understood properly from the compartementalized views characterizing today's disciplinary outlooks. For example, if you raise questions such as, 'How and why is cancer unevenly distributed over the world?' and 'How and why are carcinogenic pollution and toxic embodiment unevenly distributed over the world?', you cannot give meaningful answers from either a medical or social science perspective. And you actually need not just collaboration between medicine and social science, but to rethink epistemologies, ontologies and the overall structures for knowledge production to address such issues adequately.

AL: Speaking of epistemology, you are very invested in the theoretical and methodological openness and challenges created through interdisciplinarity. First of all, I find the reference points in your work very interesting, which include authors such as Sandra Harding (*The Science Question in Feminism*, 1986), Judith Butler (*Against Proper Objects*, 1997), Hélène Cixous (*Coming to Writing*, 1991) or Julia Kristeva (*Revolution in Poetic Language*, 1984). Could you talk more about the relationship between theory and method in your work? I am interested in this, because I feel that theory is always valued more than method, with method frequently getting too little attention in terms of how it matters and affects relationships with participants or the rest of your research. For me, the material asks you to use a certain method, but researchers often do not make the connection and do not seem to listen to the theories they employ.

NL: I originally came from a background in literary studies, where methods were not much talked about. But when I came to this very interdisciplinary TEMA department at Linköping University, I was suddenly confronted with the question of methods, especially by my colleagues with backgrounds in the social sciences. Retrospectively, I also found out that there are a lot of methods in my own original discipline, literary studies, but, as in philosophy, 'methods' is very much a 'bad word' there. So, on the one hand, there are disciplines such as literary studies and philosophy that really hate the term 'methods', and on the other, there are disciplines such as sociology, anthropology, etc., that really fetishize 'method'. In an interdisciplinary setting this can become a real clash. So, because I have committed myself to an interdisciplinary space, I have been confronted, on the one hand, with reductive aspects of a focus on methods – how they can be handled rigorously and in a policing mood from disciplinary points of departure, ending up in tunnel vision. But, on the other hand, I have also learned to respect the way in which you really need to think in-depth about the analytical strategies and thinking technologies you mobilize when you grapple with your material.

In particular, you have to think these analytical strategies and thinking technologies through extremely carefully if you want to pose and address research questions in alternative, interdisciplinary ways. It is not enough to think about theory and epistemology, you have to think about how you actually interview people, interpret textual materials, listen to music, look at what's going on in a laboratory . . . You really need to pay attention to this – to the *doing* of research. In this sense thinking about methodology and the methods you apply is very important. Some of the many students whom I encountered through teaching and supervising were very interested in theory and particular theorists, but got into problems when it came to the choice of analytical strategies. And, although I do think that it's central not to make an artificial cut between empirical work and theory, I also argue that it is very important to think carefully through which analytical strategies to choose and how these strategies resonate with your chosen theory. Sometimes I meet students in whose work theories and analytical strategies get to contradict each other, and this is of course to be avoided. Such contradictions do, for example, occur when methods are being employed in a traditional way that clashes with the untraditional theories which the student is relying on. Let me give an example: it would be contradictory to build theoretically on Donna Haraway's concept of 'situated knowledges' (Haraway 1991) and then construct an empirical analysis, where, instead of dialoguing at eye level with your research participants, you start talking about them as if you were situated 'above' them. Researchers (and not only students) are, indeed, doing these contradictory sorts of moves all the time – no one is immune to them. But if you then think about and discuss these problems, you notice how things do not fit together and this insight could be used as a first step toward exploring ways to do things differently.

Of course, you need to think through all the levels of research, and this is an area where I actually think that writing as a method of enquiry (Richardson 2000) has a lot of potential, because it is a very good tool for breaking up traditional methods and for breaking up traditional cuts and dichotomies between theory and empirics.

AL: I always find that, in terms of methods, there are things that are not discussed enough. For instance, how you present or represent your work. I could do this interview transcript in so many different ways, and each different style of editing will most likely create a different impression. You see this effect in a lot of official reports, where 'experts' tend to be edited more heavily to come across as knowledgeable and 'faultless', whereas representatives of other publics are not as heavily edited to give a more 'person on the street' feel that has an asymmetric relationship with these experts. Also, academics can be rather anxious about disciplinary boundaries despite methodological overlaps. There is also the question of whether the specific disciplinary belonging of a lot of so-called disciplinary methods is an illusion in the first place – surveys, interviews, observation and so on do not just belong to one discipline. Despite these shared methods, boundaries are often being policed in an unhelpful way, e.g. 'this is not geographical, this is not sociological, etc.', whereas what people should really be looking at are crucial questions such as: What methods does this project need? How do I cultivate the sensitivities necessary to give justice to my project? Or, as you mentioned earlier, how do I interact with my research participants – how do I situate myself?

NL: I agree very much with you on this – you cannot just stick to a disciplinary tunnel vision and ignore what is needed for a project. Instead of going 'this is not geographical, this is not sociological', you should ask: 'Is this a suitable method? Is this an unsuitable method?' You have to argue why something is a good method with an open mind – you have to cut through disciplinary boundary concerns. Heavily interdisciplinary organizations are ideally places where no discipline has the final say. If you are in a more traditional place and you do interdisciplinary research there, your colleagues might say: 'you do very interesting work' – in the sense of 'you do your work, and we do ours' – but it is still the disciplines that set the agenda and lay down which methods and forms of representing your work are allowed. But when you are in an interdisciplinary setting, like the TEMA department at Linköping University, then no discipline has the final say. I really think organizations, in that sense, can deconstruct the claims or right to police boundaries, and I think this is important.

AL: You also talk about experimentation in your work – this is something that I, too, am very interested in. You can, of course, do a lot of experiments by yourself, regardless of what anyone else is doing around you. But often, the structure and ethos of the institution that you are embedded in, has a strong influence on what you are doing and how you are experimenting. I don't think that there are any particular rules around that – I even think that, often, a more restrictive environment can make you more experimental. We also had some discussions amongst the editorial team of this book on whether there is such a thing as 'interdisicplinary methods' or whether 'interdisciplinary methods' are really a case-by-case experimentation. I wondered if you had any thoughts on that?

233

*NL:* It brings a lot of things to my mind when you frame your question like this. First of all, I agree with the people who say that methods today have become attached to and been claimed by disciplines. For this reason, too, it is important to deconstruct the disciplines. The deconstruction of the disciplines is important to get to more open-ended understandings of methods and to counteract boundary policing ways of using methods. In my book (Lykke 2010), I emphasized a distinction between interdisciplinarity, multidisciplinarity, transdisciplinarity and postdisciplinarity. In this framework, interdisciplinarity is between multidisciplinarity and transdisciplinarity – interdisciplinarity here represents a step beyond multidisciplinarity, where you stay with the disciplinary outlooks, adding them to each other. Though, when you talk about interdisciplinary methods, it can very easily be reduced to what I would call 'multidisciplinary methods'. You end up taking this method from that discipline, and that method from this discipline, and then first carry out this analysis and after that this analysis, and you then end up with a more complex result than if you had performed this from within one discipline. BUT, still, you remain within the framework of disciplines – you have not made the epistemological deconstruction that I talked about earlier. So, therefore, I would prefer to talk about 'method' in a 'postdisciplinary' or 'transdisciplinary' sense, where you do not refer back to disciplines.

At the same time, I think that it is important to keep up a respect for multidisciplinarity. There are people working on multidisciplinary grounds who are doing good work. However, I would very much like to see a book like the one that you are working on, that does not reduce interdisciplinary methods to mere multidisciplinarity in the above sense. I think that really important things are happening in the area of what I call 'trans- and postdisciplinary methods' – i.e. methods, of course experimental, mobilized in spaces and vis-à-vis research questions which are not defined within disciplinary frameworks. In such contexts, you cannot just take 'a method of inquiry', be it for example discourse analysis or surveys, and then 'apply' it. You have in such spaces to effectively rethink the whole discourse about methods as well as the connections between epistemology and methodology, ethics and representation. To in-depth question and scrutinize these connections, and to make sure that there is alignment all the time, is a totally different way of looking at methods compared to what you do in the case of multidisciplinarity, and this is, in my view, very important.

*AL:* Yes. I am also very sceptical of the 'magic bullet' approach to interdisciplinary methods. What the *Inventive Methods* book (Lury and Wakeford 2012) and this book are trying to do, for instance, is to make suggestions for different possibilities of how one could think about interdisciplinary methods in terms of shared devices or processes. These are provocations – that might be rejected, taken up, considered, re-appropriated . . . they want to be in dialogue with other provocations. I think thinking about methods is always developing – I can certainly notice it in my research and teaching – and it is always interesting what happens next. At the moment, the 'decolonizing methods' provocation from people such as Linda Tuhiwai Smith (2012), for instance, is gaining an interesting momentum and is hopefully gaining traction in wider Academia. It was something we also discussed in relation to the book, including the possibilities and difficulties in interpreting and applying this aim in relation to individual entries and the demands of internationalization.

*NL:* I think this 'search for the magic bullet' is also one of the problems. People think that they can find it, and that is like looking for *a* cure for cancer. You cannot find it in one place or by using a universalizing approach.

AL: I guess you could say that the opposite also holds true – being too particular. The two also sometimes combine in a very problematic way. I have to admit that I have become very frustrated with a lot of publications about interdisciplinarity. Perhaps I am judging to quickly or harshly – I feel like I should be more sympathetic to a spectrum of positions – but some publications, in my view, go too excessively into the details of particular interdisciplinary projects – the types of coffee cups on the table at meetings, people's body language, etc. I know that these very science and technology studies (STS) style observations serve a purpose – which has even been framed as a decolonial project – but, for me, that purpose or bigger picture often gets lost. You end up with a degree and type of reflexivity that does not put the overall framework into question. I know that you also partially come from this direction. I was curious about your thoughts on that.

NL: I have never considered myself as a 'proper' STS scholar. I think I am too little a sociologist, and too much a literary scholar here. But I have been collaborating a lot with STS scholars and, therefore, I know the field pretty well. My position on STS is that, on the one hand, the field is carrying a strong deconstructive project – decolonial is perhaps too radical a way of characterizing STS endeavours, though. All this looking at coffee drinking, choreographies and how people take things from one end of the lab to the other, is actually, as I see it, part of a very experimental form of deconstruction. As it has been pointed out by a lot of feminist critics, in not all of STS but in some more mainstream STS, analysts forgot about power when developing their methodological approaches, and ignoring intersectionality as well. An example is the kind of critical discussions that Donna Haraway has had with Bruno Latour and other STS scholars, where she is bringing out very forcefully issues of power and power differentials in terms of gender, race, class, sexuality, pointing out that these important intersections were not well dealt with in mainstream STS. And, of course, this is a major flaw in STS. I think that Donna Haraway and other feminist STS scholars, such as for example Amade m'Charek, Celia Roberts and Maureen McNeil, have let themselves be inspired by STS, but they have also integrated intersectionality and power relations.

One could say that, from the outside, analyses of power relations were not integrated well enough in STS. However, if STS methods are constructed in a more experimental way all the way through, like many feminist STS scholars have done, then these methods might become something different. In *When Species Meet* (2008), for example, Haraway takes inspiration from the ways in which STS calls for an attentiveness to all kinds of nitty-gritty details – how it matters if the dog runs like this or comes back like that, the use of the leash or other technologies – but she takes the analysis further into intersectional power relations. I did an interview with her once (Haraway 2004), and here she framed her relationship to mainstream STS like this: although she is very well known and in no way marginalized, she said that she felt that people in mainstream STS had difficulties 'digesting' her ideas and approaches. This says something about the ways in which mainstream STS threatens to congeal into *a* method, rather than stay open to issues of method, and to provocations from feminism and other kinds of activist approaches. In fact, people sometimes become quite anxious in terms of 'this is STS' or 'this is not STS', and we are then back in the old way of handling method as a fixed thing.

AL: This reminds me of something that I recently read in the *Chimurenga Chronicle* (Hardy 2013). It was an article about the black classical composer Julius Eastman and his relationship with the emerging new establishment at the time – self-proclaimed rebels such as John Cage, Earle Brown, Christian Wolff. Some promoter wanted to 'diversify' the representation of classical music in a rather superficial way by 'adding black and stir', as you might put it. He invited Eastman, who was a composer in his own right, to perform the work of Cage and other 'rebel' composers. Of course, he was angry at first, but then decided to turn this insult into a challenge. What he did was that he performed the Cage piece he was hired to do in such a way that he called out the composer on his lack of attention to power relations. Of course, he got blacklisted as a performer after that, but the challenge to the illusion of anti-establishment gestures could not be erased – he had really managed to nail the problem. Also, it is methodologically interesting in another way: even if some people only get it decades later, this performance will echo on. I thought this was fascinating, because, even as a musician, I often do not think about music being connected to these kinds of methodological possibilities, especially politically informed ones, but of course it makes sense.

NL: I think this is totally interesting. I think breaking the boundaries between 'the academic' and 'the creative', and letting methods and approaches spill over from one to the other, is very important. That goes for all sorts of creativity – visual art, music . . . my own medium is poetry and creative writing. For instance, I think that the non-verbality of music can be a strong area of expressing what you cannot deal with in other ways. Music is one of the areas that is sadly less integrated in the social sciences – you are much more likely to encounter poetry, although it is often considered 'radical' or at least 'non-scholarly'.

AL: In Geography, we have been integrating these practices more and more, but I often feel that, in many cases, it is done in quite a polite way, compared to the challenges that are coming from the music and activism crossover scene, which is challenging people through performance, but also through teaching others how to perform across established boundaries – training female DJs and producers, for instance. But sometimes, I feel, it is precisely the lack of institutional support that makes you more inclined to do something more performative, more controversial, because you need to express this lack. But there are other institutional conditions that affect performance and registers of creativity in academia. Also, you sometimes end up importing stuff that you have done outside of academia into your academic work – often it is something that you would never have been able to get funding for, because it does not meet particular definitions of academic performance of creativity. Actually, one of the things that I wanted to ask you earlier, was about funding and the problems you encounter with funding interdisciplinary work.

NL: Funding can be a really tricky issue. I think it is a place where the old disciplinary structures are pretty heavy, and there is a lot of lip service paid to interdisciplinarity in the discourses of funding agencies. But when it actually comes to distributing the money, disciplines take the front seat again. I have discussed this with a lot of colleagues, not only in gender studies, but also for instance colleagues at TEMA's other units (child studies, environmental studies, science and technology studies). And we share a lot of common ground in terms of wishing to pressure the research councils and funding agencies more broadly to recognize interdisciplinarity. From the most humble multidisciplinary approach to the most radical postdisciplinary one, there is a need to go beyond the rather empty buzzword 'interdisciplinarity'. But as pointed out with annoyance by one of my colleagues, at a recent TEMA department meeting with a reference to an interview in the *Times Higher Education Supplement* with Professor Ian Goldin, Oxford Martin School (Goldin 2016) so very much in today's academic world – from journal rankings to funding to appointments – is still governed by disciplines. A good thing about being in an interdisciplinary department is that you are not alone in the frustrations as regards the ways in which the strong disciplinary powers govern today's academia.

It is interesting to see how external funding schemes, for example, can continue to develop along disciplinary lines despite strong tendencies for universities to transition from mode 1 to mode 2, which, according to Gibbons et al. (1994) and Nowotny et al. (2001), is supposed to lead to inter- and transdisciplinarity. The old mode 1 universities, which effectively were institutions with small student populations mainly composed of elite white European men, and which had very exclusionary practices in relation to all Others, have, on the one hand, been eroded. But, nevertheless, on the other hand, some modes of funding, appointing, etc., are strictly keeping up some of the old structures, such as certain forms of disciplinarity. It is thus the case that you can get funding for a project that appears interdisciplinary, but is actually a very technocratic multidisciplinary project that does not go into the real complexities of the problem, because this possibility is blocked by old-fashioned outlooks. My experience is that a lot of funding is channelled to such projects.

AL: Yes, a lot of the time, you end up writing a really straight project, with the intention to do the interesting stuff covertly. There is then, of course, the question whether this perpetuates the system or whether the alternative results can effect a change in funding criteria in the long run. I guess this is an experiment in itself!

## References

Appiah, K. A. (2015). What Is the Point of College? *New York Times Magazine*. 8 September 2015. Retrieved 27 February 2017 from: www.nytimes.com/2015/09/13/magazine/what-is-the-point-of-college.html

Butler, J. (1997). Against proper objects. In E. Weed and N. Schor (Eds.) *Feminism Meets Queer Theory* (pp. 1–31). Bloomington and Indianapolis: Indiana University Press.

Cixous, H. (1991). Coming to writing. In D. Jenson (Ed.), *Coming to Writing and Other Essays* (pp. 1–59). Cambridge, MA: Harvard University Press. Trans. S. Cornell, D. Jenson, A. Liddle and S. Sellers from *La venue à l'écriture*. Paris: Union Générale d'Éditions, 1977.

Crenshaw, K. W. (1995). Mapping the margins: intersectionality, identity politics, and violence against women of color. In K. Crenshaw, N. Gotanda, G. Peller and K. Thomas (Eds.) *Critical Race Theory: The Key Writings that Formed the Movement* (pp. 357–384). New York: The New Press.

Gibbons, M., Limoges, C., Nowotny, H., Schwartzman, S., Scott, P. and Trow, M. (1994). *The New Production of Knowledge: The Dynamics of Science and Research in Contemporary Societies*. London: Sage.

Goldin, I. (2016). Interview in *Times Higher Education*. 3 May 2016. Retrieved 19 May 2016 from: www.timeshighereducation.com/news/multidisciplinary-research-career-suicide-junior-academics

Haraway, D. (1991). Situated knowledges: the science question in feminism and the privilege of partial perspective. In *Simians, Cyborgs and Women: The Reinvention of Nature* (pp. 183–201). London: Free Association Books.

Haraway, D. (2004). *The Haraway Reader*. New York, London: Routledge.

Haraway, D. (2008). *When Species Meet*. Minneapolis: University of Minnesota Press.

Harding, Sandra. (1986). *The Science Question in Feminism*, Ithaca, London: Cornell University Press.

Hardy, S. (2013). 52 Niggers. *The Chimurenga Chronicle*. 13 March 2013. Retrieved 27 February 2017 from: http://chimurengachronic.co.za/52-niggers/

Harney, S. and Moten, F. (2013). *The Undercommons: Fugitive Planning & Black Study*. Wivenhoe, New York, Port Watson: Minor Compositions.

Kristeva, J. (1984). *Revolution in Poetic Language*. New York: Columbia University Press. Abbreviated. Trans. Margaret Waller from *La révolution du langage poétique*. Paris: Éditions de Seuil, 1974.

Latour, B. (1993). *We Have Never Been Modern*. New York, London: Harvester Wheatsheaf. Trans. C. Porter from *Nous n'avons jamais été modernes*. Paris: La Découverte, 1991.

Lury, C. and Wakeford, N. (2012). *Inventive Methods: The Happening of the Social*. London: Routledge.

Lykke, N. (2010). *Feminist Studies: A Guide to Intersectional Theories, Methodologies and Writing*. London, New York: Routledge.

Nowotny, H., Scott, P. and Gibbons, M. (2001). *Re-Thinking Science: Knowledge and the Public in an Age of Uncertainty*. Cambridge: Polity.

Richardson, L. (2000). Writing as a method of inquiry. In N. K. Denzin and Y. S. Lincoln (Eds.) *Handbook of Qualitative Research* (2nd ed., pp. 923–948). London: Sage.

Smith, L. T. (2012). *Decolonizing Methodologies: Research and Indigenous Peoples*. London: Zed Books.

Tuana, N. (2008). Viscous porosity: witnessing Katrina. In S. Alaimo and S. Hekman (Eds.) *Material Feminisms* (pp. 188–213). Bloomington and Indianapolis: Indiana University Press.

# 4

# Haunting seedy connections

*Tahani Nadim*

> . . . So
> no my love
> whatever we've run short of
> this hasty day
> its name cannot be
> time.
>
> *Ayi Kwei Armah, Seed Time, 1988*

Seeds are potent things. So potent in fact that much effort has been poured into debilitating their self-replicating capacities through legal frameworks, gene-use restriction technology, gene guards and other termination devices. One could describe these efforts as seeking to curtail the seeds' talent for return. On the other hand, seed banks like the Svalbard Global Seed Vault want to preserve this talent indefinitely, keeping it on the cusp without end. Svalbard, also called the 'Doomsday Vault', stores seeds *ex situ* in a purpose built repository on the island of Spitsbergen in the very north of Norway.[1] This is ostensibly done for the preservation of genetic plant diversity in the face of its steady decline due to land use and climate change, or as its name suggests, sudden catastrophic annihilation.[2] The Doomsday Vault and similar facilities, as a recent report by the UN's Food and Agricultural Organization put it, 'bridge the past and the future' by securing the perpetual availability of genetic resources (Food and Agriculture Organization of the United Nations 2014: x). The adoption of similar prospects dictates the ongoing development of many museum collections, which are regarded as safeguarding the heritage of 'endangered' cultures and natures. Expressly designed to withstand even nuclear war, the Doomsday Vault has, one could say, come to terms with various ends of the world. It, in fact, banks on it.[3]

In this text I want to attend to the challenges posed by collections and archives such as the Doomsday Vault in relation to their 'hauntings' (Gordon 2008), their 'unfinished business' (Rushdie quoted in Subramaniam 2014: 21) particularly as these relate to the 'ruinations' that describe 'the ongoing quality of processes of decimation, displacement, and reclamation' of imperial formations (Stoler 2013b: 8). What or who lingers on in these collections despite their best efforts to vanquish and purify specimens and rhetorics? How can 'haunting' be configured as a method to trouble the neat returns underwritten by collections? And how to reckon with and remain responsive to the capabilities of archival materials such as seeds whose global histories and polyvalent presents customarily vault across disciplinary practices? These questions address the tensions between the Doomsday Vault – and similar attempts at more or less total archives that avow an untroubled progression of time – and the fact that, as William Faulkner put it, '[t]he past is never dead. It's not even past' (Faulkner 2011[1951]: 73). Yet, questions about the hauntings in archives and collections also direct analytical sensibilities to the situatedness and site-specific co-evolutions of archival practices and imperial formations. Looking for their hauntings makes apparent how archives and collections have emerged through specific material-semiotic arrangements whose 'stubborn attachments' (Ahmed 2007: 133) refuse to be vanquished by new technologies or alternative political paradigms. And last, questions about such 'ghostly matters' (Gordon 2008) are important to consider when engaging with the 'complexity and historicity' of collections and their organization (Bowker 2006: 121). Adding to Bowker's concerns about relevance and functionality of archives and their production of 'reconfigurable pasts' (2006: 136), the stakes pertain also to their enactments of specific futures. While this might be particularly evident in the case of the Doomsday Vault, less sensational efforts to collect and preserve seeds, such as the Mai Collection discussed in more detail below, are equally implicated in setting the terms for realities to come.

For the last three years I have been working in the Natural History Museum Berlin (*Museum für Naturkunde*, MfN). Questions about ghostly matters and imperial ruinations are urgent in an institution filled with dead animals and the exploits of over 300 years of imperial extractive activities. As a conspicuous venue of science, education and nationalism it has shaped an enduring and far-reaching vision of scientific inquiry, of citizenship and of the nation-state, particularly through the dogged pursuit of its 'acquisitive impulse' and attendant compulsions of ordering, arranging and dividing (Livingstone 2003: 29). While ethnographic and anthropological museum collections have been the focus of sustained critique for their complicities with various imperial regimes and technologies of governing (most recently Bennett *et al.* 2017), their natural history counterparts have remained fairly unfazed. Sheltered firmly on the side of Nature, they have by and large escaped the troubles that readily come with the categories of the ethnographic (e.g. 'race', 'tribe', 'culture' and so on). Such troubles are being raised by activists, scholars and, increasingly, artists and so they have become a busy arena for interdisciplinary and collaborative engagements. Here, the concern for hauntings becomes doubly demanding if we consider inter-disciplinary work as learning to question our – at times – pathological disciplinary conventions. At the same time though, a too disciplined commitment to the in-between runs the risk of losing sight entirely of the genealogies shaping our perceptions and thinking. Although some museums and disciplines might be only too happy excising their more embarrassing ghosts (e.g. human remains, craniometry), these ghosts are always already in the machine so to speak.

In 2014, I began collaborating with the visual artist Åsa Sonjasdotter on conceiving and producing an exhibition in the MfN based on troubles gathered during ethnographic and collabo-rative research. Our shared disconcertments were provoked by what we perceived, through our respective experiential and disciplinary sensibilities, as absences, obfuscations and neutralizations in museum displays and practices. Rather than formulate and present an orderly critique, however, we wanted to intervene more delicately and playfully, restoring some of the wild complexities (barely) contained in the museum through subverting orthodoxies of display and narrative. Entitled *Tote Wespen fliegen länger/Dead wasps fly further* (March–May 2015) our exhibition com-prised three artistic interventions presenting protagonists – a wasp, lunar dust and seeds – from the museum's collections and their 'factual and imagined journeys' as we called it. Sonjasdotter's artistic practice has for over ten years focused on the potato as object, archive and companion species, tracing its travels and political pasts and presents across centuries and continents.[4] So quite naturally our research first took us to the palaeobotanical collection where we searched for potato traces but came up empty. Instead we found the Mai Collection, an extant plant seed collection that is kept in rows of nondescript cupboards from GDR times, tucked away in the far corner of one of the collection rooms.[5]

> In the midst of the museum's palaeobotanical collection we find a curious body: a vast assortment of extant plant seeds, stored in glass vials, laid out in flat drawers and arranged by taxonomic order. These were put together by the palaeobotanist, Prof. Dieter H. Mai (1934–2013) from the 1950s onwards, though some parts had come to the museum from the Prussian Geological Survey (1873–1939). The extant plant seeds serve as a reference collection to allow comparison with fossilized seeds whose identity and kinship structures can thus be ascertained. Side by side sit seeds from Togo, Brazil, Cuba, Indonesia, Japan, the UK and Sweden. They come from global seed exchanges maintained by the world's botanic gardens. Seeds, like dust and the wasp, are seasoned travellers, accustomed to many means of transport, from winds to turtles, rivers and birds. The seedlings of the sisal agave (*Agave sisalana*), for example, have experienced a most intrepid sojourn, abducted from their Mexican homeland in Yucatán by German botanist Richard Hindorf (1863–1954) to Hamburg from where the surviving seedlings were sent to Tanzania in 1893, then the German colony of German East Africa. There they became the root stock for a sisal industry that exported over 90,000 tons of sisal each year and forever transformed the landscape that the plantations had occupied. Like cotton, coffee, rubber or the potato, the agave and its movements contributed to the rise of the global agro-industrial complex.
>
> In this display and wall installation we present parts of the museum's Mai Collection and follow the sisal agave plant from Mexico to Germany and Tanzania. They combine archival materials with contemporary ephemera, telling stories of collecting, smuggling and losing.[6]
>
> *Text from the information panel accompanying the seed display and installation*

There are 288 drawers that make up the Mai Collection. There the seeds stand still, arrested in development, seemingly robbed of the capacity to germinate and grow and return and disperse and grow again. The collection is rarely used and, as part of a research collection, not meant for public display. Its unspectacular appearance appealed to us and we began probing the curator (the always forthcoming and open-minded Dr Barbara Mohr), cupboards and catalogues, assembling strands of stories about seeds and their circulations. Arondekar has written beautifully about the perils that might beset researchers when encountering the (colonial) archive 'as a central site of endless promise' (Arondekar 2009: 6). She warns that not every time 'a body is found . . . a subject can be recovered' (p. 3). Developing her argument in relation to the queering of archives Arondekar's caveat remains instructive for querying the seed collection. It is not just about troubling authenticity and recovery, although these remain important sites for contestation. While the difference many of us want to make is the recovering of presences that have been rendered absent in and through archives, such a difference might also work to confirm the archival ordering, taking at face value its patterning of presences and absences. Or, as Spivak put it, such nostalgic revisioning 'would restore a sovereignty for the lost self of the colonies so that Europe could, once and for all, be put in the place of the other that it always was' (Spivak 1985: 247).

As with many of the collections and specimens in the museum, the seeds do not lend themselves to tidy narratives. Patchy documentation, ambiguous labelling and upheavals large and small render reconstructions of their histories forever speculative. Where exactly specimens were taken from, under what conditions and in what specific state are questions that for many parts of the collection (fossil and extant) cannot be answered with any certainty. Nevertheless, faithful reconstructions of collection items' institutional histories have become a prominent genre for acknowledging and re-mediating museums' complicities in imperial formations. These so-called 'object biographies', which describe the life histories of specific artefacts including their acquisition, storage, display and so on, have found uptake especially in the context of ethnological collections (Kopytoff 2013; Gosden and Marshall 1999).[7] Informed by the material turn in the social sciences and humanities, object biographies appeal to museums as they celebrate the 'power' of objects to tell bigger stories and thus confirm the continued relevance of vast collections of stuff. We learn, for example, that a 'large Egyptian boat model' was purchased by Pitt-Rivers 'from the London-based antiquities dealers Rollin and Feuardent, some time before 1879' (Stevenson 2011), or that '5 Mexican pots', also from the Pitt-Rivers Museum, have a 'duty . . . to be cultural ambassadors' and 'to stand up for Mexico' which is 'a fabulous mix of ancient and modern, Christian and pagan' (Gray 2011).

Despite their intentions to provide a richer, more diverse set of histories that could perhaps work toward unsettling dominant evolutionary narratives, object biographies still succumb to what Harriet Bradley has identified as the archive's 'assurance of concreteness, objectivity, recovery and wholeness' (Bradley 1999: 119). And so they often fail to productively sustain the disconcertment that has given rise to novel ways of engaging collections and archives in the first place. Trouble, as feminist methodologies show, is a matter of *crafting* and not *recovering*, of *inventing* and not *finding*. Arondekar suggests that '[t]he intellectual challenge here is to juxtapose productively the archive's fiction-effects (the archive as a system of representation) alongside its truth-effects (the archive as material with "real" consequence), as both agonistic and co-constitutive' (Arondekar 2005: 12). Here, haunting can become an appropriate method in evoking necessarily uneasy stories of and with museum objects. In this sense, fictitious or inappropriate entities creeping into scientific and historical orderings can bring into relief the parameters, conventions and terms that are, always unthinkingly, wrapped up with museum collections and their display.

> The sisal agave (*Agave sisalana*) is a member of the Agavoideae subfamily and known for its leaf fibre, which is valued as cordage and has been widely used in marine, agricultural, shipping and industrial settings. Sisal fibre can also be found in carpets, musical instruments, tea bags, paper pulp and alcohol. The plant's thick and spiny leaves can reach a length of almost two metres and, within four to eight years after planting, a central flower stalk will appear. This can reach a height of six metres and bears yellow flowers emitting an unpleasant odour. The sisal agave is native to Central America, where it has been cultivated since Mayan times (1800 BCE–900 CE). Until the early twentieth century it was at the centre of the henequen industry, based in the Yucatán region of Mexico, which sustained a monopoly on farming sisal in concert with North American rope manufacturers. In 1893 Dr Richard Hindorf, a German agronomist working in German East Africa (Tanzania), smuggled 1,000 young plants (some speak of 2,000) out of Yucatán in the belly of a stuffed crocodile or in the folds of a large coloured umbrella. Only 62 (or 66 or 72) plants survived the journey that took them first to Hamburg and then to Tanga, a port town in Tanzania.
>
> Plants are canny travellers that can use different means of transport: Some are gone with the wind, others prefer to be carried by rivers or turtles while yet others choose human help. With the support of Wardian cases, portable greenhouses designed by the English botanist Nathanial Bagshaw Ward (1791–1868), masses of plants reached Europe and its colonies in the 19th century in organised plant raids. In their global travels plants also often bring unexpected companions with them such as fungi.
>
> *Text from the information panel accompanying the seed display and installation*

Very few of the seeds in the Mai Collection are of cultivated plants. This is not surprising given that the collection serves to identify seed material from a time before humans, and cultivated plants are commonly 'humanly socialised' (Åsa Sonjasdotter). One of the most precious and policed practices in natural history museums concerns divisions and boundaries.[8] There are the divisions between the kingdoms of life – animal, plant and mineral – as well as the many divisions within these kingdoms (phylum, class, order, family, genus, species). These also translate into the organizational structure of the MfN as collections are divided among curatorial staff according to phyla. Another major structural division in many museums concerns the separation between non-public research collections, ordered according to taxonomic rank, and public display collections that are arranged to communicate specific stories about, for example, evolution, biodiversity, dinosaurs or the solar system.

When Åsa and I began thinking about our interventions, we early on committed ourselves to troubling the museum's divisions between present and past, nature and culture, cultivated and wild type, fact and fiction. Sharing feminist, postcolonial and environmental concerns, our intention was to return things and troubles to presence even if they have never before been (quite) present. In researching the seed collection, the extensive network of seed exchanges that involved botanic gardens and, more generally, the role of botany in the German Empire (1871–1918), the sisal agave plant appeared again and again in different written and photographic accounts, especially about German colonial rule in Tanzania.[9] It thus seemed to us a perfect protagonist for telling uneasy stories about the implications of plants in the imperial project, the role and shape of science and the continued devastations of ecologies. Since the Mai Collection did not include any sisal agave seeds we 'planted' them: The sisal seeds on show in one of the installation's vitrines were purchased specifically for the exhibition from a seed trader in Hong Kong. A live sisal plant, shipped from a nursery near Cologne, sat alongside the seeds atop a plinth. It caused much concern as potting soil can carry unwelcomed guests (pests). Introducing sisal seeds makes it possible to haunt and implicate the collection to tell a story specifically about German colonial botany but also, more generally, about the colonial and imperial circulations that linger on.

Subramaniam talks of 'interdisciplinary hauntings' (2014: 1) when describing how her combined disciplinary backgrounds – biology and women's studies – allowed her to perceive and recognize the ghosts of eugenics that continue to pervade evolutionary biology and the history of variation, of women in science and of flower colour. Her hauntings were hence, in the spirit of Arondekar, an achievement of bringing together these histories' truth-effects and fiction-effects. The 'interdisciplinary' was also borne out of encountering her subject matters as both object *and* archive, subject *and* resource. Such double vision can also be understood in terms of recognition: hauntings '[pull] us affectively into the structure of feeling of a reality we come to experience as a recognition' (Gordon 2008: 63). Hughes and Lury have written how returns and returning can be understood as a 'coming back to persistent troublings; they are turnings over' (Hughes and Lury 2013: 787). Adding to their polysemic register of re-turning, which includes putting things on their head and giving back, haunting recognition would thus mean giving shape to a 'seething presence' (Gordon 2008: 8). In placing the sisal seeds alongside the Mai Collection they become an apt figure to turn things over (with). On one hand, they provide a beginning for the story, a provision close to their nature as seeds. On the other, they serve as vehicles for shuttling across times and spaces, soldering connections that the divisions of natural history, the order of the museum, are at pains to keep from growing.

After its introduction to German East Africa, the 62 (or 66 or 72) sisal plants smuggled from Mexico were planted in the Tanga region and became the foundation for large-scale sisal production in East Africa, which at its height accounted for 47 per cent of world production.

The transformation of land into labour-intensive, agro-industrial plantations in German East Africa and other colonies was supported by a network of botanic experimental stations, such as the *Biologisch-Landwirtschaftliche Versuchsanstalt* (biological-agricultural experiment station) in Amani, coordinated through the *Botanische Zentralstelle für die Deutschen Kolonien* (Botanic Central Office for the German Colonies). The *Zentralstelle* was based at the Botanic Gardens in Berlin and was also tasked with popularizing the colonial project. The stations carried out extensive farming experiments to test the suitability of agricultural crops in the local climates and soils. These crops were obtained from the *Zentralstelle* which collected and distributed seeds and seedlings of crops such as coffee, cocoa, rubber, sisal, rubber, potatoes, pepper, cotton, tobacco and tea from Java, Brazil, India, Egypt and British Ceylon among other countries. The experimental planting was also facilitated by a lively international exchange of field reports and stories by farmers and planters.

In the pursuit of imperialist politics by means of plants many actors coincided: botanic gardens, plant scientists, gardeners, entomologists, financial institutions like the Deutsche Bank (1870–), railway companies, farm machinery, cattle, migrant work forces, plantation owners. The sisal agave was a particularly imperious coloniser, radically restructuring local ecologies. For this, the lands which had been designated German East Africa became a vast laboratory for experimenting with introduced plant species, agricultural techniques, investment practices and labour economies.

*Text from the information panel accompanying the seed display and installation*

For Gordon haunting is 'the language and the experiential modality' (2011: 2) by which to understand the ongoing ruinations of racial capitalism on bodies, social bonds and the sense we have and make of ourselves and the worlds around us. 'Haunting', she writes, 'raises spectres, and it alters . . . the way we normally separate and sequence the past, the present and the future' (ibid). In other words, it's an episode where the containment of what is past (trouble, injuries, violence) no longer holds.

Gordon's ghosts are specific. They have a certain appearance, a definitive haunting ground. Oftentimes their power (to manifest, to harm) is tied to specific places although in those places they attest to the permeability of divisions and walls. Planting sisal seeds in the collection was an act of mischief by which to conjure up spectres of injurious pasts. They were recognizable enough so as not to seem entirely out of place. Plant seeds are, in the end, part of the collection. Yet, true to their nature, they helped us cultivate a shadowy place, an otherwise barren corridor connecting *Masterpieces of Taxidermy* with *System Earth*, both part of the museum's permanent display. Following the planned path, museum visitors would move from the showcases detailing the preserving, stuffing and mounting of animal bodies to a dimly lit hallway, an artefact of the ongoing renovation works that require large sections of the building to disappear behind temporary drywalls.

On the right side of the corridor four illuminated table display cases contain sketchy stories (the ones running through this text) that string along an assortment of objects we had collated: drawers from the Mai Collection, the sisal seeds, a nappy (used to smuggle seedlings), a Kenyan banknote depicting sisal plantations, some of the few traces left of the Botanic Central Office of the German Colonies.[10] On the drywall to the left, huge bright yellow curtains signified three (non-existent) windows, their dimensions mirroring the ones in the adjacent exhibition halls.[11] Inside (outside) those windows we hung a selection of framed photographs depicting a sisal plant in Yucatán, portraits of Wardian cases (transportable greenhouses) and a monoculture sisal plantation, possibly from Tanzania. The arrangement was carefully crafted; indeed, the question of *how* to display was equally significant to the question of *what* to display. This 'how' also points to the continuation of method by other means, its lurking presence in forms of presentation that often function as 'marker for appropriate conduct' (Last, this volume). Complying with formal constraints of academic publishing or museum displays thus becomes a ready indicator for proper method. Likewise, messing with form can immediately disqualify otherwise methodical rigour and in academic scholarship, unlike in art, is a privilege of seniority.

> In reports from the Amani Institute we find references to pathogens and diseases affecting the usually so robust sisal agave. Small gouache paintings depict the damage done to its leaves and spikes by various kinds of rot and pests. After the First World War the German colonies were distributed to the Allied powers and German East Africa became a British Mandate (Tanganyika Territory). In 1917, the hacienda system in Yucatán, a continuation of the slave plantation, had been outlawed. The *Zentralstelle* was dissolved in 1920 and briefly resurrected by the National Socialists in 1941 only to be destroyed in air raids in 1944 that also eradicated most of its archive. The experiment station in Amani was taken over by the Royal Botanic Gardens, Kew, in London. From the 1960s onward, sisal production in Tanzania began its slow death, in no small part due to the rise of synthetic substitutes such as polypropylene. There the mono-cropping of sisal has left swathes of degraded land, having robbed the soil of nutrients and thus fertility.
>
> The seeds in the Mai Collection too lie in dim stagnation with little possibility to ever germinate and propagate, receiving only cursory attention from researchers and museum visitors. A register of different figurations of loss then – defeat, destruction, removal, extinctions, banishment, forgetting – that also points to the devastation of relations, interdependencies and collectives.
>
> *Text from the information panel accompanying the seed display and installation*

Despite or rather because of the museum's accumulations, it is a place filled with absences. The dead bodies that fill its halls and shelves continuously evoke what Butler has called 'constitutive outsides', the 'excluded sites' that make those present matter while containing those that don't away from sight (Butler 1993: xvii). Such outsides also haunt the neat narrative of discoveries, earth history and evolution that presents 'ideas of a knowledge at once *positive* and *comprehensive*' (Richards 1993: 6, emphasis in the original). Ruinations can be understood as the material reverberations of enforced presences, lingering absences and persistent Otherness. They are differently sensible across domains and sites and this requires, compels even, a collective effort in address and redress. For Stoler, one of the tasks of postcolonial practice is attending to the 'distinctions between what holds and what lies dormant, between residue and recomposition, between what is a holdover and what is reinvested, between a weak and a tenacious trace' (Stoler 2013a: 12). And this is also a question of form insofar as it is through form, which pertains to the how of telling/writing, that we can manage (or not) things as present *and* as not-present (Verran 2001). Thus, Arondekar's challenge of juxtaposing the archive's fiction-effects and truth-effects, of turning the archival object into a 'recalcitrant event' that refuses simple access (Arondekar 2005: 22), is not only an intellectual one. It is also always a problem of form and genre.

This is why a project such as the Doomsday Vault, that posits the seed-as-archive, not only fails practically but more worryingly perpetuates a logic of ruination. Storing all the world's seeds will not do if there no longer is any place to grow them, or if the microbial communities in the soil have been harmed, or if planting and plant-tending practices have been erased and their people destroyed or if Svalbard falls prey to the mounting geopolitical tensions in the Arctic Circle or to industrial accidents of the extractive industries stationed there. Another way of putting this is that the seed is and is not a whole, is and is not a part. The fallacy of its banking lies in always settling on one and thus forever excluding the others. This is also true for the efforts that are currently being mobilized around seeds as part of planning Africa's 'green revolution',[12] in the name of (food) security and in expanding the integration of biotechnologies, such as Bayer's recent acquisition of Monsanto, which are part of the unravelling aftermaths of colonial botany, its institutions and ecologies.

Natural history museums have to forgo the cheap chills of taxidermy and take seriously their ghosts. Haunting as Gordon notes produces 'something-to-be-done' as it is 'a contest over the future, over what's to come next or later' (2011: 3). Hauntings thus evoke truly frightening questions for museum collections imbricated in imperial and colonial practices and orders. This is because the 'something-to-be-done' might indeed entail restitutions, returns, dispersion or in any case a radical re-imagination of purpose, away from endless accumulation and infinite preservation.

## Notes

1 Despite its near-Arctic location, the vault is not quite cold enough to ensure the permafrost conditions necessary for seed storage. Sub-zero temperatures are guaranteed by the extensive cooling system run on the power infrastructure left behind by the island's extractive industries.

2 The facility was set up and is maintained by funds from the Rockefeller Foundation, Monsanto and the Bill and Melinda Gates Foundation. The first two were important actors in India's green revolution that is now, with the help of the Gates Foundation, being extended to Africa.

3 The returns promised by managing knowledge of the world's seed stocks, including old (heritage) cultivars, are considerable.

4 See www.potatoperspective.org/

5 The MfN is situated in Invalidenstrasse 43 which is only about 200 metres from where the Berlin Wall in East Berlin stood. This location is reflected in the collections which disproportionally feature specimens collected from former Eastern Bloc countries as well as affiliates (for example, Cuba).

6 Material from Botanischer Garten und Botanisches Museum Berlin-Dahlem (Botanic Gardens and Botanical Museum Berlin-Dahlem), Freie Universität Berlin (Free University Berlin), Ibero-American Institute Berlin, Image Collection of the German Colonial Society, Frankfurt University Library, Kew Gardens, Der Palmenmann™, Seed Area™, and eBay™.

7 The Humboldt Lab in Berlin, a series of interventions in the *Ethnologisches Museum* and the *Museum für Asiatische Kunst, Staatliche Museen zu Berlin* (Ethnological Museum and the Museum for Asiatic Art, State Museums of Berlin) included object biographies. There is also an object biography project at the Pitts-River Museum, Oxford. See http://web.prm.ox.ac.uk/rpr/index.php/objectbiographies/.

8 It is not surprising that the notion of the 'boundary object' (Star and Griesemer 1989) emerged from research in a natural history museum (Berkeley's Museum of Vertebrate Zoology).

9 The Museum's centrepiece, the *Brachiosaurus branci* (the world's largest mounted dinosaur) was taken from what was then German East Africa (now Tanzania) in the course of the so-called Tendaguru Expedition (1909–1911) that excavated and shipped to Berlin over 200 tons of dinosaur bones. The expedition took place after the Herero and Namaqua genocide committed by German troops between 1904 and 1907 that had cleared vast swathes of land for such extractive programmes. Many of these histories are being reconstructed in the research project 'Dinosaurs in Berlin! The Brachiosaurus Brancai as an Icon of Politics, Science, and Popular Culture' led by Ina Heumann (MfN).

10 References to this office and, more generally, colonial entanglements of German botanical institutions are scant although Katja Kaiser has been doing important work to make these histories present again (Kaiser 2015). The office, founded in 1891, together with the herbarium and library of the Botanic Gardens and Museum were destroyed in the Second World War and its experimental stations disbanded or taken over by Allied forces. The experimental station in Amani, for example, was 'revived' by the British and continued agricultural and other experiments (Conte 2002). What documents survived in relation to the office's work in Berlin are periodicals produced and distributed for the benefit of German settlers and farmers in the colonies as well as four boxes containing the estate of a German colonial botanist who had worked at the experiment station in Amani.

11 The museum was designed to work with natural light rather than electric lights, hence huge windows line the walls.

12 See the recent African Green Revolution Forum, 5–9 September 2016 in Nairobi.

# References

Ahmed, S. (2007). Multiculturalism and the promise of happiness. *New Formations*, 63(3): 121–137.

Armah, A. K. (1988, May 23). Seed time. *West Africa*, p. 926.

Armah, A. K. (1998). Seed time. In U. Beier and G. Moore (Eds.) *The Penguin Book of Modern African Poetry* (p.115). London and New York: Penguin Books.

Arondekar, A. (2005). Without a trace: sexuality and the colonial archive. *Journal of the History of Sexuality*, 14(1): 10–27. doi:10.1353/sex.2006.0001.

Arondekar, A. (2009). *For the Record: On Sexuality and the Colonial Archive in India*. Durham, NC: Duke University Press.

Bennett, T., Cameron, F., Dias, N., Dibley, B., Harrison, R., Jacknis, I. and McCarthy, C. (2017). *Collecting, Ordering, Governing: Anthropology, Museums, and Liberal Government*. Durham: Duke University Press.

Bowker, G. C. (2006). *Memory Practices in the Sciences*. Cambridge, MA: MIT Press.

Bradley, H. (1999). The seductions of the archive: voices lost and found. *History of the Human Sciences*, 12(2): 107–122. doi:10.1177/09526959922120270.

Butler, J. (1993). *Bodies that Matter: On the Discursive Limits of 'Sex'*. New York: Routledge.

Conte, C. A. (2002). Imperial science, tropical ecology, and indigenous history: tropical research stations in Northeastern German East Africa, 1896 to the present. In G. Blue, Martin P. Bunton and Ralph C. Croizier (Eds.) *Colonialism and the Modern World: Selected Studies* (pp. 246–263). Armonk, NY: Sharpe.

Faulkner, W. (2011[1951]). *Requiem for a Nun*. Vintage internat. ed. New York, NY: Vintage Books.

Food and Agriculture Organization of the United Nations (2014). Genebank Standards for Plant Genetic Resources for Food and Agriculture. i3704e. Revised Edition. Rome: FAO. www.fao.org/docrep/019/i3704e/i3704e.pdf.

Gordon, A. (2008). *Ghostly Matters: Haunting and the Sociological Imagination*. New University of Minnesota Press ed. Minneapolis: University of Minnesota Press.

Gordon, A. (2011). Some thoughts on haunting and futurity. *borderlands*, 10(2). Retrieved 1 March 2018 from: www.borderlands.net.au/vol10no2_2011/gordon_thoughts.htm

Gosden, C. and Marshall, Y. (1999). The cultural biography of objects. *World Archaeology*, 31(2): 169–178. doi:10.1080/00438243.1999.9980439.

Gray, R. (2011). Five Mexican pots: Are we potty? 5 Lonesome Pots try 5Y Analysis, Rethinking Pitt-Rivers (blog). Retrieved 1 March 2018 from: http://web.prm.ox.ac.uk/rpr/index.php/object-biography-index/7-farnhamcollection/319-five-mexican-pots/

Hughes, C. and Lury, C. (2013). Re-turning feminist methodologies: from a social to an ecological epistemology. *Gender and Education*, 25(6): 786–799. doi:10.1080/09540253.2013.829910.

Kaiser, K. (2015). Exploration and exploitation: German colonial botany at the Botanic Garden and Botanical Museum Berlin. In D. Geppert and F. L. Müller (Eds.) *Sites of Imperial Memory: Commemorating Colonial Rule in the Nineteenth and Twentieth Centuries* (pp. 225–242). Studies in Imperialism. Manchester: Manchester University Press.

Kopytoff, I. (2013). Cultural biography of things: commoditization as process. In A. Appadurai (Ed.) *The Social Life of Things: Commodities in Cultural Perspective* (pp. 64–91). Cambridge: Cambridge University Press.

Livingstone, D. N. (2003). *Putting Science in Its Place: Geographies of Scientific Knowledge*. Chicago, IL: University of Chicago Press.

Richards, T. (1993). *The Imperial Archive: Knowledge and the Fantasy of Empire*. London: Verso.

Spivak, G. C. (1985). The Rani of Sirmur: An essay in reading the archives. *History and Theory*, 24(3): 247–272.

Star, S. L. and Griesemer, J. R. (1989). Institutional ecology, 'translations' and boundary objects: amateurs and professionals in Berkeley's Museum of Vertebrate Zoology, 1907–39. *Social Studies of Science*, 19(3): 387–420. doi:10.1177/030631289019003001.

Stevenson, A. (2011). Egyptian boat: 1884.81.10, Rethinking Pitt-Rivers (blog). Retrieved 1 March 2018 from: http://web.prm.ox.ac.uk/rpr/index.php/object-biography-index/1-prmcollection/322-egyptian-boat/

Stoler, A. L. (Ed.) (2013a). *Imperial Debris: On Ruins and Ruination*. Durham, NC: Duke University Press.

Stoler, A. L. (2013b). 'The rot remains': from ruins to ruination. In *Imperial Debris: On Ruins and Ruination* (pp. 1–37). Durham, NC: Duke University Press.

Subramaniam, B. (2014). *Ghost Stories for Darwin: The Science of Variation and the Politics of Diversity*. Urbana, IL: University of Illinois Press.

Verran, H. (2001). *Science and an African Logic*. Chicago, IL: University of Chicago Press.

# 5

# Dirty methods as ethical methods?

## In the field with 'The Cultural Politics of Dirt in Africa, 1880–Present'

*Stephanie Newell, Patrick Oloko, John Uwa,*
*Olutoyosi Tokun, Jane Nebe, Job Mwaura,*
*Rebeccah Onwong'a, Ann Kirori and Claire Craig*

### Abbreviations and acronyms

Dirtpol – The Cultural Politics of Dirt in Africa, 1880–present
ERC – European Research Council
FGD – Focus Group Discussion
LAWMA – Lagos Waste Management Authority
PC – Project Coordinator (Claire Craig, Sussex University)
PI – Principal Investigator (Stephanie Newell, Yale University)
PR – Project Researcher (John Uwa, Olutoyosi Tokun and Jane Nebe at the University
of Lagos; Job Mwaura, Rebeccah Onwong'a, Ann Kirori at the British Institute of East
Africa, Nairobi)
RC – Regional Coordinator (Patrick Oloko, University of Lagos)

In 2013, 'The Cultural Politics of Dirt in Africa' project (Dirtpol) set out to understand practical as well as cultural, political and historical aspects of urban living through people's perceptions of waste management, public health, migration, public morality, environmental hazards, neighbourliness and town planning in two African cities: Lagos and Nairobi. We wanted to find out about local understandings of 'dirt' – a term we chose for the diverse numbers of African-language words, phrases and connotations it generated in translation, as well as for its own rich array of connotations – as an entry-point into people's responses to urbanization and the environment. In examining local and transnational concepts of dirt from the period of European colonial expansion in the 1880s through to the present day, the project positioned contemporary media and public health debates in relation to the two cities' long histories of intercultural encounters with the Global North and other parts of the world.[1] Interdisciplinary at all levels of enquiry, the aim was to historically contextualize and compare wider policy issues relating to public health, urbanization and community relations in African cities, as well as to position these policy issues in relation to the media and public opinion. At its core, the project asked how public opinion is shaped, including the opinions of one particularly neglected section of the 'public': children and young people. The six Project Researchers (PRs), who came from disciplines as diverse as literary studies, media studies, education, public health, environmental and biological sciences, focused their work on how particular urban spaces came to be regarded as dirty or as full of dirt, and how certain objects and subjects came to be labelled using categories related to dirt. We asked: how do ideas about dirt shape local perceptions of the urban environment, and in what ways do the media contribute to the formation of public opinion?

This project was 'lo-fi' in terms of hardware. Each team member was given a digital voice-recorder and plug-in microphone, and a laptop with the relevant programs installed, including NVivo for classifying data, plus word processing, spreadsheet and transcription software. In three weeks of intensive research training in the UK provided by NatCen Social Research and tailored to the requirements of the project, we learned about – or relearned in the case of the researchers with social science training at Master's level – qualitative research and fieldwork skills such as in-depth interviewing and focus group organization. As part of this short training period, we also attended a conference together, drew up a project timeline and work packages, and agreed deadlines and key goals. Our objective was to ensure mutual understanding within the team about research methods and ethical guidelines, as well as to build a collective identity that would help to maintain coherence when we returned to our respective countries. Training also included techniques for obtaining informed consent and for recruiting children and young people as participants.

Working together in this three-week period helped us to understand that the project depended upon collaborative processes of knowledge exchange and regular communication between team members, rather than knowledge transfer up, or down, a pyramidal management structure. In particular, given the importance to the project of African-language work across multiple language groups and urban socioeconomic contexts, and given the European Research Council's (ERC) removal from the budget of the large tranche of money requested for professional transcription and translation services, we debated the methods best suited to translation and transcription, and agreed that the labour required for these tasks necessarily also positioned the researchers as interpreters – as well as collectors – of data. The challenges of transcribing and translating with insufficient training are discussed below.

For the duration of the project we held regular fortnightly online meetings using a video conferencing package. Problems with connectivity caused frequent interruptions to these team meetings: rarely did all six PRs successfully sustain a full conference call. Even without these technical difficulties, however, meaningful communication between such a large team spread across three different regions of the world with diverse disciplinary backgrounds was not feasible through conference calling alone: six individual researchers generally had six separate sets of issues to discuss. Substantial discussions with the team leaders took place through the feedback on the fortnightly reports submitted by each PR using a template that included sections on the reasons for the success or weaknesses of particular methods, reflection on problems encountered in the field, details of individual and collective efforts to resolve these problems, and details of ongoing or new research support needs.

An extract from a report submitted by the PR for Health and the Environment in Nairobi in June 2014, during the three-month pilot period, gives a flavour of the issues raised in our fortnightly reports:

> In the last fortnight, I started all over again building new networks in Kangemi. This is because the networks I had previously built collapsed. The private garbage collector with whom I had made connections seemed to be too busy for me. Kangemi slums[2] are located along Waiyaki Way in Westlands Area, Nairobi. The slum is subdivided into smaller villages.

I focused on a village called Gichagi. The village is located close to a high-end neighbourhood called Mountain View. The contrast between the two neighbourhoods is clear. On one side of the wall, there are small houses made from iron sheets. Open sewers run along the muddy footpaths. In contrast, the other side of the wall has modern brick houses, with well-manicured lawns.

Gichagi area has many small wetlands. The open sewers, which run along the footpaths, drain into the swamps. I saw children in the wetlands and I thought they were playing. However, a local told me that they were fishing. When it rains, some areas in the slums are completely cut off and the residents have to use alternative routes. There are no *matatus*[3] plying the alternative routes. This means that during the wet season, I have to walk for almost 45 minutes to get to the village.

I had prepared to try out my topic guide through an in-depth interview, but when I reached the site things changed and I had to adapt to the situation. I met five women sitting by the road side. I introduced the project and myself. They agreed to participate in the study, but they refused to be interviewed separately. They insisted on having the discussion together. I did not have a room for a focus group discussion (FGD), but because it was not easy to get volunteers I decided to hold the discussion by the roadside where I found them. The place was wet as it had rained the previous night. A key factor in my decision to hold the discussion there was to accommodate one of the ladies, who was selling sweets and biscuits by the roadside. The other four women had young children with them. There was nobody they could leave their children with so that they could fully concentrate in the interview. Despite these potential distractions, I decided to give it a try. The FGD went well. However, the interruptions from children and customers made me forget some questions and consequently, I did not prompt and probe well. When I listened to the audio recording that evening, I realized I did not ask some questions based on the participants' answers.

After the discussion, the women promised to introduce me to other ladies in the neighbourhood but on the second day, no one was willing to talk to me. I conducted four in-depth interviews in the area. The interviews went on well and I hope they will get better as time goes by. It is a wet season in Nairobi and, as I mentioned before, it takes a long time (traffic in the city and long walking distances) to access the site so I can only manage to conduct one or two interviews in a day. Also, because of the rain, I have to leave the sites early.,

*Rebeccah Onwong'a, Report, 13 June 2014*

Researchers' fortnightly reports often included detailed observational commentaries of this kind, furnishing important information about the contexts and power dynamics shaping the collection of audio data in neighbourhoods where our ethical guidelines prohibited us from taking photographs.

Many similar obstacles to those described above – 'obstacles', that is, if the models furnished by our training programme were to be regarded as 'ideal' – arose in the environments in which the researchers worked, necessitating creative thinking and negotiation, and generating considerable commentary and reflection in the fortnightly reports. Our work with diverse urban communities helped us to think about the challenges of standard research methods in cross-cultural contexts. At one Lagos primary school, for example, socioeconomic factors prevented our plans to schedule follow-up interviews with participants several months later and also impacted on the categories the PI had proposed for the analysis of different groups of participants. In her fortnightly report, the Nigerian PR for Education and Schools wrote:

I am very conscious of the fact that I am working with pupils in a state-owned primary school. In these schools, tuition is free and books are supposedly provided for free, to an extent. It appears that most of the thirteen- to fifteen-year-olds in this primary school do not stay with their biological parents. Usually, this age group should be in the secondary school. Some of the teachers have mentioned that a lot of the pupils in the school are domestic house-helps brought from the rural areas. They also observed frequent drop-out rates due to these domestic house-helps returning to where they came from, changing guardians, or merely [being removed from school] according to the self-centred whims of their employers. If there are instabilities and relocations as indicated above, it would not be surprising to find such overage children in a primary school. It is therefore important to contextualize the data within assumed socioeconomic status dimensions.

*Jane Nebe, Report, 13 June 2014*

Overage children, such as those described above, became a 'problem' because our project wished to work with, and compare, two specific age-groups of children and young people: children aged seven to nine and young people aged 13 to 15. At the proposal stage, the British-born PI had not questioned her assumption that the first group would be found exclusively in primary schools, and the second group in secondary schools. Furthermore, as the Nigerian PR for Education and Schools pointed out, 'There have been incidences of differences between the age in the school records and the one that the pupils declare for themselves during the interviews' (Ibid.). On one occasion, 'a pupil whose year of birth in the school records was indicated as 2005 insisted that he was twelve years old' (Ibid.). Obviously, the presence of a 12-year-old in an in-depth interview or FGD designed for seven- to nine-year-olds would distort the content. In this case, and in principle, the researcher chose to accept the information provided in school registers rather than the pupil's self-declared age, and so this participant was included in the study. Such a principle, however, required sensitive handling as it risked generating feelings of embarrassment and shame in the pupils concerned if they were simply ejected from a group of participants. Jane Nebe also had to address the problem of forged signatures on some of the forms submitted by pupils who were eager to participate in the project, but could not obtain parents' or guardians' consent as signatories:

I observed irregularities in [two of] the signatures, which aroused my suspicion. After much probing and cajoling, they confessed and explained the circumstances surrounding their decision to forge the signatures. The fourteen-year-old pupil lived with different guardians, while the mother of the eight-year-old had travelled. It was not until I gave them new parental consent forms the third time, did I get properly signed forms from a guardian and a parent, respectively.

*Report, 13 June 2014*

There is a large and growing literature on techniques for research involving children, including methods for encouraging them to talk, and considerations of how adult researchers 'can understand the child's world' and 'free him/herself from the adult-centered perspective' (Kyronlampi-Kylmanen and Maatta 2011: 87; see also Thomas and O'Kane 1998; Drotner and Livingstone 2008; Stanley and Sieber 1992). With these considerations in view, we decided to adopt a competency-based method that employed children's and young people's abilities and skills, including drawing and painting ('draw-and-talk' techniques), diaries following particular urban themes, dramatizations and performance workshops, essays on set topics, recorded (voice only) group interviews and structured, recorded (voice only) one-on-one interviews.

The emphasis in our project was not so much upon children per se, as upon schoolchildren in relation to their peer groups as interpreters of public opinion. Children learn social and cultural information at school as well as at home and within the community, and schools are vectors for children's and young people's perceptions of urban identities in multicultural educational contexts. As such, the focus of our research was upon young people's perceptions, representations and experiences of urban environments and broader issues concerning health, consumption and recycling.

Children, however, are often considered secondary, or inferior, to adults (teachers, parents and gatekeepers) by researchers involved in data collection. The PR for Education and Schools in Lagos offered a compelling description of the space she opened up for young people and illustrated how our project emphasized the necessity 'to consider the children themselves as research subjects' (Kyronlampi-Kylmanen and Maatta 2011: 87):

> I had great focus group discussions with the thirteen- to fifteen-year-olds. It was like a normal peer group, but this time you have a topic for them to discuss, so it was quite interesting to watch the dynamics. You know, it's very uncommon to do this research where you allow these young people to say their mind: it's not a normal thing, especially in our context. It's not common to have them say their mind about issues. It was quite interesting to get them to *speak*, you know, speak through interviews and focus group discussions, without fear, in a safe space for them to say their mind, as they felt. That was my experience in the field.
>
> *Jane Nebe, Interview, 3 May 2016*

With the seven- to nine-year-olds, however, a different approach was required:

> Now, for those who were aged seven to nine, the option I had in mind initially was ask them maybe to draw their thoughts, and the other option was the [one-on-one] interview. I started with public schools, and I found that the option of drawing or painting their thoughts was an unfamiliar platform for them to express their thoughts or opinion about the issues I had an interest in. Even the private schools I worked with, and the middle-class public school I worked in, the idea of using paint to express their thoughts was not very effective. It was not something they tried to pick up, it was not an option that worked. So for those schools I focused on one-on-one interviews with them. I got them to tell me things from a familiar place, tell me what they like, tell me what they don't like, so it was just the face-to-face conversations.
>
> *Jane Nebe, Interview, 3 May 2016*

As these comments demonstrate, activities such as artwork projects cannot be adopted regardless of context and competency, and must be used cautiously and reflexively. One cannot assume that all children can express themselves through the medium of drawing. The PR rightly abandoned the paint-box because the medium inhibited, rather than facilitated, the children's self-expression.

We operated on the principle that children absorb and reflect upon socialization processes and political currents in their communities, and they often do so through knowledge- and information-exchange in school environments (Punch 2002; Nesbitt 2000). Working with children and young people in a wide variety of educational settings, the PRs for Education and Schools were keen to emphasize that it was necessary to adapt our formal research methods to each school environment. Methodological 'innovations' were off the agenda. 'I cannot see anything that could be considered an invention', Jane Nebe explained. Rather:

I had to plan my research work around the school calendar, and then around the school day, and around the weekly timetable, and around the school events, because when there are school events you can't do so much. Then I had the problem that the research period, when you could get the most data, was a very noisy period, during the break, you know how our children play quite noisily, and usually where I held the interviews are not far away from where the children are playing. I would always have interruptions, I would always have distractions arising from the children playing a lot. So a lot of my issues were about logistics, trying to see how I could fit myself around the children, when they had a free period or maybe something else is going on somewhere, and then you move away from where those things are. So that was the kind of adaptation I had to do.

*Interview, 3 May 2016*

We adopted the following core principles in our work with children, drawn from the work of educationists in countries in the Global South and Africanist researchers who work with children and young people as participants: careful identification of and discussion of the research with gatekeepers (see Tindana, Kass and Akweongo 2006); appreciation that the terms 'children' and 'young people' are inadequate to describe the multiplicity of perspectives, skills, interests and experiences of participants (Moses 2006: 5); training of the researchers on how to deal with the possible disclosure of information of a personal nature during the in-depth qualitative interviews; training of the researchers to recognize signs of anxiety, stress or humiliation among participants and to address these with sensitivity (France, Bendelow and Williams 2000); deletion from the database of incidental disclosures and material not relating to our structured questions; a guarantee of confidentiality to the participants and full anonymization of the results according to international and national protocols (for further discussion of this, see below).

In contrast to the often over-enthusiastic Nigerian schoolchildren, obtaining signed informed consent forms was especially difficult in Nairobi. In the final months of her work for the project, Ann Kirori, the Kenyan PR for Education and Schools, was compelled to return to many of the schools in which she had worked to retrospectively collect missing informed consent forms, arousing the mistrust of teachers and administrators who were sceptical about her reasons for requiring such a paper trail. 'I found that it was not very easy to get somebody to sign a paper', Job Mwaura, the Kenyan PR for Media and Communications, said in an interview with his Nigerian counterpart, John Uwa:

> although some participants would openly agree, others would say, 'I don't want to sign anything. Just ask me the questions. I will respond. Whatever is personal I will not answer, but whatever the answer I will not sign the consent'.
>
> *Interview, 26 May 2016*

Asked to speculate about why this problem arose, he suggested that many people were suspicious of putting their name to paper because 'they don't know what they are going to get into when they just sign that paper' (Ibid.). Other members of the Nairobi team gave up completely on attempting to obtain written consent forms to accompany the oral consent recorded at the start and end of interviews:

*Job Mwaura*: The ERC required every participant to sign an informed consent declaration. How did you go about ensuring that?

*Rebeccah Onwong'a*: Unfortunately, I did not. At the beginning of the discussion, I just asked them: 'Would you allow me to record the discussion?' If you listen to the audio recording of the discussion, you will hear the voices saying, 'Yes I agree, yes I agree, yes I agree'. That is what I relied on. You know how our people work. They are suspicious of anything and everything, so you have to reassure them. I thought signing would cause a problem, so I left it out. I know it is a requirement, but I didn't do it.

*Rebeccah Onwong'a, Interview, 28 May 2016*

Of all the ERC's ethical stipulations about research methods and procedures, obtaining written informed consent caused the most difficulties to the researchers in Kenya, even with our provision for orally recorded consent to be witnessed in writing by an independent third party and held in a secure database for sampling.

In low-income neighbourhoods of Nairobi, we encountered considerable reluctance and refusals to participate in one-on-one interviews, especially on the part of women. People were suspicious, and when in-depth interviews were secured, participants 'just said "yes", "no", and even if you probed further, you just got a "yes" or "no", and short sentences' (Ibid.). No sustained data could be obtained in low-income settlements in Nairobi using one-on-one interviews. Participants were unwilling to risk accusations of 'secrecy' by fellow community members if they entered a closed space alone with the facilitator. Indeed, the very assurances of confidentiality we offered at the outset of the interview aroused suspicion on the part of participants about the content and purpose of our questions, especially when accompanied by the appearance of an audio recorder and an informed consent form.

Viewed from the perspective of 'the field', these problems arose as a direct consequence of the tensions between, on the one hand, the ethical standards regarding confidentiality agreed in the research proposal and, on the other hand, culturally specific notions of appropriate and inappropriate behaviour and communication in the townships. As with several of the other examples given in this chapter, our research methods were often caught in-between incompatible principles about confidentiality and disclosure in the presence of strangers.

This reluctance to be isolated in a closed space with the PR also arose in particular types of focus group discussion in Nairobi: for example, after repeated instances of male domination of FGDs and women's silence in mixed groups in Kibera, the PI suggested hiring a room at a local community centre outside the entrance to the township, in which a women-only group could meet. We would use the same topic guide, familiar to everybody, but the space would exclude the more dominant members. This would be followed up by FGDs with all other interested parties, divided according to age, gender and other criteria, to be held at the same venue. Nobody would be excluded from the FGD experience, but each group would be governed by our selection criteria.

For several reasons, this idea was found to be unworkable. As one PR recalled,

It was difficult to get a room in those low-income areas where you could conduct your interviews, and the rooms that were available [for hire] were very far away, so it would be very inconveniencing to tell people, 'Now, thank you for agreeing to participate in the interview, let's go somewhere else about two kilometres away to have our discussion there'.

*Rebeccah Onwong'a, Interview, 28 May 2016*

The situation was compounded by the fact that other members of the community would accuse participants of gossiping – or worse – if they met behind closed doors in the manner the PI had suggested.

As the project progressed, the 'Dirtpol' blog, managed by the PC, emerged as a space in which we could discuss challenges such as the ones described above in a relatively informal, yet public-facing, environment. As a form of communication that did not require a scholarly, analytic mode, the blog allowed us to showcase the disparate elements of our work as it progressed, and provided a vital forum for comparing ideas about methodology and content. With a rotating structure for blogposts generated by the PRs and managed by the PC, the blog showcased examples of methods-in-action, and provided a vital toolkit for team members to use in their own fieldwork. The blog brought the team together as readers and commentators on the project as a whole, as well as creating a public interface through which 'Dirtpol' could consolidate its identity: through it, the PRs became interpreters of the research process as well as producers and interpreters of data.

In a blog posted on 28 April 2015, the PR for Health and the Environment in Lagos, Olutoyosi Tokun, offered a careful reflection on her work for the project, describing what she had learned and understood about the project as a consequence of her interactions with interviewees and other participants:

My Dirtpol journey has been an eventful one. At the start of the project it was quite challenging to get study participants to speak about dirt beyond hygiene and sanitation. I had to do some digging on social media and make a list of behaviours, situations and trends that have been described as dirty. This list seemed to help facilitate conversations. From that point the list grew longer and longer as the project progressed.

In the first few months of the project I established communication with a community known as Dustbin Estate, located in Ajegunle, Lagos. The community is so-called because it was built on a refuse dump. Interviews conducted with residents of this community have been quite revealing; we talked mostly about human interactions within the community. These kinds of discussions have become relevant considering the rate of urbanization in Lagos.

After obtaining approval from the Lagos State Waste Management Authority (LAWMA), I visited the main landfill site in Lagos and other waste management facilities operated by LAWMA. I also interviewed private waste collectors; it was interesting to know how they felt about their job and the attitude of Lagosians towards them as they carry out their duties.

Discussions with public health users and providers have been very rich. Issues such as sexuality, polygamy, the activities of traditional birth attendants in certain areas, skin bleaching, and individual beliefs that influence health-seeking behaviours have been covered. I have been able to gather very diverse opinions on these issues.

Leading up to the general elections in Nigeria that held on the 28th of March and 11th of April, 2015, a lot of data was collected in the form of election campaign materials. It was not unusual for one political party to accuse another party of playing dirty politics. Also certain campaign strategies were simply described as 'disgusting' and 'repulsive'.

I was surprised to find that the labelled recycle bins provided on campus at the University of Lagos in order to encourage the separation of recyclable waste were not being used appropriately. Hence a quick survey was conducted on campus to investigate the situation. I interviewed Environmental Health Officers, as well as students on campus, and the phrase 'The Nigerian Factor' kept coming up as the reason for the situation. It would help to know what 'The Nigerian Factor' is all about; meanwhile one of the Environmental Health Officers interviewed on campus told me all about the role of sanitary inspectors during the colonial era. We touched on the 'White man's plague' and other issues prevalent during the colonial era.

Given that the project involved six researchers and other personnel in three separate countries, to ensure the stability of fieldwork across the continent, very few novel or unconventional methodologies were encouraged in the pursuit of data. The following example of an adaptation of the conventional FGD illustrates the ways in which the research teams on this project were compelled by local circumstances to adopt alternative methods, the manner in which the blog facilitated discussion of these innovations, and the drawbacks as well as the potentialities, of such initiatives.

In a blogpost on methodology, one of the Kenyan PRs described how her FGDs in low-income areas required different skills from those we learned during our research methods training in the UK. In one low-income neighbourhood of Nairobi, FGDs could not be physically contained in the controlled environment of a room with a door, and discussions were frequently attended by more people than could possibly contribute to the session. In training, we had agreed that our ideal-sized focus group was six to eight people, but FGDs in low-income communities were never this small. They generally attracted as many people as would fit into the room provided by the local host, with onlookers poised outside the open door to enter whenever a person left. People constantly wandered in and out of FGDs, cell phones rang, babies cried, participants were summoned out of the room on business, only to return ten minutes later, and on every occasion the owner of the space, as host, was given first refusal by the group to set the tone of discussions and debates. Such space-owners were always men of status in the community, the gatekeepers with whom social convention required us to work closely to convene each meeting. Ann Kirori's blogpost captures the combination of public curiosity and individual mobility that characterised these sessions:

> On my part I would say 'participant exchange' during a focus group discussion has been my remarkable experience. What I mean by this is that: you start a FGD with ten participants and after exhausting the first topic, one or two of the participants leave the group and are replaced by another participant who is new and quickly gets absorbed into the discussion. This then means you might end up with different people from the ones you started with. In my case I only had two constant participants in an FGD of 13 people. The other 11 kept exchanging and new ones coming. The most interesting bit is that it never affected the quality of data and the discussion became more exciting as we progressed. The discussion was mainly about current and emerging issues such as Ebola, teenage pregnancy, culture degradation, people's lifestyle changes/behaviours just to mention a few.
>
> *6 November 2014*

With outsiders walking in, and insiders walking out, the already oversized focus group environment was anything but containable. As the blogpost explains, this mobile type of FGD could be relabelled a 'Participant *Exchange* Focus Group' to account for the PR's perception that, rather than introducing confusion, 'the quality of data and the discussion became more exciting as we progressed'.

The two other Kenyan PRs experienced the same type of mobile, flexible, uncontainable FGD in low-income neighbourhoods of Nairobi. Rebeccah Onwong'a described the research environment:

> We would sit in the open and the discussion would start with a group of six people. As you know, most people in low-income areas don't have anything to do; most of the time, they are just sitting along the pathways, having a good time with friends, idling, or waiting for something to 'come up', meaning a call from a friend alerting them that there is a job

somewhere. The discussion could start with six participants. And then somebody else just comes along, stands and listens to what we are talking about, and then joins. By the time you realize that there is a new addition to the group, this person has already contributed to the discussion. Then another person would join. So you can start the discussion with six participants, but end up with 21 participants! So that's the cumulative focus group discussion, and it's really difficult to control people.

I remember one time in Kangemi I found a group of six people. Some were working in a shop, and they all agreed to take part in the study. A few minutes into the discussion, other people started coming along and listening. I paused my recorder, just to tell them 'Please don't contribute to the discussion. We'll interview you later'. They would oblige, but when we were talking about sensitive or interesting issues, they couldn't contain themselves. They just joined. So it became a mess. Controlling the crowd was not easy.

I remember one particular man said, 'I have to speak before I leave. I don't have any time. I cannot wait to be interviewed another time. Just allow me to speak'. Considering how difficult it was to get participants, I let him speak. Today you might get people, but tomorrow everybody might refuse. Those frustrating days! So I allowed them to participate. For me that worked well, because then people are more free, and they talk about a lot of things.

*Rebeccah Onwong'a, Interview, 28 May 2016*

The Lagos team also reported 'welcome intruders' whose presence necessitated adaptations to the research methods developed in training, but in Nigeria there was nothing on the scale of the 'cumulative FGD' experienced in Nairobi. Interviewing his Nigerian colleague Olutoyosi Tokun in May 2016, John Uwa, the PR for Media and Communications in Lagos, remembered the blogpost about 'participant exchange focus groups' in Nairobi:

*John Uwa*: Do you remember the blog where Ann Kirori talked about participants who were not scheduled to participate and had to smuggle themselves somehow into the room?
*Olutoyosi Tokun*: [*laughs*] Passing the baton!
*John*: Have you had any experience relating to that?
*Olutoyosi*: Yes, I have been there, but it wasn't with the focus group discussion. It was with just the regular one-on-one interviews. Somebody comes along with his friend, or I'm conducting the interview in his office, and there's somebody else sitting in that office, and the person cannot help but voice his own opinion about things.
*John*: When you find yourself in situations like that, with an intruder, do you respond to it?
*Olutoyosi*: Yes. You have to. I am very open to it. The more the merrier, I think, although of course it really changes the standard. It is supposed to be a one-on-one interview, but, hey, the overall aim of the project is gathering information and opinions, and usually that kind of chipping-in helps because it also – I will use the Lagos slang here – *gingers* the original interviewee to keep speaking [*laughs*]

*Interview, 11 May 2016*

In the first annual appraisal of the project in December 2014, the ethical auditor on our advisory committee terminated the 'Cumulative FGD', 'Participant Exchange Focus Group', and other similarly flexible methods we had developed in the field. She explained that the objectives and outcomes of the research had not been explained to every individual present in one of our cumulative groups, and unless we were willing to halt the discussion every time a new person entered the room in order to re-read our introductory statement describing the project objectives and participants' rights, we could not claim to have obtained informed consent from all parties. Moreover, in the 'exchange' format, we had no signed consent forms from those participants who had left and not returned during the session. Data from these FGDs were therefore deemed unusable according to the ethical standards that we had agreed with the ERC and participating institutions at the outset of the project and discussed at length in training.

As part of the audit, further problems were identified in relation to disclosures on the blog. In one set of photographs, the PR for Education and Schools in Nairobi was celebrated by children and teachers at a secondary school in Kibera as their 'Guest of Honour' after a term of research involving creative projects and extra-curricular activities. The class and teachers dressed the PR in a school uniform for fun, and asked her to give a speech as part of their 'cultural day' on the theme of environmental conservation. The spirit of celebration in the photographs was clear, and we unthinkingly posted the images on the blog alongside the PR's enthusiastic 'end of term' report which included several teachers' and students' first names. The report also named the school. While the photographs did not name the teachers or children who featured in them, the school uniform was clearly visible, and their unpixelated faces caused a serious breach of confidentiality. As a consequence, we were advised to withdraw all research associated with the school from the data pool, even though the material had been anonymized according to protocol. The fact that the images and report had been published online meant there was a risk that individual children might be identified from the anonymized transcripts.

Reflecting on the divergence in ethical standards between the stringent ethical criteria we were required to follow and material freely available on the Internet or published in national newspapers without concern for anonymity, the Nigerian PR for Media and Communications remained frustrated at what appeared to be double-standards based on geographical location:

> We cannot snap photographs of people without their consent. And some days you have to take photos of a marketplace to portray what you are talking about, and you discover the ethical concerns of the project are prohibitive of snapping without getting consent. Am I going to get consent from the whole lot of people I see in the market?

He continued:

> Even our newspapers, our media houses, don't follow all of these consents, these *ethics*! They snap freely. What is ethical for the European Research Council may not necessarily be ethical for our media houses here. They put all manner of things in the newspapers. Why should I as a [Nigerian] researcher be bound by the European ethical standard? It prevents me from getting this important data. In our social-cultural background, some of these ethics may not really be practical.
>
> *John Uwa, Interview, 11 May 2016*

For this member of the team at least, national newspapers' regular representations of named individuals, without consent or concern for defamation, highlighted the significant disjuncture between our academic research methods and those of other professions, including journalists in Lagos.

At the time, we felt the removal of Ann Kirori's data was overly punitive, but an example from our Nigerian fieldwork illustrated the risks of disclosure carried by non-anonymized photographic material. As part of an FGD in Lagos, one of the PRs produced a picture of a well-known low-income settlement in the city, downloaded from the Internet, featuring a ragged man picking waste in the foreground. The participants were invited to comment on the obvious lack of waste management facilities in that community, and to speak about the environment in this type of urban setting in Lagos. One of the participants recognized the person in the photo-graph. This caused the termination of a focus group that had taken weeks of careful planning, as well as great discomfort to the PR and participants. After this incident, we agreed in principle that FGDs in both cities would use images sourced from locations outside the city. Even so, we remained nervous about using non-fictional material, and in Nigeria we turned instead to a rich source of stimuli for discussion and debate about urbanization and public morality: popular 'Nollywood' movies, with a plethora of DVDs on topics current to our project, such as Ebola, urban sexualities and urban popular culture.

A different type of ethical breach occurred in Nairobi when the PR for Media and Communications was pickpocketed on a *matatu*: his digital recorder was stolen, containing three un-transcribed interviews. In training, we had agreed to upload audio material to the secure online database as soon as interviews had been completed, but day-to-day circumstances had prevented this as the PR moved around the city. The interviews, each containing an oral con-firmation of the participant's name, age and occupation, were lost. In response to this incident, we could only reiterate and collectively re-confirm the principles of confidentiality and data security agreed at the outset of the project.

In institutional environments such as schools and health centres, the researchers were more likely to be able to control numbers within FGDs than in low-income areas, and to conduct confidential one-on-one interviews following agreed informed consent procedures. Even so, the quest for a secluded space away from interruptions often meant that PRs found themselves in unconventional parts of buildings. In Lagos, the PR for Education and Schools appropriated the sick-bay of one school as an interview room. 'During this fortnight', she reported,

> interruptions during the interview became a common occurrence. It appears that my activity of talking to pupils one-on-one in the sick-bay room is arousing the curiosity of both pupils and staff alike. There were times when pupils drop in for one thing or the other and I had to stop to attend to them. Some of these interruptions would be noticed in the audio recordings.
>
> *Jane Nebe, Report, 13 June 2014*

Once she sought space in a dilapidated school building adjacent to the new school, and on other occasions she used a head teacher's office for interviews that were punctuated by multiple interruptions (Interview, 3 May 2016).[4] In Nairobi, one interview with a caretaker took place at the back of a school bus awaiting the bell at the end of lessons, while the driver was – or appeared to be – asleep. Other interviews took place sitting under a tree in the open air, by the side of busy roads, or indoors in the presence of onlookers who chipped-in with comments, or interrupted in the ways described above.

A particular obstacle we faced in Lagos at the outset of fieldwork was the impact of a controversial BBC documentary entitled *Welcome to Lagos* (Prod. Will Anderson), first screened in 2010 and reproduced on a number of online platforms. For its many Nigerian critics, this three-part documentary reflected the Western media's obsession with the people of global 'slums' far more than it represented the full complexity of life in Lagos. Nobel Laureate Wole Soyinka described the documentary as 'jaundiced and extremely patronising. It was saying, "Oh, look at these people who can make a living from the pit of degradation"' (cited in Dowell, *Guardian*, 28 April 2010).

In spite of the multiple ways in which this documentary was a boon to our project as an example of Eurocentric mediations of Lagos, unfortunately the BBC had also chosen to represent the Lagos State Government as an urban bully, treating its most vulnerable people as trash, with heavy-handed slum-clearance tactics and schemes to shape Lagos into a global 'mega-city' by removing the majority of its low-income inhabitants. The programme clearly positioned itself on the side of the small people – represented as citizen-survivors – whose shacks were bulldozed by the Lagos Task Force, and who were driven into further extremes of poverty and survival where human relationships snapped under pressure.

The memory of *Welcome to Lagos* was still strong among officials four years after the broadcast, as Olutoyosi Tokun approached the LAWMA and Jane Nebe approached the Ministry of Education for permits to conduct their research in the city. The BBC had, in the view of officials, exploited the trust of the Lagos State Government who granted the film crew access to Olususun landfill and other key sites. As one commentator in the *Guardian* vividly stated, in the programme 'our dirty linen were yanked from our very loins and aired on the international veranda' (Nwaubani, 6 May 2010). Understandably, officials were not going to make the same mistake a second time. As a consequence of the word 'dirt' in our project title, suspicious managers interrogated the researchers again and again about their intentions, and were reluctant to grant access to municipal dumpsites and government schools. As Olutoyosi Tokun recalls:

> Actually this suspicion originated in the BBC's *Welcome to Lagos* documentary . . . When I went to LAWMA to get approval, it was such a difficult assignment for me. It took several months for them to approve that I access their refuse dumps, because it is exactly the same research environment that *Welcome to Lagos* was based on, because some of it was based on the refuse dump in Lagos and I was also going back to that refuse dump with an international organization. It helped that the University of Lagos was part of the project, and they gave me all the support necessary to secure approval from LAWMA . . . Nobody wants a bad name. As much as things are a bit different in Nigeria, we want to promote the good aspects of the country. We don't want to be misrepresented. Somebody called it 'the single story' – Chimamanda Ngozie Adiche[5] – that's what she calls it. We want at least a holistic, a fair, representation of what's happening in Lagos.
>
> *Interview, 11 May 2016*

Arriving in Lagos with a research project containing the word 'dirt', we had become caught up in the very media discourses we wished to research. Government officials were 'reading' us through their negative interpretations of the BBC's international media text. In an effort to allay officials' concerns, we emphasized the ways in which our objectives differed from the BBC's; we highlighted the involvement of the University of Lagos as a research partner; and we agreed to send copies of our final reports to the ministries at the completion of the project in 2018. Jane Nebe observed:

> Because of this BBC documentary it was not surprising that people within the Ministry [of Education] were very sceptical of such research. What they felt the BBC documentary did was to hype the negatives and ignore the positives, or talk about the positives in a way that was not really hyped as positives. The scepticism was really there.
>
> *Interview, 3 May 2016*

In spite of the examples given above of participants' reluctance about particular research methods, people from all socioeconomic backgrounds were generally willing to participate in our research once the themes and objectives of the project had been explained. The voluntary nature of participation and the right of participants to withdraw at any stage were emphasized at the outset of interviews, and information was provided about the anonymization process and the ways in which the data would be stored and used. The PRs emphasized repeatedly that this project could not contribute directly or materially to changes in the urban infrastructure, nor could we influence government policy. People's complaints and requests could not be taken forward to officials or ministers: rather, we would circulate our findings at workshops and conferences, and in academic papers, as well as making use of data in our interviews with public health and waste management professionals. The most our project could hope for, we explained, was a 'trickle-up' effect. Even so, volunteers were plentiful in the FGDs run by local researchers in multi-linguistic settings where African languages were the dominant medium for discussions.

Asked about what motivated participants, the PRs offered a number of explanations ranging from the practical to the intellectual. In particular, the topic of material benefits arose in the PRs' reflections. We were not allowed to offer cash payments for participation, and this generated considerable discussion between the PI and the PRs. According to our ethical agreements,[6] cash payments for participation were forbidden. In Lagos, however, our agreement in principle to cover participants' transport costs for their journeys from various parts of the city to the 'Dirtpol' offices at the University of Lagos, to a maximum of N4,000 (approximately £14.00 in 2016), rapidly became a euphemism for cash payments. Meanwhile, in Nairobi the PR for Media and Communications undertook in-depth interviews at restaurants over drinks and lunch charged to 'Dirtpol' expenses, and our FGDs in low-income communities were always concluded with the distribution of soft drinks.

In Lagos, John Uwa expressed a strong sense of injustice at the contradictions in this prohibition of overt payment. During our training in the UK, he pointed out, we had seen examples of research for which participants were compensated in cash for attending interviews:

*John Uwa*: . . . but in Africa they say it's unethical to pay. I have very serious qualms over this.
*Olutoyosi Tokun*: [*interjects*] When money is given as part of recruitment, as an incentive, it leads to some sort of bias, so the people that need the money are eventually recruited into the study. That might be why it's unethical. But if this incentive does not influence the selection of the participants, it's okay, because the person has given his time anyway and didn't demand anything, so just to appreciate his time and his effort, you give him some money.
*John*: Yes, you are right. I never give money before an interview, and sometimes I don't tell them I am going to give them anything. Some people expect it, you know, you have these NGOs going out interviewing people and they give them money, so when they see you come like that they say, 'Ah, what are you going to give?'

*Interview, 11 May 2016*

At least one of the Kenyan PRs strongly agreed with this view:

Some of the ethical issues, I found them very ridiculous! Why would you pay somebody in Europe after the person has given you data, and not in Africa? Are we not the same? That was very ridiculous. I think some of these ethical issues should be contextualized. They should not just look at Europe and decide, these are the people you are going to give money to and leave these others. We are all human beings. We all have needs. We all want to be appreciated after such an activity. So it was also very difficult for me to tell somebody, 'Yes, I'm coming to interview you, but you are going to pay for the snacks, for the refreshment that you are going to take, because I don't want to influence the responses that you are going to give.' That was totally wrong. Nobody would accept to come for such an interview!

*Job Mwaura, Interview, 26 May 2016*

The European examples cited in the PRs' recollections related to examples of market research and data gathering for government departments that we had examined during the training period in the UK, for which cash payments had been made to participants. We, however, were disallowed from offering any financial incentives to participants deemed 'vulnerable' according to the ERC's language for the residents of 'developing' countries. The team was confronted with the status of the project as a scholarly undertaking following ethical standards designed for academic data collection across international contexts, in which the economic and power inequalities between Global North and Global South were understood to introduce potential biases to data. We could not adopt the principles used in European research; but, as the PRs insisted, we also could not refuse to give participants refreshments and snacks at the end of an interview, nor to reimburse their travel to and from our venues. As such, few of our respondents participated 'for free. They would expect you to give them something at the end of the interview' (Job Mwaura, Interview, 26 May 2016).

All the researchers commented on the effectiveness of using local languages in one-on-one interviews and FGDs. 'When you speak their language, they tend to let their guards down, then I'm able to connect very well with them and probe more and more', John Uwa said of his Lagosian participants:

> They do not think the recorder can interpret their language, so they come out and say all manner of things, especially on some very topical issues. They are very careful: they don't want to be seen as racist, or chauvinistic, but when you begin to speak their language, they open up and are able to tell you that the Igbos are dirty in very emphatic terms, or that the Yorubas sleep with shit around them. Most of the English interviews I've had, I've not heard anyone say that the Yorubas can eat with shit around them, or sleep with shit around them, or use their potty to eat regular meals.[7] But when you begin to speak other languages you see all the discriminations. All the sentiments, begin to play out. African language is a very wonderful thing!
>
> *Interview, 11 May 2016*

Across our research contexts, African-language interviews generated a fluency of opinion, whereas English often produced stilted, overly formal responses. 'One way to get some good responses was to converse with these respondents in a language that they were comfortable with', Job Mwaura commented in Nairobi: 'When I broke the formality completely, such as when they were able to talk in Sheng or in their mother-tongue, then those sensitive issues became a little bit easier to talk about' (Interview, 26 May 2016).

A problem arose in transcribing this rich data, however. While all the researchers were fluent in more than one African language, they had no training in how to write these languages. 'The major challenge I had was transcribing', Jane Nebe remarked:

> because I needed to be able to find expressions as accurate as possible, and as near as it was possible to be in the English language equivalent. We were asked to transcribe in the original language and then do the conversion, the translation to the other language. This was really difficult, you know, because Nigerian Pidgin and Yoruba are languages that you don't really get to do a lot of writing in. It's normal you speak them, you talk, you discuss with them. Writing them was another situation entirely.
>
> *Jane Nebe, 3 May 2016*

This posed especial difficulties for the transcription of Yoruba interviews, in which diacritical marks were essential to the meaning of words, but also for the transcription of flexible languages such as Nigerian Pidgin and Sheng, in which words and meanings change continuously. 'There is always a possibility that you might mistranslate. There's always a possibility that you might over-translate. It's possible you might under-translate', Jane Nebe added, 'so what I always do is ask questions':

*Jane Nebe*: Sometimes there are words that have contextual meanings, there are phrases that have contextual meanings, that if you don't understand that context, you might take the literal translation, but that is not what it means. What comes to mind is when someone says 'eh-*heh*'. Now, eh-*heh* has many meanings. It's a Pidgin word, eh-*heh*. You can never transcribe it, too, you cannot really transcribe the word. If someone says eh-*heh*, it could mean 'yes', eh-*heh* could mean 'yes continue', eh-*heh* could mean . . .

*John Uwa* [*interjects*] '. . . you have been caught, you have been caught! Have you seen? It's over!'

*Jane*: Exactly so, yes, it means up to ten different things if you sit down to analyse what eh-*heh* means. But when I write it in my transcripts, people might not understand what the eh-*heh* means if I do not try to put in a comment to say this person was trying to say, 'yes I agree with what you are saying', or a person can say eh-*heh* to mean 'you don't mean to say, I don't believe what you just said is true'. So you see, some of these things are not just pure language, but also contextual meaning as treated with these phrases. Some of them are slang. Some of them are proverbs. So you really need to explain these things as footnotes, as endnotes, and even sometimes in brackets, especially when you are translating from the familiar language of data collection to the English language.

*Jane Nebe, Interview, 3 May 2016*

In response to these challenges, in our transcripts we developed the referencing systems mentioned above, including footnotes for the cultural translation of terms translated literally in the text and square brackets for the inclusion of non-verbal expressions and for phrases in the original language where the English translation was only an approximation.

Project management by the PI and RC revolved around the effort to maintain intellectual and methodological coherence across the regional and disciplinary boundaries of our research. Of special concern was how to ensure sufficient methodological consistency to enable comparisons between data produced by PRs with parallel portfolios in Lagos and Nairobi. For the 'Public Health and Environment' researchers, for example, who were responsible for building relationships in the field of public health, environmental strategy and waste management, Nairobi and Lagos yielded such different sets of local topics, sites and concerns that the work of the two researchers rapidly diverged. For different reasons, the two 'Media and Communications' researchers were also unable to build comparable archives from their work, because their portfolios involved sampling a wide range of material from local newspapers, radio and television documentaries, as well as from popular music, popular films, television soap operas, online blogs, local publications and other popular urban media. They sampled numerous different media on different days, and sorted data according to project themes, but the sheer quantities of material meant that methodological consistency was difficult to maintain between Lagos and Nairobi. The 'Education and Schools' researchers were able to produce comparable data by sharing topic guides, and by communicating regularly about research methods and strategies. Of particular relevance to these researchers in schools, and generating conversation between them, was the question of how to reduce the power-relationship between themselves and the pupils, who were often overly deferential to the researchers in the context of schools' hierarchical institutional structures.

In total, over 200 interviews were recorded by the team, including translations into English from Yoruba, Nigerian Pidgin, Gikuyu, Swahili and Sheng.[8] In both cities, we met with public health providers and users, waste management professionals, residents from diverse neighbour-hoods, school children, teachers, other school staff, media producers and media consumers. Clearly this project could not have been undertaken by any individual researcher, but, as the Project Coordinator reflected, understanding the conceptual underpinnings of the project *as a whole* posed ongoing problems for individual team members. Alongside the PI and RC, 'our team was composed of research assistants, collecting data, but they were not just going out and harvesting data. The project was much more demanding than that because it was so conceptual' (Claire Craig, Interview, 1 June 2016). While individual work packages were anchored to definable topics and fields, the question of how to develop an overarching understanding of the project remained open and unresolved for the duration of the project. As a consequence of the vagueness of the overarching project title, each of us developed a different understanding of the project, and the resulting data are heterogeneous. Nevertheless, through our interviews and FGDs with diverse urban communities, we have established that urban encounters and identities – relationships with others, as well as the implementation of environmental and public health policies – can be understood differently when filtered through concepts relating to dirt in its local and global manifestations, rather than concepts relating to hygiene and cleanliness. Current discourses around the spread of diseases and urban planning are underpinned by cleanliness as a desirable goal, but a focus on the antithesis of cleanliness potentially takes us much further in comprehending urban processes and relations.

Our project attempted to address practical questions relating to research methodologies alongside conceptual questions about urban experience. The interviews and media materials collected by the researchers – and the PRs' blogposts and reports – helped to build a database of public opinion about environmental policy, waste management, urban morality and daily routines in East and West Africa, as well as a commentary on research methods in African urban contexts. At the same time, as the project matured, we tried to identify and reflect on the presence of what the Project Coordinator eloquently described as our own 'cultural skeletons' within the data and interpretations we produced (Interview, 1 June 2016).

One methodological outcome of this project – unanticipated in the initial proposal – was our realization not only that adjustments to standard social science methods have ethical implications, as discussed above, but that the attempt to implement standard social science methodologies might actually *produce* ethical problems. Whether conducted in the Global North or the Global South, the agreement to abide by international ethical standards is fundamental to academic research, but the researchers on this project found that the concept of 'confidentiality' is culturally variable. It can produce risks to participants if assumed to be universal. Assurances of confidentiality might, in some fieldwork contexts, be interpreted as 'secrecy' rather than a guarantee of safety, and thus pose a danger to the participant's well-being. A further set of ethical considerations might thus be produced in the very process of implementing international ethical standards for scholarly research.

## Notes

1 Strong Arab-Islamic religious and cultural networks can be found in Nigeria and Kenya. Kenya's centuries-old history of trade and religious exchange is focused largely on Mombasa, with its pronounced Arab-Islamic urban influences and ancient trading networks with the Indian subcontinent along the famous spice routes. Flows of trade in the nineteenth century, including the slave trade, generated global networks in Nigeria and Kenya that far exceeded European colonial routes and boundaries.
2 While in the UK for training, we had debated appropriate terminology for low-income urban settlements, but in our various research contexts we sometimes reverted to local terms such as 'slum' and 'ghetto'.
3 Privately owned minibus for public transport.

4 Risk assessment was almost impossible to implement in the field due to such rapid, last-minute changes of venue.

5 Adichie is author of the global bestselling novel, *Half of a Yellow Sun* (2006). In an interview in July 2009, she famously described negative Western constructions of Africans as the creation of a 'single story', saying 'show a people as one thing, as only one thing, over and over again, and that is what they become' (www.ted.com/talks/chimamanda_adichie_the_danger_of_a_single_story/transcript?language=en).

6 The project underwent ethical screening and approval by committees at three separate institutions: the ERC, the University of Sussex and the University of Lagos.

7 In residences with no interior bathroom, some households keep a potty in their living quarters for night-time use, preferring this to the option of going outdoors to the communal toilet at night. This practice has given rise to stereotypes and abusive jokes about Yoruba domestic hygiene. In an unrecorded interview with John Uwa, one elderly woman living in Lagos insisted that the Ijebus (Yorubas from a part of Ogun state) use their potty as an eating bowl: in her opinion, this is why they are called '*Ijebu oloorun*' (smelly or dirty Ijebu). This idea was strongly refuted by a participant of Ijebu origin in a separate FGD. On another occasion, while at a car mechanic's, John Uwa's request that an apprentice dispose of his engine oil in a container, rather than on the ground, attracted the attention of the boss: 'is he not a Yoruba boy?', the manager said within earshot of the employee: 'That is how Yoruba people behave! Yoruba people can live with shit in their house, eat where there is shit, and even eat with the same plate they use for shit, so I am not surprised at his stupidity'. Such negative cultural stereotypes are generally expressed in African-language encounters rather than in formal FGDs or English-language interviews.

8 We did not have language coverage for Arabic speakers, which affected the communities with whom we could work in Lagos and Nairobi.

## References

Dowell, B. (2010). Wole Soyinka Attacks BBC Portrayal of Lagos 'Pit of Degradation'. *The Guardian*, 28 April.

Drotner, K. and Livingstone, S. (Eds.) (2008). *International Handbook of Children, Media and Culture*. London: Sage.

France, A., Bendelow, G. and Williams, S. (2000), A 'risky' business: researching the health beliefs of children and young people. In A. Lewis and G. Lindsay (Eds.) *Researching Children's Perspectives* (pp. 150–162). Buckingham: Open University Press.

Kyronlampi-Kylmanen, T. and Maatta, K. (2011). Using children as research subjects: how to interview a child aged 5 to 7 years. *Educational Research and Reviews*, 6(1): 87–93.

Nesbitt, E. (2000). Researching Eight to Thirteen Year-Olds' Perspectives on their Experience of Religion. In A. Lewis and G. Lindsay (Eds.) *Researching Children's Perspectives* (pp. 135–149). Buckingham: Open University Press.

Nwaubani, A. T. (2010). Nigeria's Anger at the BBC's Welcome to Lagos Film. *The Guardian*, 6 May.

Punch, S. (2002). Research with children: the same or different from research with adults? *Childhood*, 9(3): 321–341.

Stanley, B. and Sieber, J. (1992). *Social Research on Children and Adolescents: Ethical Issues*. London: Sage.

Thomas, N. and O'Kane, C. (1998). The ethics of participatory research with children. *Children and Society*, 12(5): 336–348.

Tindana, P. O., Kass, N. and Akweongo, P. (2006). The informed consent process in a rural African setting: a case study of the Kassena-Nankana district of Northern Ghana. *PMC*, 28(3): 1–6.

Note: Five of the PRs describe their work here:
Ann Kirori: www.youtube.com/watch?v=GbBf8njMrNE
Job Mwaura: www.youtube.com/watch?v=Qa8ixy8G4iI
Jane Nebe: www.youtube.com/watch?v=REdCmyD5Yws
Olutoyosi Tokun: www.youtube.com/watch?v=G11zCWiD7FY
John Uwa: www.youtube.com/watch?v=Rwtyy9gIxEI

# Section 5
# Valuing and validating

# 1

# Valuing and validating

## On the 'success' of interdisciplinary research

*Mike Michael*

## Introduction

Although in this collection we have asked authors to discuss interdisciplinary methodology through particular verbs, one cannot help but, or one is tempted to, touch also upon nouns. Specifically, the nouns that attach to the verbs valuing and validating can be presented as contrasting pairs which connote rather different practices, interests and arrangements. Thus, for Valuing we can compare value and valuation, and for Validating we can compare validity and validation. At base, the former term of each pair implies an 'external' or 'objective' relation by which some thing or some activity is judged against pre-given standards: value against market forces, validity against scientific or epistemological criteria. Conversely, the latter term of each pair – valuation and validation – points to an intersubjective or interactional relation in which some thing or some activity is assessed by means of standards and criteria. Thus, validation, say of a person's work within an organization, might entail shared negotiation of criteria as to what is of worth; and valuation of a property might involve the situated weighting and juggling of criteria in the process of moving through a house, say, to come to a proposed money figure.

Of course, this contrast is hardly absolute. As social scientific studies of audit culture and standardization have often noted (for example, Bowker and Star 1999; Power 1999), external criteria are mobilized in the process of locally negotiated valuing, and the 'fruitful' application of external criteria often entails their situated negotiation. Nevertheless, this contrast hopefully sets up a frame to address the complex, involutionary processes that enter into the doing of Valuing and Validating. At the very least it should hint at what we might call a 'topology of valuing' in which 'valuing' and 'validating' in relation to interdisciplinary methods operate at several interwoven, intersecting and interacting levels. Such a topology might simultaneously entail institutional encouragement, systemic devaluation, ontological invention, professional and political accountability, and embeddedness within genealogies of practice (and much more, as we shall see). It should be obvious that I will not be dealing with more formal methodological accounts, especially those concerning variants of validity, such as scale, probability or reliability. There are plenty of text-book definitions and applications in the literature. In keeping with the interdisciplinary ethos of the volume (and as hinted in the foregoing), I wish to engage with 'validity' as a complex notion that can connote the more or less successful accomplishment of some form of agreement about the 'facts of the matter' that emerges in different ways in the process of doing interdisciplinary work.

In what follows, I intend to discuss these issues in a little more detail by thinking them through what has been called translational research. As a means by which research in one discipline is transferred to another which then translates it into practical intervention (say the movement of laboratory-based research into clinical application), *prima facie*, translational research looks like a relatively uncontentious site. After all, it would seem that here interdisciplinarity functions with a 'lightness of touch', given, on the one hand, the absence of a need to fashion new interdisciplinary tools, and, on the other, the existence of infrastructural supports (in co-habitation within research centres, governmental backing and financial resourcing).

Despite a willingness to trial alternative approaches, however, things do not run so smoothly, as rather different values – epistemic, cultural, educational – are in place, which militate against even this apparently 'easy' form of interdisciplinary collaboration. Nevertheless, some groups do 'hold' and below I examine how this might work through the redistribution of value, or a re-patterning of valuing. This examination will serve as a basis for an expanded discussion of the complex role of valuing and validating that draws directly on the chapters that comprise this section. As we shall see, valuing and validating range across a number of concerns: data, participants, project, case, research question, methods, profession and institution. They also function on a number of registers – epistemic, ontological, inter-personal, institutional, temporal – not all of which are easy to reconcile. The aim of this section, therefore, is primarily to illustrate the complex interweaving of valuing and validating within a range of interdisciplinary practices and projects, rather than to provide a systematized approach. To the extent that there is a common theme across these illustrative cases, it concerns the 'matter of success' which is taken up in the concluding remarks. What can count as a 'successful' interdisciplinary project, finding, team or researcher?

## On the valuing and validating in translational research

Translational research (also routinely referred to as 'bench to bedside' research within the medical field) is the term applied to the movement or translation of 'basic research' (often produced in the laboratory) into the clinical domain where it can serve as the basis for innovations that are therapeutically, diagnostically and/or preventively useful. According to Watts (2010), translational research has become a key policy priority, shooting 'up the (UK medical research) agenda' (n.p.) and being championed by the likes of the Office for Strategic Co-ordination of Health Research, Medical Research Council, and the Academy of Medical Science. It is also central to the funding initiatives of the European Commission's 7th Framework Programme and US National Institutes of Health (the latter calling for bids for Centres of Excellence of Translational Research).[1]

However, despite all the enthusiasm for this useful translation of knowledge across disciplines, there are underlying problems, not least with regard to the ways in which the different disciplines envision and enact such fundamentals as 'knowledge' and 'scientific rigour'. As various authors have documented, these divergences hinge in large part on the values that pertain within the respective 'epistemic cultures' (Knorr Cetina 1999) of, in this case, lab sciences such as embryonic stem cell research and clinical sciences such as treatment of Parkinson's disease (for example, Cribb, Wainwright, Williams, Farsides and Michael 2008; Wainwright, Williams, Michael, Farsides and Cribb 2006; Wilson-Kovacs and Hauskeller 2012). Thus, Michael, Wainwright and Williams (2005) traced how, for instance, lab scientists derogated the 'experimental techniques' of clinicians (who would often work with a few cases rather than extended sampling), while the clinicians derogated what they saw as the lab scientists' obsession with scientific rigour at the expense of the immediate needs of patients.

In these two research communities – those of the lab and the clinic – we might say there is a valuing of different values: 'proper experimental procedures' versus 'commitment to patient well-being'. At the same time this distinction also determines a validation of certain sorts of cultural practices. For instance, within lab science culture, what is most valued is scientific excellence, and academic 'voice' is attended to on the basis of the quality of one's science. By contrast to the meritocratic culture of the lab, within clinical science, hierarchy on the basis of seniority (and presumably 'experience') structures who can have voice. It is not surprising, then, that when a research centre comprised of both clinical and laboratory scientific teams have research meetings, attendance is affected by the disciplinary type of research that is being presented (Brosnan and Michael 2014). Not only is the substance of a topic in need of translation (determining the 'value' and 'validity' of this or that finding), but so too is the *form* of discussion and debate (who is 'permitted' to have voice).

We can approach this divergence in terms of Annemarie Mol's discussion of ontological politics. In *The Body Multiple* (2002), she argues that a disease such as atherosclerosis manifests itself in different ways depending on the practices that enter into its enactment. Atherosclerosis *is*, therefore, something different when studied microscopically by a pathologist as compared to when a physician manually gauges the temperature of a patient's feet. What atherosclerosis 'is' differs across these two cases. As Mol (1999) notes these different versions do not necessarily need to 'cohere' – though practically and routinely they do, simply because a patient must be treated. Sometimes, however, they can diverge and create tensions, sometimes co-exist in parallel quite happily; sometimes they can be coordinated and managed.

What this suggests is that values are 'emergent' and, by extension, 'valuing' is contingent. Values do not sit 'above' practices, animating them, so to speak. Rather, what is of value within and across specialisms or disciplines can vary with the practical issues at stake. Sometimes this will revolve around the immediacy of patient needs; sometimes it will attach to the work regimes of different hospital or disciplinary specialisms. In the case of translational research, the divergent, multiple ontologies of clinical and lab practitioners seem to be managed in a variety of ways. For instance, neglecting writings from 'other' disciplines can be put down to sheer lack of time (the arduousness of lab work means that one must read instrumentally, and that means one does not have the luxury of reading work from more distant disciplines). However, this does not necessarily imply an unbreachable divide between the ontologies of lab and clinic. Brosnan and Michael (2014) note that researchers themselves see the eventual 'calibration' of lab and clinical ontologies taking place in the future. This 'promise' of future ontological articulation is embodied in the research Group Leader who straddles lab and clinical practice. It is the Group Leader who, by virtue of their vision of prospective translation, symbolizes the continued 'adhesion' of the research group. As Brosnan and Michael suggest, this future, yet present, conception of the research community can be seen as an instance of Barry, Born and Weszkalnys's (2008) interdisciplinary logic of accountability, namely 'the idea that (interdisciplinarity) helps to foster a culture of accountability, breaking down the barriers between science and society, leading to greater interaction, for instance, between scientists and various publics and stakeholders' (Barry et al. 2008: 31). In the present case, this translates into the processes by which scientists and clinicians enact themselves as accountable to one another. The former perform their work as relevant to clinical problems (as identified by the clinicians); the latter make their clinical work accountable to the lab scientists who can provide the underlying organic bases for that clinical work. But this accountability is 'done' in a somewhat diffused way, and, crucially, such accountability is enacted via the figure of the Group Leader.

The point of this example from medical research is to show how in the various modes of articulation across disciplines touched upon here, valuing and validating play complex, shifting and contingent roles. From this albeit brief discussion of translational research, we can tentatively derive the following pattern: there is, on the one hand, a valuing of scientific rigour, practical application and (future or promissory) interdisciplinary collaboration, and, on the other, a validating of one's own epistemic culture, research practice, research relations and research leader. Embroiled in all this are mediations of ostensibly abstracted scientific (for example, sample size, use of control groups) and moral (for example, the imperative to find ways of alleviating the suffering of humans) values that might or might not govern the prevailing norms of the research team. In the next section, we will pursue these involutions further by considering the insights provided by the chapters in this section.

## On the varieties of valuing and validating

In relation to the contributions to this section, we find that the meanings of valuing and validating have proliferated extravagantly. In a quick summary of these many relations in interdisciplinary work we can crudely abstract the following:

- the value of the singular observation (as opposed to a large sample) that simultaneously evokes its conditions of possibility and their analysis – Latimer and Munro;
- the value of disruption in a performative methodology that emphasizes surprise and change – Akama and Pink;
- the value of using designerly practices to derive problems (rather than find solutions) – DiSalvo;
- the value of attending to the impact of the scale (and quantity) on interdisciplinary research processes (the problems of too large or broad a collaboration) – Fukushima;
- the value of 'care' and affect – understood as compromise – in overcoming epistemic and cultural differences in interdisciplinary research teams – Michael;
- the value of keeping things open to enable collaboration while acknowledging interdisciplinary divergences – Calvert;
- the value of accommodating different disciplinary explanations as a way of enriching the object of study – Gisler;
- the value of disjunctions between research problems as framed by social scientists versus practitioners in problematizing the value of social science analysis – Irwin and Horst;
- the value of pursuing the multiple qualifications of 'an' object (such as bottled water) in shaping the range of methods that was adopted – Hawkins;
- the value of an empirical 'example' in its complex relationality to this or that generality – Driesser;
- the value of deploying interdisciplinary dissent to enable engagement with disciplinary presuppositions, not least with regard to the possible politics that flow from such presuppositions and their interrogation – Tironi;
- the value of (cultural, individual) difference within and across projects for interdisciplinary work (and its validity) which draw in past projects (on the proposed uses of Information and Communication Technologies) and project futures (for the uses of ICTs) – Graham;
- the value of designed devices as a means of enacting interdisciplinarity capable of doing 'speculation' – Wilkie;
- the value of designerly fictions in enabling speculative ethnography – Galloway;
- the value of indigenous knowledges and practices which reconfigure 'interdisciplinarity' toward a means of revivifying and sustaining indigenous worlds – Elder and Potskin.

Needless to say, I cannot in this introduction do full justice to the contributions that follow. However, the point I wish to pursue concerns the variety of valuings and validatings that is evidenced in these entries. Rather tentatively, we can re-group them under the headings: 'Doing data/object of research', 'Accounting/practice', and 'Relations/reproductions/repairs'. While these are certainly not empirically mutually exclusive, they can heuristically serve to highlight different moments of valuing and validating. Provisionally, and with some laxity, we might say that: Elder and Potskin, Hawkins, Graham, Driesser, and Latimer and Munro fall under the auspices of 'Doing data/object of research'; under the label 'Accounting/practice' can be included Akama and Pink, Gisler, DiSalvo, Tironi, Wilkie, Galloway, and Irwin and Horst; 'Relations/reproductions/repairs' encompasses the work of Michael, Fukushima, and Calvert. I do not address all these studies but simply draw on those that seem to be most useful in exploring these aspects of valuing and validating.

## 'Doing data/object of research'

'Doing data' addresses the values entertained in the derivation and generation of data. Gay Hawkins speaks of how her object of study, 'bottled water' – as a complexly re-qualified and re-qualifying, multiple reality-provoking entity – necessitated a range of methods. That is, the object somehow validates the methods that are used. In Stengers and Latour's (2015) introductory account of Souriau's 'different modes of existence', an object of art (or any object that is made, for that matter) places a demand on its maker to 'work it out'. But this is a fraught process, filled with the possibility of error and failure. Bottled water is just such an object that, in order to be made to exist properly – or successfully – demands that its (social scientific) makers apply their skills, technique and capacities, but always hesitantly, carefully, so that the data (about plastics, water, contaminants?) thus derived 'successfully' make that object.

Conversely, the 'work' is not simply crafted by its maker, but also, in return, crafts its maker (see also Sybille Lammes, this volume). In Joanna Latimer and Rolland Munro's chapter, it is the object (or event) apprehended as an ethnographic moment that 'makes' the researcher – that asks questions of her. Specifically, this singular case of an ethnographic moment precipitates a bringing to the fore, and re-valuation, of one's theoretical and analytic situation such that one sees anew that object, along with its conditions of possibility. In this process, the researcher is herself reconfigured, and made to 'work successfully'. However, we need to expand this image of a single work confronted with its individual maker (and vice versa). For each carry with them a pattern of pasts and prospects. For Connor Graham, in researching the potential uses of ICTs (or in crafting an account of such uses), it is important to set these within an array of (among other things) past futures of ICTs and past projects researching such futures. In other words, on our present reading, the 'success' of the object of a research study, and the researcher themselves need to be situated in relation to the past successes (and failures) of ICT futures and studies of such futures. Finally, for Tuur Driesser, we also need to address the value of the research in terms of its processes of exemplification – that is, in relation to the 'example' per se. Accordingly, the example does not relate to the general unproblematically; rather, they need to be situated in ambiguous spatial and temporal relation to one another, not least when the example – what can count as an example – is itself discipline-bound.

In taking all these accounts together, there seems to be a valuing of something that approximates the 'success' of a research study. However, this hard-won success points in several directions at once – to the 'object' of the study, the researcher, the histories of researcher and researched, the examples and cases that 'make up' the research. Indeed, we might say that any perceived or actual success is validated by the configuring of these different elements. But even if a particular study does emerge, or conclude, successfully, it often tells us little about how to proceed in relation to future 'objects of research'. Each object of study, according to these contributions, makes its own peculiar demands upon research in relation to 'being worked out'.

What are the implications of this account for the doing of interdisciplinary research? On the face of it, these are pessimistic: the object is 'worked out' through distinct and discrete disciplinary techniques that are generative of distinct 'works' (what above was called multiple ontologies). However, we can pause and reflect here on the fact that a work of art, in order to be 'successfully worked out', might 'call on' the artist to pull in techniques from multiple disciplines – painting, sculpting, video, performance, for instance. In a parallel way, in order to be 'successfully worked out', the object of research might 'demand' a range of methods and techniques that normally sit in different disciplines. This raises, for present purposes, the issues of what counts as 'success' in general, and what counts as success for an interdisciplinary object in particular – what then is the relationship between ontology/ies and the heuristic? We will approach these concerns through a discussion of, respectively, 'accounting/practice' and 'relations/reproductions/repairs'.

## 'Accounting/practice'

To propose that a research object demands certain (combinations of) methods to be 'successfully worked out' is to connote a sense of completion. If it has succeeded then the object reaches a form of closure or triumph or attainment; and we have findings or claims to be delivered. And yet, some of the contributors to this section seek anything but closure when they judge the value of their own research. By contrast, success is associated with openness, tension, doubt, uncertainty, different and better problems. Thus Manuel Tironi speculates on the value of dissent (that never materialized) among different practitioners (for example, architects, social scientists, government officials) in opening up the possibility of democratic potential; Yoko Akama and Sarah Pink describe a range of methods for surprising and disorienting researcher-participants about their respective objects of research, by introducing uncertainty and possibility into those objects; and Alan Irwin and Maja Horst, insist on the need to attune to the problem of many problematizations (related to different disciplinary specialisms) which as a meta-problematization, as it were, may be productive (leading to a useful questioning of presuppositions) or destructive (leading to tension and impasse); and Anne Galloway, in developing, disseminating, and decanting responses to her fiction of the PermaLamb, cuts across design and ethnography in order, in Haraway's (2016) phrase, to 'stay with the trouble'. In these four examples, we might detect a value placed on something we could call 'a-success' in which success is grounded in 'unsuccess', unity in disunity.

Put another way, there is a series of general collectively enacted disruptive practices in play in these three examples: dissent, disorientation and meta-problematization. All these could be discounted as a negative upshot of interdisciplinary collaboration, however, what we witness are accounts which (also) valorize these disruptions in terms of their potential productivity. The suggested notion of a-success not only connotes the utility of disruption, but also the lack of certainty that can be attached to such utility. The productivity of disruption remains potential – there is no guarantee (as Irwin and Horst are all too keenly aware).

We can tentatively draw out some more general implications of this discussion by considering the practice of triangulation. Triangulation routinely refers to the deployment of different methods that aim to derive a variety of data about a given topic (for example, interviews and observations, or interviews and survey data) that, by virtue of corroborating one another, serve to increase the likelihood of validity of the findings (Denzin 1978). However, Silverman (2006) –among others – has been highly critical of this technique not least because the different methods entail different theoretical backgrounds and different contexts of application, which renders suspect the comparability of their data. However, it might well be more fruitful to explore the productivity of their differences: what does a disparity in interview and biographical data and non-participant observation tell us about, say, a scientist's characterization of science? Where Merton's (1973[1942]) biography-based work derived the norms of science (which stressed communality universality, disinterestedness and organized scepticism), Mulkay's (1979) observational approach noted the lack of relevance of norm-following in the reward of scientists (rewards, such as further funding, were based on the usefulness of scientists' findings rather than on their adherence to norms). In the process, the very notion of norms was challenged. Even for the scientists there was a value in the opening up of the problem of norms – which they, in light of the triangulation across Merton and Mulkay, either continue to see as structurally important (Merton; also see Mitroff 1974) or as 'ideological' (according to Mulkay), or waver between these statuses. In sum, the practice of triangulation is accounted for in terms of opening up, of problematization; and the validity of triangulation lies less in verification or consolidation and more in its value as a means of generating problems.

## 'Relations/reproductions/repairs'

In the previous two parts of our segmentation of methods of valuing and validating, we have focused on the complexities of interdisciplinary success by focusing, respectively, on the object of research itself, and on the types of accounting for success (or, specifically, a-success). In this final part of the introduction, we turn to a consideration of the relations among research practitioners drawn from different disciplines. How are these relations engendered, reproduced and repaired? We have already seen in the discussion of translational research that such relations can be very loosely woven: in the above example, the relations between laboratory and clinical researchers were essentially vicarious, being mediated by the Group Leader. Nevertheless, this yielded success insofar as the research group itself enjoyed longevity, and did not break apart. This suggests a strongly social configuration of methodological differences, and so these next contributors explore some more features of interdisciplinary relations.

For Jane Calvert, the management of difference among collaborating social scientists, designers and synthetic biologists was facilitated by an emergent metaphor, that of 'the wedge in the door'. This evoked an opening (albeit fragile) through which dialogue among different researchers could potentially proceed. Mike Michael addresses some of the more overtly affective dimensions of interdisciplinary relations. Here 'compromise' among interdisciplinary colleagues is possible because of a certain 'care' toward one another (which might be based in socializing and liking) where there is a sort of tolerance of, or indulgence toward, each other's disciplinary commitments and quirks. By comparison, as Masato Fukushima tellingly warns, successful interdisciplinarity might also be affected by the sheer volume of participant researchers, that is, by the number and density of relations. As he notes, in addition to the problems of managing and mediating cooperation across a multitude of researchers, there are tensions born of the simultaneous push for larger research teams and the chronic and systemic pursuit of individualized recognition. If Calvert and Michael hint at the ways in which researchers from different disciplines might validate each other, Masato problematizes the very possibility of such validation where teams are large and dispersed.

However, lest we think that Calvert's and Michael's chapters simply point to means of 'successful collaboration', we can take note of the following. Calvert describes a metaphor that serves to render broken or problematic relations 'reparable'; indeed, even if difficulties persevere/ persist, there is an in-principle possibility of future collaboration. One might say that a variation of an 'imagined community' (Anderson 1983) seems to be at play here, in which the various members of an 'interdisciplinary community' remain prospectively inter-linked by virtue of a series of (wedged) open doors or agreed terms of discourse. However, arguably, such a metaphor also militates against collaboration. That is to say, it can deflect or diffuse any anxieties about the lack of collaboration by evoking the *potential* of collaboration: 'we're not collaborating now, but we could if we wanted to!' In Michael's case, the prominent role afforded socializing (drinking, gossiping and joking in the pub) might end up being precisely what is privileged in the collaboration. In addition to this socializing being potentially a mode of exclusion, not least a gendered one (who in the team is left out because of carer responsibilities?), it can also end up being an 'end' in itself. More specifically, for the sake of sustaining these socializing practices (one might call them a particular type of friendship), the difficult negotiations that would benefit an interdisciplinary study might be avoided; certain elements of an 'imagined community' are prioritized over others. In sum, we can detect the obverse to both Calvert's and Michael's accounts: the successful repair and reproduction of relations might end up, ironically, negatively impacting upon the success of 'working out' the research object.

Finally, when re-considering Fukushima's chapter, it becomes clear that 'multitude/ multiplicity' takes a variety of forms. Not only is there the issue of the sheer numbers of relations, but there is the issue of the range of types of relationship. Over and above intra-/interdisciplinary relations, one can imagine other sorts of relations at play (for instance, gender, national/international, intra-/extra-institutional, early/mid/late career, fully/mainly/partially employed on/ committed to the interdisciplinary project). To focus only on 'career phase' and 'commitment' to the project, it is not difficult to identify bottlenecks in communication within teams where late career or part-committed collaborators do not see the relevant (interdisciplinary) project as sufficiently high priority to warrant their full, or even partial, attention. Needless to say, this can be very frustrating for junior researchers or those fully employed on the project. This in turn suggests that it is not simply the 'research object' that is in the process of being 'worked out successfully', so too are careers, salaries, promotions, future relationships, longer-term research programmes, and so on and so forth. Another way of framing this is to extend Fukushima's tacit contrast between quality and quantity of interdisciplinary research. So far we have primarily been thinking about the different types of quality of interdisciplinary relations, and how these are variously inculcated, managed and secured. However, quantity also plays its part: the energy and enthusiasm (or, conversely, indolence and indifference) with which an interdisciplinary project is pursued, how this energy and enthusiasm is distributed across an interdisciplinary team, affects the 'success' of both the object and the relations of study.

## Concluding remarks

This section introduction has attempted to situate 'valuing' and 'validating' in relation to the complex practicalities of doing interdisciplinary work. Drawing on discussions of translational research and on the chapters that make up this section of the book, the essay has implied that valuing and validating take on a variety of trajectories through the processes of interdisciplinary research, tying together such elements as the object of study, the generation of better problems and productive uncertainty (a-success), the quality and quantity of social relations and affects, and the careers and future plans of researchers. Out of this 'topology of valuing' – that comprises an interdisciplinary collaboration – might or might not co-emerge a 'successful' object of study aligned with its 'successful' collective maker (research team). As Stengers and Latour (2015) repeatedly note, such is a tempestuous process that is as likely to yield failure as success (though, of course, both these terms are multiply interpretable). That said, as we noted above, success is not simply a matter confined to the relation between maker and made, researcher and researched. It is also embedded within larger dynamics. On this score, and with the 'topology of valuing' in mind, let me end with an anecdote.

Many years ago, I was involved in a small interdisciplinary project (sociology, social psychology, cognitive psychology, anthropology, science and technology studies). Despite being a naive early career researcher, even I could recognize that there were fundamental problems with the research design. As the research project unfolded, it also struck me that we were simply not doing enough – either in terms of collecting adequate data, or working sufficiently hard on that data. My sense was that, in spite of a few publications, the project had failed to deliver on its promise and was, in a word, unsuccessful. However (and here my then-naivety was fully on display) I failed to reckon with the context of the project. As it turned out, emerging political and policy elements of the research field meant that the project found itself placed at the 'cutting edge', and was regarded as, more or less, a triumph. For me, it remains a missed opportunity, if not an outright failure, though I have not resisted benefiting from its success in the intervening years. The moral here is that both object of study and researcher emerged 'successfully', in part because together they could be said to comprise an 'object of politics' crafted in an other's successful political project . . . in which sense values shape research methods more than we might always anticipate or understand.

## Note

1  http://grants.nih.gov/grants/guide/rfa-files/RFA-AI-12-044.html#_Part_1._Overview

## References

Anderson, B. (1983). *Imagined Communities: Reflections on the Origins and Spread of Nationalism*. London and New York: Verso.

Barry, A., Born, G. and Weszkalnys, G. (2008). Logics of interdisciplinarity. *Economy and Society*, 37(1): 20–49.

Bowker, G. C. and Star, S. L. (1999). *Sorting Things Out: Classification and its Consequences*. Cambridge, MA: The MIT Press.

Brosnan, C. and Michael, M. (2014). Enacting the 'neuro' in practice: translational research, adhesion, and the promise of porosity. *Social Studies of Science*, 44(5): 680–700.

Cribb, A., Wainwright, S., Williams, C., Farsides, B. and Michael, M. (2008). Towards the applied: the construction of ethical positions in stem cell translational research. *Medicine, Health Care and Philosophy*, 11(3): 351–361.

Denzin, N. K. (1978). *Sociological Methods*. New York: McGraw-Hill.

Haraway, D. (2016). *Staying with the Trouble: Making Kin in the Chthulucene*. Durham, NC: Duke University Press.

Knorr-Cetina, K. (1999). *Epistemic Cultures: How the Sciences Make Knowledge*. Cambridge, MA: Harvard University Press.

Merton, R. K. (1973[1942]). *The Sociology of Science: Theoretical and Empirical Investigations*. Chicago, IL: University of Chicago Press.

Michael, M., Wainwright, S. and Williams, C. (2005). Temporality and prudence: on stem cells as 'phronesic things'. *Configurations*, 13(3): 373–394.

Mitroff, I. I. (1974). Norms and counter-norms in a select group of the Apollo moon scientists: a case study of the ambivalence of scientists. *American Sociological Review*, 39(4): 579–595.

Mol, A. (1999). Ontological politics: a word and some questions. In J. Law and J. Hassard (Eds.) *Actor-Network Theory and After* (pp. 74–89). Oxford and Keele: Blackwell and The Sociological Review.

Mol, A. (2002). *The Body Multiple: Ontology in Medical Practice*. Durham, NC: Duke University Press.

Mulkay, M. (1979). *Science and the Sociology of Knowledge*. London: Allen and Unwin.

Power, M. (1999). *The Audit Society*. Oxford: Oxford University Press.

Silverman, D. (2006). *Interpreting Qualitative Data: Methods for Analysing Talk, Text and Interaction* (3rd ed). London: Sage.

Stengers, I. and Latour, B. (2015). The Sphinx of the work. In Etienne Souriau (Ed.) *The Different Modes of Existence* (pp. 11–90). Minneapolis, MN: Univocal.

Wainwright, S., Williams, C., Michael, M., Farsides, B. and Cribb, A. (2006). From bench to bedside? Biomedical scientists' expectations of stem cell science as a future therapy for diabetes. *Social Science & Medicine*, 63(8): 2052–2064.

Watts, G. (2010). Lost without translation? *British Medical Journal*, 341: c4363 doi: http://dx.doi.org/10.1136/bmj.c4363 (Published 18 August 2010).

Wilson-Kovacs, D. and Hauskeller, C. (2012). The clinician-scientist: professional dynamics in clinical stem cell research. *Sociology of Health & Illness*, 34(4): 497–512.

# 2

# Compromising

*Mike Michael*

## Com-pro-mise

It goes without saying that interdisciplinary collaboration can be a fraught business. Of course, there are various modes by which different disciplines might more or less happily pattern their methods, practices and concepts in relation to one another (see Barry, Born and Weszkalnys (2008) for an important statement on this). However, there are also cases when the process of collaboration requires delicate negotiation, and even then there is no guarantee of a happy, productive outcome. An example of this is the tension between laboratory scientists and clinicians who both ostensibly share an interest in neuroscience and yet place value in very different forms of knowledge and experimentation, and epistemic culture (Brosnan and Michael 2014).

In my own recent experience of collaboration with designers, what at one level appeared to be an excellent working relationship with common points of intellectual reference, revealed itself to be far more complex. For instance, my accounting of relevant design traditions was seen by my designer colleagues to be lacking accuracy let alone nuance, while in my view their engagement with users was less than it could have been (in terms of its generation of empirical material). Now neither of these critical views is to be seen to be 'correct'. The point is, rather, that they indicate the tensions that arise because of the value placed in/on divergent priorities: a 'proper' history of design that appropriately situates the project, as opposed to a quick sketch that gets us to the point of discussing the social theoretical implications of the project; an unsystematic engagement with the users of the design in order to gather clues as to the design's role in their everyday lives versus a systematic collection of data so that we can access and analyse the range of users' relevant views and practices.

Despite these differences in accent and interest, a compromise between social scientists and designers was achieved. But to say this obscures a complex set of features. If we break up compromise into its three component parts, we can note that, etymologically at least, we have 'mise' which translates as 'putting', 'pro' which translates as forward, and 'com' which translates as together: in other words, a 'putting forward together'.

At first glance, this might imply a sort of unproblematic, seamless, integrated 'co-projection', but things are a little more complicated. 'Putting forward together' does not necessarily connote either integration or seamlessness. Things can be put forward together in parallel, after all. That is to say, compromising might mean acknowledging one's collaborators' views and practices but not necessarily valuing them, or valuing them only contingently. How then might we think about this process of valuing disciplinary others? In what follows, it is suggested that we can do this by addressing the affective dimensions of compromise. Here compromise might be understood as a 'staging of care' in such a way that participants collaborate in ways that accommodate and value their differences as well as their similarities.

## Threat and promise

Surprisingly, it turns out that promise is an antonym of threat. Compromising can be reconsidered in this light: it can be said to connote the common overcoming of threat. In the case of my collaboration with the designers, the threat I perceived was one in which their seemingly piecemeal and opportunistic use of data challenged my conceptions of empirical research. What could they usefully say about users' use of their designs, in particular, such artefacts as biojewellery? (For a fuller discussion of the details of biojewellery and my responses to it, see Michael (2012).) It was only when I came to realize that the usefulness of their designs lay, indeed, in their use (as opposed to the subsequent social scientific commentary on that use) that I came to grasp the limitations of my own assumptions about the value of social science. As a result, there was a shift – not total by any means – toward a re-viewing of a nexus of values running through my attachments to social scientific research. To clarify, these disciplinary (in my case, the sociology end of science and technology studies) attachments were not only to specific theoretical and methodological content. Attachments were also drawn to the 'ethos' of doing social scientific research per se.

For instance, there was value placed in the clarity of a research question: a particular problem (in my case a discrete technoscientific controversy on stem cell research, say) is identified and investigated empirically. In contrast, for the designers I was collaborating with, while there were certain design and technical problems to be solved, the 'research problem' seemed to emerge through the doing of the research. From my perspective, the political point of my research was to illuminate and champion the value of the framings and knowledges of more or less disempowered groups (as against those of predominant expert institutions). By contrast, while the designers were certainly interested in the ways in which people interacted with their designs, they were as much concerned with displaying those designs and the processes that went into their making.

The designers' rather different positions on 'research' were threatening, not least because they did not initially make sense. In what follows, I discuss how the designers' practices did gradually begin to make sense to me, and how, in the process, my own sense-making machinery began to change. Stengers' discussion of cosmopolitics will prove particularly useful in addressing the parameters of this process of change.

## The cosmopolitics and the compromises of compromise

In both Bruno Latour's (e.g. 2004) and Isabelle Stengers' (e.g. 2005a) discussions of cosmopolitics, a key aim is to fashion a 'good common world'. However, what this good common world looks like, and how it is composed, are not straightforward. While Latour and Stengers focus on the realm of politics, the various points they raise translate to the processes of interdisciplinary collaboration (which is of course politically charged).

In the foregoing, I have placed emphasis on the *practice* of research, where practice takes in such dimensions as the clarity of research questions and the enactment of particular constituencies, as well as technical methodological procedures. For Stengers, and for Annemarie Mol too (2002), practices enact realities or ontologies. The different disciplines' divergent practices – Stengers (2005b) calls this an ecology of practice – yield divergent ontologies. For Mol, this ontological multiplicity lies at the base of what she calls 'ontological politics' and can be managed or coordinated in various ways so that sometimes one ontology takes precedence over others, sometimes a composite ontology emerges, and sometimes ontologies simply run in parallel, avoiding conflict because both hold to a promise of future coordination (on the latter, see Brosnan and Michael 2014).

If Mol documents a selection of forms of ontological politics, Stengers explores the potential parameters of cosmopolitics. The aim is definitely not definitively to determine what counts as the best procedures for engendering coordination (and thus the good common world) through which disciplines can work together. For Stengers, this would amount to the application of the parameters developed by another practice or ontology (say that of politics), or, in the present case, another discipline (say, that of studies of interdisciplinarity). Rather, with each new interdisciplinary venture new forms of coordination need to arise through the very processes of coordination in their specificity. As the ethnomethodologists would say, there is no outside to cosmopolitics – if there were, this would simply be another element of the cosmopolitical process. Thus, in finding a compromise among disciplines, one must not attempt to impose that compromise by installing the means that realize it (e.g. particular sorts of workshops, or the co-writing of working papers). These means will import their disciplinary baggage and be as likely to undermine as to mediate collaboration. This certainly seemed to be the case in my collaboration with designers. The compromise that emerged derived as much through more informal means such as a mutual sense of irony and self-deprecation, and a pleasure in beer-fuelled socializing.

For Stengers the danger is that there is too quick a 'compromise' – that disparate practices, multiple ontologies and different disciplines move to a 'common ground' too easily. To put this another way, we need to 'compromise over compromise' – to be circumspect in proceeding lest we miss 'something more important' as Stengers (2005a: 1001) puts it. This means paying attention to that which does not make obvious sense within an unfolding collaboration (the 'idiotic' in Stengers' terms). In my case, this initially took the form of acknowledging – rather than denying or deferring – the lack of sense in the designers' practice (from my social scientific perspective), and a slowing down to rethink my efforts at compromise, that is, at facilitating a putting forward together. Specifically, I was very enthusiastic about co-writing working papers, keeping to a particular timetable, and having regular 'serious' team meetings all as a way of enabling collaboration. As things stood, these efforts hardly took root.

For instance, team meetings always struck me as chaotic affairs as colleagues wandered off mid-discussion to get cups of tea, or print something off, or look at something online, or abruptly leave for some other meeting. To say the least I found this frustrating. But let me slow down and ask how does one do compromise under these conditions? First, one could simply dismiss this as an indication of bad, unprofessional habits. Second, one could embrace the designers' way of conducting meetings as an interesting and intriguing way of negotiation. Both are too quick and easy. The alternative is to ask 'how is this designerly meeting practice put together?' (and thus one can ask 'what sustains my view of good meeting practice?'). In brief, given that the designers occupied the same studio space, a meeting is simply a continuation of their ongoing, routine and dispersed interactions. There is nothing especially special about a meeting (hence its annoyingly dispersed and dissipated character). From my perspective as a sociologist with his own office, among other sociologists with their own offices, a meeting carries far more weight as a discrete and important event. In sum, the com-pro-mise of a meeting needed, itself, to be subject to com-pro-mise – a putting forward together about how to put forward together.

However, this did not happen. It is only in retrospect that I grasped these differences. Where then did the 'compromise of compromise' take place? I suspect it was in the pub, drinking, gossiping and joking. Stengers (2005b) remarks that a staging needs to happen in which participants are placed in a position of 'equality' to one another, such that both their views about the issue at stake, their views on how to proceed in tackling it collaboratively, and the contingency of these views is made available to all. This is no easy matter – it requires great efforts of reflexivity and overt openness. The suggestion – albeit precarious – is that this proceeds affectively and requires a certain staging (in the present instance, this was provided by the pub). Here we can introduce the notion of care in the sense Puig de la Bellacasa (2011) articulates it. Accordingly, care 'signifies . . . an affective state, a material vital doing, and an ethico-political obligation' (p. 90); thus to 'care' for a thing (in the present case, the object of design and study) is to recognize that it is emergent through both one's own and others' practices. As a corollary, one needs to care – with due circumspection – for those others who care for that object, whatever their (disciplinary) background, and despite disagreement in what counts as care, and what precisely one is caring about. Having said that, one also needs to 'care about care' – care is not innocent in that it can be used to moralize, or to belittle, or to arrive at too easy a compromise. In the present context, this means that the disparate views on how to derive and process 'data' (as discussed above) came to be 'taken care of' in the sense that I came gradually and affectively to value the designers' particular 'idiosyncratic' approach to 'data' collection and processing. As noted this was partly mediated by socializing in the pub.

## Conclusion: compromising compromising

Lest this account seem too seamless, it is worth reflecting on some of the limitations of compromising. First, compromising cannot be operationalized as a 'method' per se – this is because the idea of what a 'method' of compromising 'is' is itself is up for grabs: it emerges practically, locally and iteratively. Second, compromising as discussed here can appear to be a process of endless, perhaps indulgent, 'negotiation'. Of course, it usually isn't, not least because there are 'external factors' (research council deadlines, publication schedules, reputational structures) that mitigate this tendency. Third, and by the same token, external factors can also militate against any compromising at all. Economic constraints and organizational hierarchies, for instance, can mean that there is no compromising, rather a falling into line with a privileged set of practices and goals (my example discussed above might be said to reflect the particularly privileged conditions of an academic setting). Finally, and conversely, forms of compromising can be encouraged to enhance 'creativity' and generate novelty in response to market conditions (e.g. Thrift 2005). In this case, compromising is itself subordinated to broader institutional strategies.

Having made all these critical points, compromising hopefully remains a promising device for addressing the complexities of interdisciplinary work. If not quite a method, then it is at least methodological in the sense that it can serve to sensitize research to the variety, situatedness, informality, affectivity and iterativeness of the means by which collaborations across disciplines might be conducted.

## References

Barry, A., Born, G. and Weszkalnys, G. (2008). Logics of interdisciplinarity. *Economy and Society*, 37(1): 20–49.

Brosnan, C. and Michael, M. (2014). Enacting the 'neuro' in practice: translational research, adhesion, and the promise of porosity. *Social Studies of Science*, 44(5): 680–700.

Latour, B. (2004). *Politics of Nature: How to Bring the Sciences into Democracy*. Cambridge, MA: Harvard University Press.

Michael, M. (2012). 'What are we busy doing?' Engaging the idiot. *Science, Technology and Human Values*, 37(5): 528–554.

Mol, A. (2002). *The Body Multiple: Ontology in Medical Practice*. Durham, NC: Duke University Press.

Puig de la Bellacasa, M. (2011). Matters of care in technoscience: assembling neglected things. *Social Studies of Science*, 41(1): 85–106.

Stengers, I. (2005a). The cosmopolitical proposal. In B. Latour and P. Webel (Eds.) *Making Things Public* (pp. 994–1003). Cambridge, MA: MIT Press.

Stengers, I. (2005b). Introductory notes on an ecology of practices. *Cultural Studies Review*, 11: 183–196.

Thrift, N. (2005). *Knowing Capitalism*. London: Sage.

# 3

# Deriving

*Carl DiSalvo*

In most all of its most familiar uses, to derive means to obtain. But to derive is not simply to come into possession of something, rather it is to acquire something by means of drawing it out from something else, and often in the endeavour of drawing it out, transforming it. We derive by extension. We derive by modification. What we derive is myriad; the term has common usage in linguistics, in mathematics and in the sciences. It also has common usage in descriptions of the everyday, for instance we say that we derive calm from meditation, excitement from sport and pleasure from sex.

Deriv-ing, then, is a process. It is a process of extension and modification through which some thing is produced. Most often, this process is assumed to be logical. That is, there is assumed to be an ordered sequence of steps, a clear articulation by which we achieve that which we achieve. For instance, we derive the term *designerly* by affixing the term *designer* with a standard suffix that changes the lexical category and meaning of the term. This general process (of adding a suffix) can be repeated time and again in linguistic contexts for producing new terms (even when, as in the case of *designerly*, the outcome is awkward, it is still sensible). Such logical sequences and articulations are similarly the case in the production of, that is, derivation of, mathematical proofs, chemical compounds and so on.

Contemporary research in the humanities and social sciences is increasingly engaging in practices of making as part of the scholarly endeavour. Making is, in fact, not a singular thing or even a stable category, but rather shorthand for a swathe of practices including design, art, craft, multiple forms of media (video, photography, sound), as well as the so-called maker movement (whatever that might turn out to be, or not; see Anderson 2012; Pye 2015; Sennett 2008). Why this interest in making in the humanities and social sciences? What is it that making purportedly provides?

What making provides interdisciplinary research in the humanities and social sciences with is novel means of deriving knowledge. This occurs through a process – making as a mode of derivation. But that process presents a challenge because of the unfamiliarity of the logics, sequences and manners of articulations used. Rather than attempting to engage with all kinds of making, I will confine this brief discussion to the domain of design.

Designers have a fraught relationship with process, which affects those who work with designers as well as those who study design. Within design, there is a fetishization of process.

Perhaps one of the most manifest historical perspectives on this fetishization is the Design Methods movement (see Jones 1992; Cross 2001) and the canonical example of Christopher Alexander (Alexander, Ishikawa and Silverstein 1977). The Design Methods movement sought to formalize processes for design. In effect, the Design Methods movement sought to formalize processes of deriving effect (or affect) through particular actions and forms. So, Alexander's pattern language could be seen as a means of arriving at 'a good city' or 'a good neighborhood' or 'a good street corner' by the extension and modification of a series of structural elements and material configurations. Design Thinking could be taken to be a contemporary version of this impulse; Design Thinking purports to provide designers and non-designers alike with a means of deriving invention, if not innovation, from a specified manner of engaging contexts – any and all contexts (Brown 2008). The promise of Design Thinking is that we all might be able to derive, to draw forth in inspired and meaningful ways.

And yet, while process – the means of derivation – is an ongoing interest to practising designers and design scholars, it is also a point of contestation. At the same time that many designers fetishize process, many also concomitantly resist any idea that the design process might be scientific, formulaic, fixed or prescribed (see, again, Cross 2001). This is not to say that the derivation is not logical, simply that it (purportedly) follows another logic or logics.

Lury and Wakeford (2012) use the term *inventive methods* to capture a range of these new ways of deriving knowledge in the humanities and social sciences, many of which involve making, and some which are decidedly designerly. I want to argue that what makes design such an exciting (and sometimes maddening) approach is that it is itself an inventive method. The logics of design are continuously being re-invented and re-articulated as the contexts and purposes of design change. If, as Buchanan (1985) claims, design has no fixed subject matter but, rather, like rhetoric is a means of invention, then each design calls for a re-imagination of what design is capable of doing and how. Making doorknobs is not the same as making buildings which is not the same as making logos which is not the same as making interfaces and yet it is all, comfortably, design.

Within the context of product and service innovation Kolko (2010) has argued for understanding the inventiveness of design (how it derives) as a process of abductive reasoning; Buchanan (1985) offers a conceptualization of design as a modern form of rhetoric, a means of situating design as a liberal art concerned with the invention of arguments for how we could and should live; and with speculative design in the context of public understanding of science and technology Michael (2012) invokes the trope of the idiot to explain the practice and purpose of design. Each of these is correct – each of these is a proper means of deriving given its particular context and purpose. This is both the resilience of design and the source of quandaries with regard to research. Design is itself a series of inventive logics.

In its context of origin – serving industry through the creation of goods – the fact that designers could approach the same situation and conditions time and again, and time and again arrive at different ends via a similar process, was of benefit. It filled the marketplace with a plethora of variation (whether or not this was beneficial beyond the marketplace is another issue). Within the context of research, this variation can be troublesome. Yes, it does produce an abundance and diversity of interpretations in both content and form (taken together as the materialization of the derivation). This abundance and diversity, however, might trouble some who look for coherence in scholarship.

One approach would be to standardize, or at least catalogue and typify, designerly approaches to knowledge production. And indeed, there are those who pursue that vector. Another vector is to shift our perspective on what we value in methods and what we mean by deriving or achieving knowledge.

Perhaps one value of designerly methods in the humanities and social sciences is the extent to which the endeavour of designing enables us to derive problems. The interpretations, the meanings produced through design are not means of settling concerns, but rather of materializing them and their significant factors. In this way, the derivation takes on its double meaning of both the process of deriving and the source from which something is derived, its origin. Rather than designerly methods producing the ends of inquiry, maybe they produce its starting point. Deriving is not merely a process, but an experiential endeavour, an event that enables productive problem-making (Wilkie 2014). There is a quizzical aspect of this. Design, in this context is not about usability, usefulness or even desirability. What we derive from design, as mode of making-as-inquiry, are not solutions, but rather productive glitches, difficulties and complications.

## References

Alexander, C., Ishikawa, S. and Silverstein, M. (1977). *A Pattern Language: Towns, Buildings, Construction.* Vol. 2. New York: Oxford University Press.

Anderson, C. (2012). *Makers: The New Industrial Revolution.* New York: Random House.

Brown, T. (2008). Design thinking. *Harvard Business Review*, 86(6): 84.

Buchanan, R. (1985). Declaration by design: rhetoric, argument, and demonstration in design practice. *Design Issues*, 2(1): 4–22.

Cross, N. (2001). Designerly ways of knowing: design discipline versus design science. *Design Issues*, 17(3): 49–55.

Jones, J. C. (1992). *Design Methods.* New York: John Wiley & Sons.

Kolko, J. (2010). Abductive thinking and sensemaking: the drivers of design synthesis. *Design Issues*, 26(1): 15–28.

Lury, C. and Wakeford, N. (Eds.) (2012). *Inventive Methods: The Happening of the Social.* New York: Routledge.

Michael, M. (2012). 'What are we busy doing?' Engaging the idiot. *Science, Technology & Human Values*, 37(5): 528–554.

Pye, D. (2015). *The Nature and Art of Workmanship.* London, New York: Bloomsbury.

Sennett, R. (2008). *The Craftsman.* New Haven, CT: Yale University Press.

Wilkie, A. (2014). Prototyping as event: designing the future of obesity. *Journal of Cultural Economy*, 7(4): 476–492.

# 4

# Disrupting

*Yoko Akama and Sarah Pink*

*Design+Ethnography+Futures* is a research programme and initiative that explores techniques of disruption to undermine the certainties that inform conventional ways of knowing – for both researchers and participants – in design and ethnography (Pink and Akama 2015). *Design+Ethnography+Futures* attempts to move beyond conventional discourses in design that validate outcomes and resolutions by incorporating user perspectives in the tight locus of the formal design phase. It likewise goes beyond the objectifying tendencies of ethnographic studies of, or about, persons or phenomena. Rather than aiming to do 'better' design ethnography, we seek to create an opening where a hybrid interweaving is underpinned by movement towards a shared conceptual foundation. And like our engagements with uncertainties of the future, this is a disruptive strategy because it challenges and undermines what it is that we habitually thought we already knew and did.

The idea of disrupting, as we develop it here, refers to an improvisatory process that 'corrupts' ethnographic and design principles when they are brought together. Disruption is implicitly pursued, for example, in critical and speculative design that takes the form of fictional and absurd future-projections to ask difficult and ignored questions and to provoke and disturb common understandings (Dunne and Raby 2013; Michael 2012). In Design Anthropology, anthropologists are bringing design into their research to dismantle ethnography to reconstruct something new (Murphy and Marcus 2013), using interventionist, material, speculative and provocative approaches. However, the coupling of design and anthropology is not neat and seamless. Rather, it is characterized more by a disruptive interdisciplinary endeavour whereby it necessitates the abandonment of competence and specialism to

> enter a terrain beset with fears of inability, lack of expertise and the dangers of failure. The transformational experience of interdisciplinary work produces a potentially destabilising engagement with existing power structures, allowing the emergence of fragile forms of new and untested experience, knowledge and understanding.
>
> *Rendell 2013: 119*

As we discuss, this abandonment of certainty gives rise to discomfort, yet also shows possible ways to produce avenues of knowing about things not yet encountered, possible alterities and potential futures. As such, we place value upon uncertainty and put the not-yet-made at the centre of inquiry: ethnographers/designers are substantively engaged in processual worlds where they work with emergent qualities and people who are sharing their journey into the immediate future.

## Departures: where we are coming from

Design anthropologists have begun to reorient the ethnographic focus towards the future through their engagements with design (Gunn and Donovan 2012). Underlying this movement is a critique of the single, specific and sometimes authoritarian ways that research can shape and restrict our participants', as well as the researchers', understanding. Like Gatt and Ingold we have a concern about the predictive and prescriptive orientations to future-making that are promoted in design which seek 'to conjecture a novel state of affairs as yet unrealized and to specify in advance the steps that need to be taken to get there' (Gatt and Ingold 2013: 145). Instead, we aspire to building capacity in relation to ever-changing circumstances, to improvise and open up passages (Akama, Stuedahl and Van Zyl 2015). In other words, we are interested in inviting people along on a journey through which they come together to discover, question, converse and reflect upon how we 'are' in the ways we understand the world and manifest our understanding, and to propose how 'we' become with one another (Akama 2012). This implies that future-making necessarily starts with how we step into the future together. 'Becoming with' (borrowing from Haraway (2008)) is an important aspect of our research, indicating a companionship in embarking on this endeavour and a commitment to travel along together, rather than being in fear of making 'others'.

There is also a double movement happening here whereby the things we make (both objects and relations) and do together, act upon us as we are making and engaging in these very things (Willis 2006). This ontological process is akin to a hermeneutic circle: the world we make is, in turn, making us, undermining what we thought we were, inscribing how we are being and becoming with others. By inviting an approach that embraces uncertainty, allowing the emergent to define what we know and share, our being and becoming with will be situated and embodied but also constituted in relating and co-shaping new entanglements (Light and Akama 2014).

## Strategies for disruption

From December 2013 to May 2014, we ran a series of workshops with a variety of researchers from different disciplines (i.e. not just from design or anthropology). These ranged in format and duration, from two full days to three hours, with 12 to 30 researchers and postgraduate students from various universities. Each workshop had a guest facilitator to frame a theme that was of interest to the participants.

On embarking on this series, one of the challenges was to investigate how to facilitate exploration, not with regard to what we already knew, but in relation to what we did not yet know. This required a form of 'disruption' on the part of the instigators (also co-facilitators of the workshops), which valued unintentionality, as opposed to intentionality, and created opportunities for uncertainty instead of aiming for certainty. This was uncharted territory and a confusing experience for many, including us as instigators. Activities were randomly suggested on the basis of serendipity, chaos and whimsical ideas in order to experiment with what might happen.

It is no surprise that several workshops resulted in some of the participants leaving partway; in addition, many showed signs of discomfort. For example, the *Myths of the Near Future* workshop led by Katherine Moline (see Akama, Moline and Pink 2017) to co-explore socio-technical interaction sought to disrupt participants' relationships with their mobile phones: as such, a familiar, practical, precious and innocuous technology was turned into a research tool. Moline invited the participants to swap their phones with strangers during the workshop to see what the interaction revealed about each other and themselves. This disruptive strategy aimed to interrogate social norms and inculcated habits, and as researchers, encouraged us reflect on how we give 'permission' to strangers to have access to our personal lives, and how the other person might attempt to understand through an 'unfiltered' medium. Several participants opted out of this exercise, choosing instead to report how they use their mobile phone. However, most participants were keen to embrace the risks. Their casual attitude towards storing personal information on their smart phones was only registered when someone else accessed it. One participant remarked, 'I know my phone collects and aggregates data of all kinds about me, but rarely see it revealed even in this basic way'. For this participant, this activity revealed a number of things about those interpreting personal phone data:

> I learned a lot about what not only my phone 'looks like' (and what I look like) through the specific lens of photos and my sporting apps, but [also] about how we understand phones to show things about people, all of this was also specific to the person doing the looking.

A disruptive process prompted participants to articulate new feelings about aspects of smart phones that they might previously have taken for granted.

Another workshop, led by Helen Addison-Smith, interrogated implicit ethics that could be revealed when invited to ingest plants and animals that were still alive. Again, this approach attempted to 'disrupt' our habitual attitudes and interrogate the blurry line between what we believe and what we actually do in practice. The act of placing unfamiliar things into one's mouth brought forth intimate and confronting ways of knowing, exploring and questioning. Several participants were disgusted with some foods that were shared, but all were willing themselves to participate to see what they might discover. For example, trying to eat pungent, sticky and slimy *natto* (Japanese fermented soy beans) led to rich conversations that encompassed admitting our own contradictions, recognizing emotive responses and cultural differences, and addressing practices that lay beneath attitudes to life and death, consumption, production and waste.

In another activity, led by Elisenda Ardèvol and Debora Lanzeni, the participants were invited to re-make the workshop room into a space to foster innovation. What was once a sterile, impersonal office space in a university became a chaos of coloured threads. Ripped and torn materials were attached, hooked, twisted, tangled and wrapped around furniture and walls. To move around the room, the participants had to duck under coat-hangers, shuffle along the wall, avoid tripping over obstacles – changing our movements, relationally and dynamically in the room – releasing the participants from conforming to normative ways of socially interacting in a formal setting.

These 'disruptions' were a provocation to us all (participants and facilitators) to 'let go' of our preconceptions, forgo the need for resolution, and step out of our respective disciplinary certainties to consider alternative approaches and experiment with what might emerge out of an assembly of ideas, people and things. They were designed deliberatively to attune, sensitize and become mindful towards the emergent unknowns. It often triggered genuine surprise and reconfigured ways of knowing collaboratively.

*Design+Ethnography+Futures* endeavoured to explore disruptive emotions and strategies because they provoke questions of ourselves that we might not like to examine. It revealed how neither the present nor change are simple to understand or to design for – thus showing how such work complicates any ambitions to use such research processes to inform 'better' outcomes, something that we had already rejected as a possibility from the outset. Instead the implication is that since the present is messy, and complex to research (see Law 2004), we need to be valuing processes of change, and possible futures that are equally messy. They cannot be reduced to simple predictable contexts in which we might design neat future-oriented solutions. These disruptive strategies are not suggested for readers to repeat as methods or approaches, but instead are intended to serve as exemplars in valuing surprise, emergent unknowns and other uncertainties, in ways that are productive and generative, and that might catalyse further thinking on change making.

For more information on *Design+Ethnography+Futures*, see http://d-e-futures.com/

## References

Akama, Y. (2012). A 'way of being': Zen and the art of being a human-centred practitioner. *Design Philosophy Papers*, 1: 1–10.

Akama, Y., Moline, K. and Pink, S. (2017). Disruptive interventions with mobile media through Design+Ethnography+Futures. In L. Hjorth, H. Horst, A. Galloway and G. Bell (Eds.) *The Routledge Companion to Digital Ethnography*. London and New York: Routledge.

Akama, Y., Stuedahl, D. and Zyl, I. V. (2015). Design disruptions in contested, contingent and contradictory future-making. *Interaction Design and Architecture Journal*, 26: 132–148.

Dunne, A. and Raby, F. (2013). *Speculative Everything: Design, Fiction, and Social Dreaming*. Cambridge, MA: The MIT Press.

Gatt, C. and Ingold, T. (2013). From description to correspondence: anthropology in real time. In W. Gunn, T. Otto and R. Charlotte-Smith (Eds.) *Design Anthropology* (pp. 139–158). London and New York: Bloomsbury.

Gunn, W., Otto, T. and Charlotte-Smith, R. (Eds.) (2013). *Design Anthropology: Theory and Practice*. London and New York: Bloomsbury.

Haraway, D. J. (2008). *When Species Meet*. Minneapolis and London: University of Minnesota Press.

Law, J. (2004). *After Method: Mess in Social Science Research*. London: Routledge.

Light, A. and Akama, Y. (2014). Structuring future social relations: the politics of care in participatory practice. In *Proceedings of Participatory Design Conference 2014* (pp. 151–160), Windhoek, Namibia, 6–10 October.

Michael, M. (2012). De-signing the object of sociology: toward an 'idiogic' methodology. *The Sociological Review*, 60(S1): 166–183.

Murphy, K. and Marcus, G. (2013). Epilogue: ethnography and design, ethnography in design, ethnography by design. In W. Gunn, T. Otto and R. Charlotte-Smith (Eds.) *Design Anthropology: Theory and Practice* (pp. 251–268). London and New York: Bloomsbury.

Pink, S. and Akama, Y. (2015). *Un/certainty*. Design+Ethnography+Futures research series, RMIT University, Melbourne, Australia.

Rendell, J. (2013). A way with words: feminists writing architectural design research. In M. Fraser (Ed.) *Architectural Design Research* (pp. 117–136). London: Ashgate.

Willis, A.-M. (2006). Ontological designing. *Design Philosophy Papers*, no. 2: 1–11.

# 5

# Dissenting

*Manuel Tironi*

July 2010. Five months had passed since the 8.8 magnitude earthquake and the consequent tsunami that ravaged Constitución, a coastal city in southern Chile. In mid-April an interdisciplinary team of architects and social scientists – myself among them – was put together to conduct a participative design process for the reconstruction of Constitución. After three months of public hearings, community forums and charrette exercises, the process culminated with a city-wide referendum to decide on the construction of the plan's most ambitious project: the 20-hectare anti-tsunami park on Constitución's riverfront.

The project had been contentious. To begin with, it entailed the forced relocation of several families and the capacities of the park to mitigate a giant wave were quite unclear (Tironi and Farías 2015). And importantly, the architects from the reconstruction team were against the very idea of a referendum. It was not responsible, they claimed. A form of protection had to be built if lives were to be saved from future tsunamis, hence the park was not an alternative but a policy and moral obligation. Sociologists and anthropologists did not concur. We were not against the park but against its imposition in the name of an uncontested principle. Democracy comes with its own risks, we argued; too many technoscientific atrocities have been carried out in the name of security, progress or the common good, especially in states of exception. After days of intense conversations and negotiations, an agreement was achieved. Architects and social scientists worked through a hybrid arrangement. The referendum would be conducted, but instead of a blank ballot with simple 'yes' and 'no' options, each alternative would be accompanied with a pedagogical description of its pros and cons. The people would be in charge of deciding the fate of the park, but after being responsibly informed about what comes along with each choice. The referendum was carried out in July 2010. The 'yes' option won almost unanimously.

This story can be narrated in different ways. First, it can be a tale about how responsibility, deliberation and 'life' were signified and ordered differently by architects and sociologists in Constitución. A story, in other words, about how interdisciplinary ventures provoke multiple – and sometimes clashing – modes of valuing. In Constitución, however, an agreement was reached. Despite their differences, architects and sociologists worked around an accord. And this is the second possible account: one about the practices deployed in interdisciplinary projects to carefully tinker a solution, a temporary settlement in which different commitments are attentively knitted together.

291

*Figure 5.5.1* Ballot for Constitución referendum

Tale #1 and tale #2 are good examples of how conflicts over values in interdisciplinary research are often invoked in science and technology studies. When it comes to reflecting the complexities of interdisciplinary valuation, we usually turn either to conflicting epistemologies, i.e. the encounter of valuing realities that do not link up (tale #1) or, conversely, to the choreography of practices and materials unfolded to reach a tentative solution (tale #2). The first tale is about multiplicity and difference (Mol 2002; Law and Singleton 2005); the second about diplomacy and cosmograms (Latour 2004, 2007).

But there is also a tale #3. This account emerges when the script changes and the protagonist shifts from the researcher, either troubled by conflicting value systems or engaged in cosmogramming a common world, to the actual people of Constitución and the political potentialities of tales #1 and #2. Sociologists and architects disagreed, but to what extent did their disagreement facilitate the possibility of political explorations, problematizations, and emancipation? Further, did the interdisciplinary accord reached by sociologists and architects create the conditions for invoking or at least imagining a technopolitical otherwise by those that suffered and endured the tsunami? Attuned in the speculative register of the 'what if?' (Pignarre and Stengers 2011), tale #3 is not about the epistemic conundrums of experts but about what the situation could have offered to the people of Constitución as they strove to persevere as etho-political subjects in late liberal Chile.

In what follows I want to play around tale #3. I want to imagine a situation in which architects and sociologists, instead of achieving an accord, *dissent*. Dissent comes from the Latin *dissentire*, or to feel (*sentire*) differently. Interestingly, then, dissenting implies not just a discursive or cognitive quarrel, but a conflict that unfolds through and upon affective, sensuous and pre-reflexive doings. Dissenting is not a dispute over opinions and perspectives, but a clash over modes of sensing, engaging and inhabiting the world (Tironi and Sannazzaro 2017). I am particularly interested in the political affordances of these affective disagreements and in their capacity to crack open new forms of collective life. Interdisciplinary projects, I argue, render workable dissenting situations that need to be carefully accounted for in political explorations. Not the modelling of dissent into a formal method, but dissenting – or the moment of feeling differently – as a situation that may be methodologically empowered for the creation of a political otherwise. With the help of Jacques Rancière and Isabelle Stengers I want to think about the political capacities that might be unleashed when value mismatches in interdisciplinary projects are not worked through but enhanced as moments of democratic expansion.

## Counting and ordering in Constitución

The referendum was celebrated as an historic success for both Constitución and post-disaster planning in Chile at large. A total of 800 neighbours voted and the 'yes' option won with an impressive 84%. The political technology devised by architects and sociologists performed as expected: it offered citizens the last word on the park while rendering the project irreversible and incontestable. Architects and sociologists cheered the result as a methodological accomplishment of their interdisciplinary compromise.

But the referendum was also a way of closing down a debate whose threads were multiple and entangled. The planning team had the mandate to design the final plan in 90 days. So while participation was placed centre-stage the process could not afford too much deliberation. Constrained by time and results, the participation apparatus deployed in Constitución enacted citizens and participation in very particular and often conflicting modes (Tironi 2015). In June, an open meeting was conducted to discuss the social and technical plausibility of the park. The format of the meeting, however, ended up congregating the project's concerned groups and 'stakeholders', namely those whose properties faced potential expropriation, and who, by virtue of their economic loss, assumed a political superiority over the park's fate. So, despite the democratic objectives imprinted on the participatory machinery put forward to deliberate about the park, the discussion did not open a space for a minority politics as Deleuze and Guattari (1987) would have it, in which other ways of seeing and feeling could be invoked. The referendum, with its large scale, anonymity, legal age constraints and the power invested in it by the force of Western liberal democracy, not only sealed the fate of the park but also validated a way of defining what the public sphere is, who can participate in it, and how.

If architects and sociologists had disagreed on the value, content and role of the referendum, the park would not have been built. Or it would, but differently. Or maybe the disagreement could have precipitated new discussions about the politics and ecologies of the ocean in Constitución, new affective encounters between the people of Constitución and their material surroundings. Or perhaps it could have congealed new procedures for thinking about and deciding on the future of the city. Perhaps an interdisciplinary disagreement could have created the conditions for what French philosopher Jacques Rancière (1999) calls political emancipation.

Indeed, for Rancière politics is – needs to be – always emancipative (Rancière 1999). It only emerges when those who have not been taken into account, or have had their right to define what it means to (ac)count usurped, emancipate themselves from the arrangements that others have prepared for their proper democratic inclusion. Politics does not appear when a place is assigned to those that have been excluded but when the assignment itself is questioned and desta-bilized. So, politics is not the recognition of 'the community' as an excluded part in post-disaster decision-making, or the consequent integration of laypersons into planning via the referendum or other political technologies. Such a gesture assumes that society's parts – their characteristics, needs, and capacities – are already known and that hence those traditionally marginalized can be properly included by participation apparatuses. Politics, for Rancière, is when the counting of the parts, the knowledge about their nature, and the procedures for inclusion are contested in the name of a radical definition of equality: a debate not about the inclusion of the excluded but regarding what 'debate', 'inclusion' and 'exclusion' actually mean. For Rancière, the political moment par excellence is when two people speak but do not understand each other because they have not agreed on what it means to speak. Politics, the possibility of radical equality, resides in the incommensurable and frictional moment of profound disagreement – something that the referendum and the participatory machinery put forward by sociologists and architects diluted.

## Sorcery

Could an interdisciplinary venture be thought differently? What procedures, what material practices could architects and sociologists have assembled to invoke their 'feeling different' as a means for emancipation in Constitución? To translate interdisciplinary dissent into a method might be tricky. Methods are complex apparatuses that provoke into being the entities being observed, elicited or intervened (Law 2004). But the question remains: what would an inter-disciplinary commitment to take dissenting as a political potentiality look like? It could very well look like sorcery, if we follow Isabelle Stengers. Stengers is interested in the political disloca-tion forced by witches. Defining their art as magic, asserts Stengers, is already a 'magical' act 'that creates an unsettling experience for all those who live in a world in which . . . the art of magic has been disqualified, scorned and destroyed' (2005: 1002). The touchstone of this art is the capacity to convoke a force – the Goddess for neopagan witchcraft. But this force is not convoked in the search for a solution but to catalyse a regime of thought and feeling 'that bestows the power on that around which there is gathering to become a cause for thinking' (2005: 1002). Ultimately, the efficacy of this force relies on its capacity to 'transform each protagonist's relations with his or her own knowledge, hopes, fears and memories' (2005: 1002).

Stengers is hence interested in those concrete experiments and modes of *empowering* a situation; practices and rituals to confer the problem at hand, with all its complexities, the power to redistribute roles, shift questions and instantiate new affections. Empowerment is an *activation*. And as such it can be extended, at least speculatively, to the interdisciplinary challenge undertaken in Constitución. By attending to the valuing of differences and facilitating their full antagonistic deployment, sociologists and architects, as sorcerers, could have empowered a situation with the capacity to provoke new relations. They could have crafted a space for being 'in the presence of' their values and their consequences.

So, for example, as a political experiment 'in the wild' sociologists and architects could have debated the referendum in Constitución's main plaza, allowing everyone to pitch in. An open-air assembly without restrictions on who should talk and what should be said; an ad-hoc *palaver*[1] devoid of any teleological expectation: an interdisciplinary quarrel over the referendum as an object of public debate and starting point for imagining collectively not just what is needed to protect Constitución against oceanic forces, but also how collective decisions about the city will be made.

Alternatively, sociologists and architects could have used probes (Michael 2012), testing devices to speculate about possible futures and through which a zone of political indeterminacy could be opened. The inventory of possible futures to be tested would have needed to go beyond typical scenario planning – epitomized by the restricted Constitución with/without an anti-tsunami park figuration. For what is at stake here is how the dissent between sociologists and architects over democracy might have congealed different political futures. What type of world is enacted if human 'life' is valued over all other principles and secured univocally as a moral mandate? What does it entail for the political future of Constitución if consensus, even over material immunization against life threats, is prioritized? Sociologists and architects could have designed a game in which a decision opens a new set of options (and closes others) to articulate indeterminate political trajectories. A sort of strategy game or sandbox (Guggenheim, Kraeftner and Kroell 2013), played collectively, in different neighbourhoods, or installed in Constitución's main square for anyone to play. Such a probe, based on the interdisciplinary clash between sociologists and architects, could have ignited a discussion about the political future of Constitución that the participation machinery set up by planning experts did not address.

*Palavers*, probes, sandboxes. The main point is that interdisciplinary dissent over the referendum, offered and extended to those that will have to cope with future tsunamis, might have engendered a moment of idiocy (Michael 2012) or precarity (Tsing 2015) with significant political effects. If experts had slowed down their attempt at reaching an agreement, if they had let their interdisciplinary disagreement flourish untamed, they might have allowed the people of Constitución to decide not just about the feasibility of the park, but also about the political ecology of things, affects and representations at stake.

## Note

1 *Palaver* [from *palavra*, 'word' in Portuguese] is a negotiation system utilized in pre-colonial Africa in which no principle for participating is imposed and all forms of knowledge are accepted. The term is often cited by Stengers as an alternative to liberal decision-making.

## References

Deleuze, G. and Guattari, F. (1987). *A Thousand Plateaus: Capitalism and Schizophrenia*. Minneapolis: University of Minnesota Press.

Guggenheim, M., Kraeftner, B. and Kroell, J. (2013). 'I don't know whether I need a further level of disaster': shifting media of sociology in the sandbox. *Distinktion*, 14(3): 284–304.

Latour, B. (2004). *Politics of Nature: How to Bring the Sciences into Democracy*. Cambridge, MA: Harvard University Press.

Latour, B. (2007). Turning around politics: A note on Gerard de Vries' paper. *Social Studies of Science*, 37(5): 811–820.

Law, J. (2004). *After Method: Mess in Social Science Research*. London and New York: Routledge.

Law, J. and Singleton, V. (2005). Object lessons. *Organization*, 12(3): 331–355.

Michael, M. (2012). Toward an idiotic methodology: de-signing the object of sociology. *The Sociological Review*, 60(S1): 166–183.

Mol, A. (2002). *The Body Multiple: Ontology in Medical Practice*. Durham, NC: Duke University Press.

Pignarre, P. and Stengers, I. (2011). *Capitalist Sorcery: Breaking the Spell*. Basingstoke and New York: Palgrave Macmillan.

Rancière, J. (1999). *Disagreement: Politics and Philosophy*. Minneapolis, MN: University of Minnesota Press.

Stengers, I. (2005). The cosmopolitical proposal. In B. Latour and P. Webel (Eds.) *Making Things Public* (pp. 994–1003). Cambridge, MA: MIT Press.

Tironi, M. and Farías, I. (2015). Building a park, immunising life: environmental management and radical asymmetry. *Geoforum*, 66: 167–175.

Tironi, M. and Sannazzaro, J. (2017). Hulliche Energy: Experiments in participation and ontological disagreements in a wind farm. *Revista Internacional de Sociología*, 75(4): e080.

Tsing, A. (2015). *The Mushroom at the End of the World: On the Possibility of Life in Capitalist Ruins*. Princeton, NJ: Princeton University Press.

# 6

# Exemplifying

*Tuur Driesser*

To give an example is always to give more than the example itself, to hint beyond the concrete at something larger, more general. Yet, there is no easy relationship between the example and that which it exemplifies. A single example cannot be generalized to the structure, theory, rule or argument without running into objections of external validity and academic rigour. Instead, as will be argued in this chapter, the example has its own logic, reworking established oppositions between the particular and the general, opening up new creative and imaginative relationships between them. Exemplifying, as a method, makes use of the capacity of the single instance to produce other forms of knowledge, by making intelligible complex contexts. It opens up its own space/time in which the part and the whole become known together. This essay will discuss this logic of exemplarity and the methodological implications of exemplifying. In doing so, it will consider how the example and the process of exemplifying relate to issues of interdisciplinarity.

## The example

Examples are controversial devices in conceptualizing the relationship between universality/generality/whole on the one hand and particularity/singularity/part on the other. Their relationship to the general is complex and paradoxical: they at once produce and are governed by the rule that defines them. Likewise, the rule simultaneously exceeds and is defined by any concrete instances in the process of exemplification. Indeed, what is at stake in the process of exemplification, according to Lowrie and Lüdemann (2015: 2), is the unstable tension between 'flat and hierarchical methods of ordering'. Too much towards the horizontal and the example loses its capacity to extend beyond itself; it is left on its own. Too much towards the vertical and the example becomes just an instance, generalizable, fully defined or governed by its rule. As Massumi (2002) describes, 'the success of the example hinges on the details', but at the same time, 'at each new detail, the example runs the risk of falling apart, its unity of self-relation becoming a jumble'. Similarly, writing about the use of examples within anthropology, Højer and Bandak (2015: 6) argue that 'the example does not invert the vertical analytical movement but rather points to a "lateral" rethinking of the relation between the particular and the general, ethnographic material and theoretical reflection'. Thus, 'to give an example is a complex act' (Agamben 2009: 18;

see also Meskin and Shapiro 2014), that requires a balancing of the horizontal and the vertical, the flat and the hierarchical.

For Agamben (2009), too, the significance of the concept of the paradigm, used interchangeably with the notion of the example, lies in the value of a concrete case. As such, it must be considered within the whole trajectory of Agamben's philosophy and his longstanding concern with dissolving the dichotomy between the universal and the particular, the general and the singular (Meskin and Shapiro 2014). The paradigm's role within this trajectory is that it breaks with the 'binary logic' – the 'total abandonment of the particular-general couple as the model of logical inference' – and should be located more appropriately within a 'force field traversed by polar tensions', in a domain of analogy rather than of logic. In other words, the example does not so much abandon the distinction between the general and the singular, but rather their hierarchical dichotomy (see also Harvey 2002).

What is at stake then, according to Agamben, is not a question of the transformation of the particular into the general, of a case into a theory, but the relationship between 'a singularity (which thus becomes a paradigm) and its exposition (its intelligibility)' (2009: 23). This exposition – which should be understood in the sense of an exhibition, a demonstration, rather than an uncovering – of the case as paradigm involves the demonstration of the conditions under which it can come to be known as such. These conditions are not external to the case, where the question would be if the case meets the conditions to qualify as paradigm, but are 'immanent' (p. 31) to the example.

In the example's force field, generalization is no longer of interest: 'as in a magnetic field, we are dealing not with extensive and scalable magnitudes but with vectorial intensities' (Agamben 2009: 20–21). Generalization presupposes a knowledge external and prior to the particular, against which the case in question can be measured. Højer and Bandak (2015: 11) describe this distinction between generalizability and intelligibility/knowability in terms of evidence and exemplification:

> With exemplification, the question of veracity and validity – that is, of finding proof ('What is this phenomenon proof of?') – turns into a question of how to produce imagination and potentiality ('What can this example evoke?'). The move from evidence to exemplification is thus a move from the passive provision of evidence from an already established viewpoint in a disciplinary tradition ('We know what we are looking for but can we find it?') to the active making of convincing connections *from within* the example ('Can we find other things by (imaginatively) using what we have found?').

With the example, attention turns to a concern with intelligibility, or knowability. While generalizability presupposes the existence of a general rule of which a particular case can be found more or less representative, for example, the rule '(if it is still possible to speak of rules here)' (Agamben 2009: 21) does not exist outside of the instances that are governed by it.

The relation between the paradigm and its intelligibility is exhibited, exposed, through what Agamben has previously called the example's 'exclusive inclusion' (Agamben 2002). This is where the 'para' comes in: the case is placed beside itself, beside the rule that decides what it is a case of: 'it shows "beside itself" *(para-deiknymi)* both its own intelligibility and that of the class it constitutes' (2009: 24). In this showing-beside, the phenomenon is temporarily suspended from its 'normal function' (ibid.), excluded from the class to which it belongs in order to make clear the grounds for its inclusion. In the process of exemplifying, the example 'defines the intelligibility of the group of which it is a part and which, at the same time, it constitutes'. It is this exclusive inclusion that makes possible the simultaneity or coincidence of the case with its class that distinguishes the paradigm's intelligibility from both the representative case's generalizability and the particular case's uniqueness. That is, what is made intelligible does not precede the paradigm: it is produced at the very moment of the latter's suspension; or more accurately, not the class as such, but the paradigm's mode of belonging, its way of relating to it is what is 'exposed' (Meskin and Shapiro 2014: 428) by its exclusive inclusion, its being placed alongside.

## The now of knowability

The exclusive-inclusive character of the example also brings into view the temporal dimension of exemplifying. In response to Agamben, Weber draws attention to Benjamin's concept of the 'now of knowability' (Weber, in Agamben 2002) – the moment, or time/space in which the example's knowability is developed. This now, in its exclusive inclusion, its suspension of the normal function, is a 'cut' – a moment of separation in which 'what is involved . . . is not so much the act of [knowing] as the virtuality of . . . becoming-[knowable]' (paraphrasing Weber 2008: 50–51):

> Such a now is an *Augenblick*, the *glance* of an *eye* whose sight is always split between what it is and what it sees. In such an instant, what becomes possible is not simply knowledge as reality, but knowability as ever-present possibility.
>
> *original emphasis*

What is important for Weber is the distinction between knowability and knowledge, and the importance of the former in its own right.

> Such 'knowability' is not, for Benjamin at least, simply a preface to its realization as full-fledged knowledge. It has its own dignity, precisely as potentiality, and above all, it has its distinctive structure. It is this structure alone – which is that of *awakening* as distinguished both from consciousness and from unconsciousness – that explains how and why knowability, whose manifestation is inseparable from its vanishing, cannot be reduced to the positive knowledge it both makes possible and relativizes.
>
> *2008: 168–169*

The now of knowability – the exclusive inclusion, the suspension from normal function – is therefore not simply a snapshot, holding still the object of investigation. It is characterized by a structure of awakening, which should be understood, not in a revolutionary sense of becoming conscious, of waking up, but as a process of positioning, of gaining a sense of direction in both time and place in the world.

The temporality of the paradigmatic method, therefore, is not about the trajectory from example to knowledge, but about the now of knowability which has its own space and time; which has a movement of itself or rather, which is movement in itself. This is what Weber (2008: 171) writes about the spatial and temporal ('the one conditions the other') characteristics of awakening:

> The (person) awakening never wakes up in general, but always in and with respect to a determinate place. The locality in turn is never closed upon itself or self-contained, but opened to further relationships by the iterations that take place 'in' it. To be sure, such iterations are never infinite, they will always *stop*, but that stopping will never amount to a conclusion or a closure. Rather, it will be more like an interruption or a suspension. A cut.

In everyday use, examples – if successful – are grasped instantly, with it being immediately understood what it is that the example exemplifies. If unsuccessful, the example might fall flat, failing to make anything intelligible, or (perhaps even worse) exemplifying something entirely different from what was initially intended. As a method, in contrast, the process of exemplifying is tasked with expanding the now of knowability. Paying attention to this now of knowability means asking not how the case makes known that which is interesting about it, but how, as a paradigm, it facilitates a particular mode of knowing – as a form of positioning, determining direction, an extension beyond itself. It is this now that seizes on the example's ability to 'proliferate, connect, and absorb', in which 'exemplification multiplies, makes connections, and evokes (the one becomes many)' (Højer and Bandak 2015: 12).

The example might lack external validity in the traditional sense but, as a method, exemplifying has its own way of validating – one that is no less rigorous than that under the criteria of generalizability. With the rejection of the latter, and the focus on intelligibility or knowability, the example is never just out there to be found, but always has to be made. In response to a question about the limits of the paradigm, Agamben (2002) speaks of 'the ability of the author to find and create the good paradigm'. Find and create – in other words, the giving of an example is about uncovering as much as it is about constructing. An example is not first defined before being analysed: 'in the paradigm, intelligibility does not precede the phenomenon; it stands, so to speak, "beside" it' (Agamben 2009: 27). Methodologically, an example is not selected to prove a particular point that is known in advance. It involves a somewhat messy or blurry process in which what is exemplified only becomes clear gradually. In exemplifying, both the general and the particular, are constructed, uncovered and exhibited together, slowly and iteratively. Indeed, the importance of the example is that it suspends and problematizes its relation to what it exemplifies, so that the unexpected, the incompatible and the unresolved can become of value to the researcher. This is a process of validation in which both that which needs to be validated and the criteria against which the validating will need to take place hang in the balance.

## Exemplarity and interdisciplinarity

One way of conceptualizing the role of examples in the formation of disciplines is through Kuhn's work on paradigms. Here, disciplines are established and develop through those 'shared examples' of 'concrete problem-solutions' that students and researchers find in textbooks, laboratories and scientific literature (Kuhn 1970: 187). Even in the absence of formal rules (i.e. vertical), these examples can still guide research by way of the lateral movements they enable. In the same vein, examples of interdisciplinary research, problem-solutions and problem-formulations can enable and inspire interdisciplinarity, even when formal rules are necessarily hard or maybe even impossible to articulate.

Another approach is suggested by Lowrie and Lüdemann (2015) who link different configurations of 'flat and hierarchical methods of ordering' to specifically disciplinary commitments. In other words, what defines a discipline is not so much, or not only, its shared examples, but also its mode of exemplification:

> Many of the disputes in the history of thinking about examples turn on the competition between flat and hierarchical methods of ordering, depending on whether the radical singularity of the thing or person exemplified or its conceptual subsumption is privileged.
> *Lowrie and Lüdemann 2015: 2*

The way in which this balancing is eventually played out, it is argued, differs per discipline, with each having a different preference towards the flat or the hierarchical. Therefore, while

Højer and Bandak (2015) argue that exemplification is distinctly anthropological, it must be added that this is a specific mode of exemplification. In the same way that disciplines have different generalization strategies (Moriceau 2010), so too do they value different exemplification strategies. In this way, exemplification becomes another fault line along which disciplines can relate to or differentiate from one another.

Can interdisciplinarity develop its own mode of exemplifying? This seems unlikely, and probably even undesirable, as the endless possibilities of interdisciplinary constellations would defy any stable articulation of a specific mode of exemplifying. There is no one interdisciplinarity, no interdisciplinary discipline. But perhaps it is precisely because of this diversity that interdisciplinarity can draw inspiration from the example. For Harvey (2002: 209–214) the formulation of a theory of exemplarity is not only a methodological or philosophical project, but also a political one. Its complex reworking of the general/particular opposition undermines binary modes of thought which, she argues, are at the heart of the colonial/'patriarchal' system. Recuperating a logic of the example then serves to shatter the violence of binary oppositions that lies at the heart of the violence of the exclusions of otherness. In this framework, the example itself becomes an example for emancipatory modes of thought. Interdisciplinary research, too, can follow the example's example of rejecting binary oppositions; not in a bid to dissolve all difference, but to enable movement across. Here, interdisciplinary research expands its own 'now' in which that allows disciplines to extend beyond themselves, for problem articulations and solutions to multiply and proliferate, facilitating new orientations, and provoking imaginative associations. In other words, the example itself as exemplary for interdisciplinarity.

## References

Agamben, G. (2002, August). *What is a Paradigm?* (Transcribed by Max van Manen). Lecture, European Graduate School.

Agamben, G. (2009). *The Signature of All Things: On Method*. New York, NY: Cambridge, MA: Zone Books, distributed by the MIT Press.

Harvey, I. E. (2002). *Labyrinths of Exemplarity: At the Limits of Deconstruction*. Albany, NY: State University of New York Press.

Højer, L. and Bandak, A. (2015). Introduction: the power of the example. *Journal of the Royal Anthropological Institute*, 21(S1): 1–17.

Kuhn, T. S. (1970). *The Structure of Scientific Revolutions* (2nd ed.). Chicago, IL: University of Chicago Press.

Lowrie, M. and Lüdemann, S. (2015). Introduction. In M. Lowrie and S. Lüdemann (Eds.) *Exemplarity and Singularity: Thinking through Particulars in Philosophy, Literature and Law* (pp. 1–15). Oxon, UK: Routledge.

Massumi, B. (2002). *Parables for the Virtual: Movement, Affect, Sensation*. Durham, NC: Duke University Press.

Meskin, J. and Shapiro, H. (2014). To give an example is a complex art: Agamben's pedagogy on the paradigm. *Educational Philosophy and Theory*, 46(4): 421–440.

Moriceau, J. L. (2010). Generalizability. In A. J. Mills, G. Durepos and E. Wiebe (Eds.) *Encyclopedia of Case Study Research* (pp. 420–422). Thousand Oaks, CA: Sage Publications.

Weber, S. (2008). *Benjamin's -abilities*. Cambridge, MA: Harvard University Press.

# 7

# Explaining

*Priska Gisler*

Imagine an interdisciplinary research presentation setting: a curator of a well-known museum in a European capital gives an introductory talk on the occasion of the opening of a new exhibition. The work of art she comments upon has been produced in the context of a mixed-methods artistic and social-scientific research project. In her talk the curator describes the work on display as part of an ongoing research project and, in addition to the artist, she credits next the research team. She then goes through a whole range of aspects that make the artwork socially meaningful and artistically valuable. As a kind of explanatory overlay to the video-installations of the research-artist presented in the exhibition, the curator relates her interpretations of some of the broader social, historical and cultural debates underlying the project. Also, she ties them elegantly to the production-process of this work as well as to some points about the position of the artist within the art-scene.

The sociologist, a member of the same research team, is astonished. How is it that the artist herself remains silent about her own findings, and refrains from talking back to the curator who is elucidating research results without having produced them? The sociologist remembers the various moments during the research process when the team was discussing vividly their object of study, adding to each other's thoughts, mutually crafting relevant propositions. One particular contentious situation comes to her mind. She had asked – after the artist had already shown a work based on the project in a show – to discuss the display to find together some answers to questions that the work had raised for the sociologist.

Attending the opening of the show, the reaction of the artist lingers in her mind. She recalls her artist colleague declaring, back then, that she was never going to explain her own work. The sociologist was astonished, not so much because she expected the artist to come down with a causal explanation of the work, but because she had expected some elucidation of what the work meant or how one might develop it further. This moment still bothers her while she listens to the introductory talk of the curator: the then refusal of the artist to account for her work, to contribute to a discussion about the results she had produced. If the exhibited work was the product of artistic research, if it could be treated – as they had decided – as a publication, then surely it needed, in some way or the other, to be part of the effort to derive a deeper understanding of ideas, actions, social realities? She is sure the artist would not oppose this goal – but why had she refused to 'explain' then? How was the artistic research work in the show contributing to knowledge production, and what was the role of the curator in delivering a series of interpretations?

In this interdisciplinary situation, the sociologist starts to reconsider her own notions of describing, interpreting, accounting and – especially – explaining and understanding. She is reminded of the sociology she was taught during her university studies, a sociology relying on a Weberian notion of a science 'concerning itself with the interpretive understanding of social action and thereby with a causal explanation of its course and consequences' (Weber 1978: S. 3). Definitely, she now feels the need to reconsider the role and status of (an) explanation in the context of a collaborative, interdisciplinary artistic–social-scientific research project, and to rethink the differences among the team concerning the idea and practice of 'explaining'. This is why she re-reads Max Weber and finds quite revealingly that '(t)hus for a science which is concerned with the subjective meaning of action, explanation requires a grasp of the complex of meaning in which an actual course of understandable action thus interpreted belongs' (Weber 1978: 9).[1]

The refusal of the artist poses a problem for the sociologist, not least insofar as their research project has 'worked' as common endeavour. How, she reflects, is explaining and understanding – or better in Max Weber's terms: an 'explanatory understanding' – going to happen if not in and through the video-installation, which both of them, the sociologist and the artist, regard as a publication of their research results? But, the artwork does not do what a classical social sciences research paper does: a scientific article already contains explanation. The grasping of an actual course of understandable action does not necessarily happen in a strict rationalistic sense in which reasons for, or causes of, social facts are neatly laid bare. Much more importantly, explanatory understanding seems to be done in the way publications are structured (observation, theoretical approach, hypothesis or assumptions, modes of operationalization, presentation of the empirical data, acceptance or not of the outcomes, discussion of the results). And any claims about a contribution to understanding are tested against criteria of validation.

*Nolens volens*, the interdisciplinary research process had, then, found itself engaged in a debate about how to grasp the complex of meanings. Perhaps the artist's refusal to explain the work, the sociologist now thinks, can be comprehended better by following how the artistic field organizes explanatory understanding in ways different to what is found in social scientific traditions. Indeed, after a while, the sociologist becomes aware that the art-field more generally takes on the task of 'explaining' the complex meaning of a work of art. A video-installation as research publication thus gains its value and validity through particular forms of circulation within the art-field.

That it is not the artist explicating her own work is not unusual in the field of the fine arts. On the contrary, the framing and contextualizing of artworks in galleries or museums is usually done by others, be they curators, art critics or collectors. Framing is delicate work. While many artists tend, as they say, to offer a series of possibilities or to provoke a debate, they usually try to avoid closure for their work.

Although the interdisciplinary research team is used to discussing the research project in a confidential setting, this openness of discussion is, as it were, externalized once the (artistic) results are publicized. This reflects the fact that from the artistic point of view there is not one 'actual' insight, conclusion or even explanation that serves as a starting point for debate through which to gain connectivity to others. On the contrary, to offer a limited account would be to risk narrowing down the debate about the artist's work and thus limit its circulation. In fact, it would make it rather difficult for the artwork subsequently to be questioned by different art critics or, later, to be bought by collectors. The curator is allowed to try out many sorts of interpretative explications; indeed, curators are experts in coming up with multiple explications, though the expectation is that they do so in a 'laudable' manner, pointing to artistic skills and deeds done, and highlighting the insights that can be gained. Crucially, as soon as the curator has spoken, the explication 'inhabits' the artwork. Every following comment, question, remark will do the same: they each enrich the work and allow it to take on a life of itself. The artist – by claiming to hold back her own explanation – effectively obliges others to do the talking. By directing most of the attention towards the work itself, ironically, she enables its mobility. It is the broader art-field that, through the production of written material (e.g. the catalogue, the flyer in the exhibition, press articles by critics, etc.), demonstrates – validates – how well the piece fits in an exhibition, in a museum, and evaluates the suitability of the work to be positioned in a wider social and artistic context. In the accumulation of views from the curator, the critic, the collector or buyer, the museum and the 'art world', a distributed network of explanatory understanding emerges.

To summarize: after exploring the nature of artistic and social-scientific disciplinary practices that have been brought together in the interdisciplinary context, the sociologist becomes aware that the division of labour applied (in the production of artistic or social-scientific knowledge) plays a role in the ways the explanatory understanding of research outcomes is organized. An artist might generate material (not data, as the sociologist would tend to say) by observation, interviews, dreams, memories, etc. The material is at some point condensed in an artistic work that, hopefully, then will be invited to a show and presented to an audience. The curator offers framing: interpretations, analyses, and additional information. The positioning she suggests, subsequently, will be continued by art critics, buyers, collectors and on and on and, hence, will play part and parcel into the transformation of knowledge and the circulation of meaning.

Comparing the artistic work in the exhibition – as a result, a research publication – to a social scientific publication, some commonalities and differences emerge. Social scientists tend to collect and interpret data, they seek to narrow down the significance of the material and work out an analysis as well as some conclusions before they present their results at conferences and in papers. In journals or in public the aim is to withstand – with one's own explications – a critical interrogation by the community. Intrinsic to this process, the social sciences have developed a system that integrates a question, a theoretical approach, a methodology and an interpretation in one (paper, presentation, etc.). This involves a chain of reference, as described by Bruno Latour (1999) that allows the reader to trace step by step the movement from empirical engagement to final textual account, that is, to bridge the distance between the word and the world.

Seen through the lens of the interdisciplinary setting, though, for the arts a different mechanism becomes evident. The video–installation in the exhibition is – as in sociological work – a result of structuring procedures. However, unlike in sociological work, these remain largely invisible, locked within 'the actual practices which generate accounts' (Smith 1974: 262). Instead, other 'external' structuring procedures are available, not least evidenced in the museum event in which the curator finds, more or less successfully, the words to communicate 'explanatory understandings' to the public. More broadly, an array of interested others – among them most prominently collectors, gallery owners, museum directors – function like a scientific community, whose 'boundary work' serves to stabilize the value and 'validity' of the artwork (Gieryn 1999). Thus, it is only together and cumulatively that this 'community' contributes to an explanatory understanding. In this distributed manner, Weber's 'grasp of the complex of meaning' is possible and gives some background insight into what might be at stake, what are the 'matters of concern' (Latour 2004).

The preceding account allows us to delve deeper into the term 'explaining' in relation to valuing and validating. So far we have depicted a curator doing the talking for the research work produced by an artist-researcher who refrains from explaining her artistic research result, while leaving the sociologist in this interdisciplinary setting confused. The sociologist contrasts against this her own practice, according to which a publication should offer its own explanatory understanding. Clearly, the voices of colleagues, friends, fiends in the same scientific field, will evaluate her scientific contribution. Step by step they interrogate the research process: Is the research question relevant? Was the method adequate? Do the results seem reasonable? Such validating procedures will point to numerous shortcomings, and propose amendments and alternative ideas about how better to treat, or understand the subject. Accordingly, the social scientific author will react and respond. By comparison, the artist will remain silent, inhibiting herself from giving 'explanatory understandings' of what can be seen, while the curator proliferates 'explanatory understandings' not least because drawing attention to, and sustaining interest in, the artistic work is linked to the standing and status of the exhibition, or even her museum. She continues with her own account, mediating the work on display: 'The mediation procedures directly enter into the constitution of the object as it becomes known', writes Dorothy Smith (1974: 264). The curator does not only provide a complex of meaning. By carefully selecting certain aspects and not others, she contributes to its composition (Michael 2004: 10). She co-constitutes it as well and, through enabling circulation, she adds to its valuation and validation.

What are the implications for the valuing and validating of interdisciplinary research methods? It is certainly a tricky task for an interdisciplinary research team to find ways of circulating their material, and the 'explanatory understandings', across or *trans* artistic and social-scientific disciplinary fields. In this case, for example, the research team has to bear in mind the differences in social science and art worlds in which there is, respectively, critique, debate and closure versus silence, distribution and proliferation. Put differently, how can artistic researchers 'attempt to attain clarity and certainty' and how can sociologists accept that 'no matter how clear an interpretation as such appears to be from the point of view of meaning, it cannot on this account claim to be the causally valid interpretation' (Weber 1978: 9)?

Ideas of ways to mediate between the two fields of the social sciences and the arts are still in a formative state, although developments seem to be increasingly evident. When scientists give tours and present their work in art exhibitions, when artists and social scientists 'mirror' each other in self-reflective interviews, or where members of interdisciplinary teams reveal some of their practices and articulate their heterodox experience in mixed (artistic research and scientific) publications (e.g. Dombois, Gisler, Kretschmann and Schwander 2012), formerly well-defined fields might suddenly yield a terrain for new assemblages.

At the opening of the show, the sociologist and the artist stand next to each other. The mediation of explanatory understanding will continue for them as long as they are bound in their common, interdisciplinary research context, personally and institution-wise. Arguably, today's valuation of projects by funding institutions (with their variegated interest for research 'output' and 'impact') increasingly encourages a *bricolage* of research activities. Indeed, there seems to be a growing expectation that different genres of output are necessary for the successful completion of a research project. This strategic refocusing suggests a shift in relation to what is validated as a result. In an ideal world this might imply an interest in broadening modes of enquiry to develop multifaceted ways of seeing – but as yet it is difficult to judge.

## Note

1 In the German original: '"Erklären" bedeutet also für eine mit dem Sinn des Handelns befasste Wissenschaft soviel wie: Erfassung des Sinnzusammenhangs, in den, seinem subjektiv gemeinten Sinn nach, ein aktuell verständliches Handeln hineingehört" (Weber 1972 (German version), p. 4).

# References

Dombois, F., Gisler, P., Kretschmann, S., Schwander, M. (2013). *Präparat Bergsturz*. Band 2, edizioni periferia, Luzern und Poschiavo.

Gieryn, T. (1999). *Cultural Boundaries of Science: Credibility on the Line*. Chicago, IL and London: University of Chicago Press.

Latour, B. (1999). Circulating reference: sampling the soil in the Amazonian forest. In *Pandora's Hope: Essays on the Reality of Science Studies* (pp. 24–79). Cambridge, MA and London: Harvard University Press.

Latour, B. (2004). Why has critique run out of steam? From matters of fact to matters of concern. *Critical Enquiry*, 39 (Winter): 225–248. Retrieved from www.bruno-latour.fr/sites/default/files/89-CRITICAL-INQUIRY-GB.pdf

Michael, M. (2004). On making data social: heterogeneity in sociological practice. *Qualitative Research*, 4(1): 5–23.

Smith, D. E. (1974). The social construction of documentary reality. *Sociological Inquiry*, 44(4): 257–268.

Weber, M. (1972). *Wirtschaft und Gesellschaft*. Tübingen: Mohr Siebeck.

Weber, M. (1978). *Economy and Society: An Outline of Interpretive Sociology*. Berkeley, Los Angeles and London: University of California Press.

# 8

# Generalizing

*Joanna Latimer and Rolland Munro*

The aim in this chapter is to open up issues around generalizing in social science research. We counter-point a tendency to evidence explanatory hypotheses through the adding up of specific 'confirmatory' observations. Our radically different approach to understandings of 'world-making' through sustained contemplation of the situated instance. What we are putting into question is the modernist insistence on attaching a valuation of 'one' to any incident or event: to denote an instance as merely a 'specific case', a 'single example', a 'one-off'. This is not just because the one-off can be dismissed as irrelevant, to be treated as either something of little or no account or as the deviation that proves the rule. Rather, as ethnographers in two different fields, we both cherish the value of the *exemplar*, the moment, incident or instance that sums up what is going on and which seems to throw everything else into light.

One of the aporias of science is the multiplication of generalizations that are required to keep pace with a never-ending process of division within research. The analytic mode results in splitting phenomena, events and substance into ever-finer parts, a process that was supposedly put into reverse by the use of logic for its re-synthesis into meaningful theory. Yet, as David Hume pointed out a century or so after Descartes revitalized interest in the analytical approach of the Greeks, logic is unable to add back the meaning that analysis has subtracted. This is because formal logic prohibits any move to generalize from the adding up of parts. While it is possible to begin with a so-called universal like 'red' and proceed to find examples that appear to have the quality of redness, no amount of observations of swans permits whiteness to be the essence of swans.

The subsequent implosion of knowledge in the West has thus entailed a loss in meaning, a loss that has gone hand in hand with the fashion for metrics: measurement having long since been employed to reinstate what logic fails to add back. Inevitably, however, since the potential for measurement multiplies as fast as division works its possibilities, meaning becomes a receding horizon. As Dave Beer (2016) argues, at work, in the body, and in leisure and consumption, uncertainty grows apace with the pressure to measure everything. It is as if people are seeking stabilities in identities by bolstering the ever-growing detail of life with ever-greater refinements in terms of numbers. Whatever counts counts; and the story of attempts by colleagues at the University of Edinburgh seeking to remove Higgs (of Higgs Boson fame) from his post because of a lack of publications for the research audit is salutary for all those magnifying themselves by watching their citations grow on sites such as Research Gate.

Pressing questions remain, nonetheless, over how, and when, the single instance is admissible. This is an issue that dominates more popular debates about generalizability; and rightly so in the face of governmental policies being driven by demands for more and more quantitatively based evidence. Examples of older people being left unattended on hospital trolleys for hours should be shocking and unacceptable, but such failings should not be interpreted as meaning the NHS is an institution that must be privatized. So, too, the call for quantitative evidence also affects the politics of everyday life: while we might feel sympathy with a person who was bullied at school in ways that soured their later life, we can all feel irritated when someone draws too broadly on an unfortunate experience of their own, such as a failed marriage, to argue marriage should be avoided by everyone else.

This brings us back to Hume's problem of induction. The positivist movement, which arose in the late nineteenth century and gathered pace in the early twentieth century, attempted to get round the embargo on turning observations ('This planet has life on it') into law-like theories ('All planets have life on them') by a reversal in direction that insisted on all 'meaningful' generalizations having to be stated in a theoretical form that permitted testable predictions to be drawn from them ('If all planets travel in an ellipse, then the Earth's orbit is elliptical'). This opened the way for Karl Popper's argument that, since science could never prove its law-like theories to be true, it had to look for contrary instances that falsified the theory. Indeed, increasing attention is now given to looking for the contrary case, the unpredictable outcome, as what has become popularized as 'black swans' (Taleb 2007).

World-making – in contrast to the illusion that there is a single, discernible universe 'out there' to which all pronouncements and propositions must refer – is directed towards knowing how each of us is drawing our ideas and experience together in ways that stabilize the 'buzzing, blooming confusion'. As Latour (1987: 101) explains, what matters in science is not the truth of each single proposition, the conventional view about it, but whether or not things *hold together* more generally:

> A sentence does not hold together because it is true, but because it holds together we say it is 'true'. But what does it hold onto? Many things. Why? Because it has tied its fate to anything at hand that is more solid than itself. As a result no one can shake it loose without shaking everything else.

Attempts to name this most elusive of human phenomena vary: complementary to the idea of world-making, anthropologists attend to cultures as shaping ideas, mores and beliefs, Kuhn referred to paradigms as generating theories in science, while Foucault talked about truth regimes and explained how discourse could discipline thought. So, too, the less totalizing notions of perspective and standpoint have taken on an extra weight of meaning, with (the later) Goffman also identifying 'framing' as exercising this kind of force.

The notion of world-making seems especially appropriate, therefore, for research in social science, where a key aim is to comprehend ways in which world-making is *organizing* what we or others are engaged in (Latimer and Munro 2015). Insofar as this requires a shift away from a focus on *what* we think, towards imputing what might be governing our thought, it might be that a more fecund alternative rests on the notion of *working backwards*. While a partial step towards this is captured in the pragmatist Charles Peirce's notion of abduction (a method distinct from induction and deduction in its positioning of generalizing as a heuristic), the roots of its more profound influence are to be found in Heidegger. Instead of finding himself always looking for 'something', asking say if it existed or not, Heidegger's (1959) deconstruction of metaphysics proceeded by questioning and re-questioning 'Why is there always something?' In persisting in his contemplation of this single question, he was eventually able to break open the stranglehold that Western metaphysics imposed on the existence/non-existence of things and ask the very different question 'Why is there something, rather than nothing?'

In reversing the usual mode of direction of generalizing, researchers in social science can ask: What are the conditions of possibility for a particular event to have taken place? This is to wonder: Can we re-visualize the world to make sense of what someone has done? Or ask: What kind of moral order would permit a person to engage in head-hunting (Rosaldo 1993)? This kind of questioning would also encompass coming to understand the very *absence* of events; for example, as Mary Douglas (1975: 4) noted, allowing a menstruating woman to cook would cause the cosmos of the Lele to collapse. Consider Marilyn Strathern's 'glimpse' of men half-running, with their path out of her field of vision:

> On my part, I shall never forget my first sight of mounted pearl shells in Mt Hagen, in 1964, heavy in their resin boards, slung like pigs on a pole being carried between two men, who were hurrying with them because of the weight, a gift of some kind.
>
> *Strathern 1999: 8*

As Strathern says, she will never forget this image – the 'dazzle' of what she saw stayed with her over decades. Yet what is the status of such a 'glimpse'? Strathern (1999: 6) argues that seeing the men half-running with the weight of the mounted pearl shells – and it would prove that it had to be men because only men could offer this kind of gift since women were not allowed to handle pearl shells – was for her an *ethnographic moment*. For Strathern it was the relation that joined 'immersion' in the field of observation (concrete apprehension of facts) with 'movement' in the field of writing (abstract analysis). As she notes, 'What makes the ethnographic moment is the way in which these activities are apprehended as occupying the same (conceptual) space' (1999: 262).

Addressing the particular can also be an approach that constitutes the instance as an 'event' (Deleuze 1992): an irruption, bifurcation, a moment of irreducible dissonance, a transition that opens. We can note, then, when and how an account or a performance is treated Goffman-like (1963) as 'infraction' or even as one of Garfinkel's (1996) 'breaches'. Such moments can indicate what is held by some participants to be most precious in the culture concerned; and so work to preserve, or even intensify, the sacredness of what may 'pass' as acceptable and what may not. Alternatively, their value might also lie in their instantiating a moment of resistance, a moment in which the usual modes of ordering and backgrounding suddenly let the implicit come into view, including all that makes the truth hold.

So particulars need not be about representing through discounting specifics into numerals (or their verbal equivalents: one, many, most). This brings us back to the question of how does a Strathern make their examples trustworthy? As revealing something, rather than saying nothing? There is, to be sure, value in the 'surprise' finding, the contrary example that puts into doubt all that has been known before. But the true value of the illuminating example comes not only from evading the trap of making the example hold as a particular 'representation' of a universal truth or 'social fact', or from the rigour of the researcher's cross-checks and balances alone.

Good examples are much more profound: they are about grounding the part in the fabric of relations and associations, connections and disconnections that makes the particular possible. For example, the particular of what Riles (1998) calls 'a figure seen twice' is about bringing into the light the conditions in which the figure is being made possible. This is why Bauman (1989) is opposed to naming men like Hitler or Stalin as 'monsters'. As he suggests, such easy nomenclature overlooks how the mores of administrative theory shape and adiaphorise social relations in ways that enable the effacement of particular forms of personhood, and heinous acts to be carried out. The division of labour into specialist roles ensures that those who gather the names have no direct relation to the soldiers who put the people on the trains; with neither having any connection to the guards who run the camps or the people (sometimes prisoners themselves) who are given the task of pressing prisoners into the gas chambers.

The notion of world-making is also highly germane to interdisciplinary research. Inasmuch as proliferating specialization through the division of labour governs the growth of knowledge, it also affects the methods by which researchers from different schools or traditions can come together and collaborate. In this respect, it is not only the world-making of those being studied that should be the focus of research in attempting to understand what the devil it is they are up to. By the same token, researchers are encouraged today to be more reflexive about their own orientations and pre-judgements. This kind of exemplification is unfolded by Helen Verran (2001) in her description of the disconcertment she felt in Nigerian classrooms when observing one of her student Nigerian teachers conducting maths and science classes: the intense feeling that something in someone else's world-making is out of order, a sense that opens up the possibility of a different kind of world-making and confounds any possibility of a generalizing logic.

Certainly, this opening up of different kinds of world-making becomes a pressing concern in collaborative projects that draw together researchers from diverse disciplines and different countries. And, critically, attention to this issue can promote collaboration, dialogics and mutual reflexivity. Indeed, rather than take the positive road of looking to make predictions on the basis that there could be a robust general theory for different cultures, it is almost now a commonplace to attend to the possibility of multiple worlds, multiple logics, multiple identities. Yet attempts to generalize about the diversity of what is underpinning thought and action may not entirely allow for a different kind of multiplicity in world-making. This is to remember that extension takes place through relations of all types. Hence world-making not only takes place within the connections people make as they interpret things and relate ideas to their experience, worlds also get shaped and pre-figured through the agency of non-humans and within boardrooms of power and behind the scenes in clandestine meetings among people of influence. As such it seems preferable to assume that such matters as *Zeitgeist*, belief systems and perspective only *contribute* to world-making, but never act alone.

Ahead of closing, therefore, we want to go one step further and press another facet to generalizing that even these more radical forms of research overlook. As authors we have taken care to introduce into our own research the notion of motility (Munro 1996; Latimer 2013), allowing people to be much more motile, both in the self they perform and in the worlds they endlessly construct. So rather than identifying how many people act or speak in accordance with institutionalized routines and repetitions, or deviate in well-trodden ways, we have tried to explicate how *switches in extension* take place from moment to moment. Extension here could include attachment to tools, narrative tropes, situations, or involve other persons. Importantly these materials of extension body-forth specific meanings, ways of seeing or doing the world, by people as they go about their everyday lives. Such switches in extension thus alter the world, even if only momentarily or fractionally, and do so without disposing of existing or alternative possibilities, as is implied by concepts such as mutation and evolution.

Indeed, the potential to exercise motility may underpin power relations. Our drawing attention to how different worlds are kept in play, therefore, is not simply to point to the making and unmaking of identities (and worlds) as fluid (as if anything goes). On the contrary, our emphasis on motility is to help researchers attend to how, when and where switches in extension take place. Noting such matters helps illuminate the complexity at stake in how, when and where stabilities are being accomplished and re-accomplished. This is to adduce the precise moments and places when switches are made, including identifying how motility can become common-place in accomplishing power relations at work (Munro 1999). And it is through this attention to how people and things become attached and detached that the stabilities or 'conditions' underpinning world-making can not only be made visible, but observed to change from time to time however momentarily. For example, Latimer (2013) in her study of the partial alignment of medicine and the new genetics, helps reveal how medical dominance is accomplished and re-accomplished not through purification, or the enactment of the scientific method alone, but also through switches in extension. In terms of motility sustaining medicine's dominance, those practising medicine can be observed to switch their attachment – one minute they perform the medical gaze as a pure clinical moment through which people and their parts are objectified, in the next they reattach to persons as human beings.

In emphasizing these different facets of world-making, we have contested the orientation to generalize along a single line of investigation, severing relations in order to add up specific examples that are taken to be similar in one way or another. To generalize in social science requires researchers to remain open to the manifold of relations and the far-reaching possibilities in world-making that connection (and disconnection) may take. So let us ask again: What value can we give the particular? Should we treat the exception as idiosyncratic, merely reinforcing the case for the normal by limiting their value to a form of deviance? Or do such matters draw attention instead to how people and their worlds are 'motile' – not only in manifesting different identities as they say or do things, but in effortlessly shifting their world-making from moment to moment as they alter their habits and routines from place to place?

## References

Bauman, Z. (1989). *Modernity and the Holocaust*. Cambridge: Polity Press.

Beer, D. (2016). *Metric Power*. London: Palgrave Macmillan.

Deleuze, G. (1992). What is an event? In T. Conley (trans.) *The Fold, Leibniz and the Baroque* (pp. 86–94). Minnesota: University of Minnesota Press.

Douglas, M. (1975). *Implicit Meanings: Selected Essays in Anthropology*. London: Routledge.

Garfinkel, H. (1996). *Studies in Ethnomethodology*. Cambridge: Polity Press.

Goffman, E. (1963). *Behaviour in Public Places: Notes on the Social Organization of Gatherings*. New York: The Free Press.

Heidegger, M. (1959). *An Introduction to Metaphysics* (trans. Ralph Mannheim). New Haven, CT: Yale University Press.

Latimer, J. (2013). *The Gene, the Clinic and the Family: Diagnosing Dysmorphology, Reviving Medical Dominance*. London: Routledge.

Latimer, J. and Munro, R. (2015). Uprooting class? Culture, world-making and reform. *The Sociological Review*, 63(2): 415–432.

Latour, B. (1987). *Science in Action: How to Follow Scientists and Engineers through Society*. Cambridge, MA: Harvard University Press.

Munro, R. (1996). The consumption view of self: extension, exchange and identity. In S. Edgell, K. Hetherington and A. Warde (Eds.) *Consumption Matters: The Production and Experience of Consumption* (pp. 248–273). Sociological Review Monograph. Oxford: Blackwell.

Munro, R. (1999). Power and discretion: membership work in the time of technology, *Organization*, 6(3): 429–450.

Riles, A. (1998). Infinity within the brackets. *American Ethnologist*, 25(3): 378–398.

Rosaldo, R. (1993). *Culture & Truth: The Remaking of Social Analysis*. Boston, MA: Beacon Press.

Strathern, M. (1999). *Property, Substance & Effect: Anthropological Essays of Persons and Things*. London: Athlone.

Taleb, N. N. (2010). *The Black Swan: The Impact of the Highly Improbable* (2nd ed.). London: Penguin.

Verran, H. (2001). *Science and African Logic*. Chicago, IL: University of Chicago Press.

# 9

# Interdisciplines, and Indigenous research and methodologies

*Catriona Elder and Jonathon Potskin*

The process and experience of doing research in the context of Indigeneity[1] is by default inter-disciplinary. Research focused on Indigenous peoples and cultures is a capacious and emerging area; that said, it is perhaps, instrumentally, described as Indigenous Studies or Indigenous research. As with other interdisciplinary fields, Indigenous Studies or Indigenous research comes with a prescription to rethink the questions asked, the theories chosen and the methods deployed in more traditional fields. What is particular about doing Indigenous research relates to the choice of Indigenous methodologies and/or epistemologies. Aspects of this choice relate to researchers as Indigenous persons, though the issues also relate to non-Indigenous researchers as well. As a co-authored chapter, written by an Indigenous person and a non-Indigenous person, the voices emerging across this narrative will tease out some of the ongoing tensions between the taking up of Indigenous epistemologies in Indigenous Studies, and the reproduction of colonial practices in an interdisciplinary space.

## Indigeneity and Indigenous research

The notion of Indigeneity or being Indigenous is different from its allied concepts of race and ethnicity (Elder 2007). Though all three can be understood in terms of powerful constructs that order understandings of communities and individuals in relation to bodies, cultures and differ-ence, Indigeneity has attached to it the additional issue of land and belonging (Behrendt 2012). As with race and ethnicity, Indigeneity is shaped by a long Western history of inequality, colonialism and oppression (Langton 1993). However, Indigenous peoples are understood, and understand themselves, in terms of their inalienable connection with, and belonging to a par-ticular geographic space (Brennan, Davis, Edgeworth and Terrill 2015). In Australia, this space is known as 'Country' (Roe 1983). Larissa Behrendt (2012) notes Indigenous people are of the land, they are formed through it and by it. Similarly, it is their custodianship of this land that enables it to continue to be productive (2012). There is a responsibility built into the relationship between Indigenous people and their land.

Indigenous worlds and Indigenous knowledges are ordered in terms of a different epistemo-logy and ontology compared to many other Western disciplines. In Australia, Indigenous epis-temologies are understood in terms of The Dreaming (see Morrisey 2015). Some of the more recent work undertaken in Indigenous Studies maps these epistemologies and translates them into frameworks for academic research (Martin 2007). A key shared characteristic of Indigenous epistemologies is that they are interpersonal by nature. In the Canadian context, Margaret Kovach explains that a 'Nêhiyaw epistemology is relational . . . so while I speak of knowledges (e.g., values, language), it should be assumed that they are nested, created, and re-created within the context of relationships with other living beings' (Kovach 2009: 47).

Much of the existing work on interdisciplinarity focuses on the institutional context in which research takes place. In this framework, the crucial question is often about the relationship of that which is interdisciplinary to the disciplines (Burawoy 2013). When assessing Indigenous Studies in relationship to traditional disciplines, as with gendered and raced approaches to scholarship, Indigeneity has enabled or sometimes forced a rethinking of key modes of Western knowledge (Rigney 1999). Paul Sillitoe (2007: 152) refers to these types of interdisciplinary challenges as a 'reordering of knowledge' and makes clear that they can be understood as 'threatening' to established disciplines. As Joe Parker and Ranu Samantrai (2010: 11) have noted 'interdisciplinarity can open up a plurality of ethics so that ethical knowledge practices may in the words of Andre Glucksmann "make appear the dissymmetries, the disequilibriums, the aporias, the impossibilities, which are precisely the objects of all commitment"'. *Indigenous Studies and research does precisely this.*

Indigenous research can be understood as the research that is conducted by Indigenous peoples worldwide, and is research that is 'for us, by us', that is, it is for Indigenous people and by Indigenous people. Shawn Wilson (2008) argues that '[t]he development of an Indigenous research paradigm is of great importance to Indigenous people because it allows the development of Indigenous theory and methods of practice'. Wilson's approach is useful when applied to disciplines. For example, he notes in relation to the discipline of psychology, that the emergence of Indigenous psychology enables Indigenous peoples to be the ones that decide what is 'normal' and 'abnormal', or even if this distinction needs to exist. This said, the development and/or recognition of a general Indigenous epistemology (as opposed to specific disciplinary interventions) can do much to shape and encourage interdisciplinarity.

Almost all modes of Indigenous Studies or research have a common form of experience whereby Indigenous people experience connection *to place* and *through relationships*. Place is often formed through two different experiences: the first is our place of origin, the second, our place within society. The role of relationships in Indigenous Studies and research suggests that we, Indigenous peoples, are connected to, and also within, our family, community, land and culture. This framework constitutes a distinct epistemology.

One of the issues that becomes apparent when thinking through Indigenous epistemologies is that many of the divisions set up in Western knowledge systems – especially those pertaining to the connection between the natural world and humans – do not have any traction. Some of the logics that drive new studies in Posthumanism are already part of Indigenous knowledges. For example, as the earlier quote from Margaret Kovach notes, Indigenous knowledges are relational. Bagele Chilisa (2012) makes a similar point and argues that in Indigenous epistemologies there is an emphasis on the I/We relationship as opposed to the Western I/You relationship with its emphasis on the individual. Drawing on the work of Nomalungelo Ivy Goduka, Chilisa recounts that among the Bantu people of southern Africa, the principle that captures the philosophy of Ubuntu, is expressed in the concept of being. This is, '*nthu nthu ne banwe*' (Ikalanga/Shona version). An English translation that comes close to expressing this principle is 'I am we: I am because we are: we are because I am' or 'a person is because of others' (Chilisa 2009: 413).

## Colonialism and Indigenous research

Aileen Moreton-Robinson has explained that for the past few centuries, shaped by colonialism, scholarly research to do with Indigenous peoples and cultures has emerged from an 'epistemic fixation with our [Indigenous] cultural difference' (2016: xv). This approach with its obsession with racial difference has shaped both the questions and answers, as well as the processes and the outcomes that are possible in Indigenous focused research. Karen Martin notes that this type of scholarship produces 'salvage research' (2007: 27). Further, the result of this mode of research has been an enormous body of knowledge that tends to centre Indigenous peoples as 'objects of study' (Moreton-Robinson 2016: xv). Salvage research emerged from the disciplines of anthropology, archaeology, linguistics and history. In many cases the outcomes were complicit with colonialist notions of the vulnerability of Indigenous cultures in the face of Western modernity and a need to protect or scoop up what was left (McGregor 1997).

## Instrumental interdisciplinarity and Indigenous research

More recently these same 'Western' disciplines have often been important in the setting up of interdisciplinary teams of researchers whose aim is to find answers to what are seen as problems for nations in relation to Indigenous peoples. In Australia, from about the 1930s, this has been referred to as 'The Aboriginal Problem'. This drive to problem solve is often described as drawing on, or creating, a type of instrumental interdisciplinarity (Kann 1979) where the state, through the work of the academy, and allied institutions, seeks to solve problems that frustrate the nation. This type of instrumental interdisciplinarity can also be understood in terms of notions of accountability (Strathern 2004). In this case interdisciplinarity is deployed as a practical coming together of the life sciences, and the social sciences or humanities, to make science seem to be more responsive to public ideas of its usefulness (Strathern and Rockhill 2013).

In the cases of Australia and Canada the reason for these instrumental or accountability approaches, deployed as they are to solve racial problems, is that they might contribute to the emergence of knowledge that goes some way to undoing the results of postcolonial injury (Kowal 2015: Chapter 1). The aim is to bring together scholars and practitioners from a range of fields to solve problems understood as being large scale and causing systemic life course disadvantages for Indigenous peoples in relation to non-Indigenous peoples. In Australia, the colloquial expression to describe the difference between these social indicators is 'the gap'. And so the aim in terms of accountability and instrumental interdisciplinarity is to 'close the gap', particularly in health indicators (Australian Human Rights Commission 2017).

In this sense, interdisciplinarity is often strongly tied to social justice. Kann (1979) was a scholar who quite early on noted the place for a mode of critical, rather than instrumental, interdisciplinarity. More recently Jacobs and Frickel (2009) have commented that the driving force of some interdisciplinary fields is akin to a social movement. They suggest the energy comes from 'collective action . . . in the face of resistance from others' (2009: 57). Some of the early and very powerful Indigenous Studies and research had this critical edge. This research involved a postcolonial critique of the Western knowledge regimes that underpinned salvage research.

As powerful as this resistance- or social justice-focused scholarship was, it meant that from the 1980s onwards much of the next generation of Indigenous Studies scholarship and research was framed as a response to, or a critique of non-Indigenous research processes rather than being able to focus on the development of innovative and productive research about Indigeneity. Again, to cite Moreton-Robinson (2016), the original Indigenous Studies research tends to refuse the 'density' of Indigenous experiences and instead the critical response tended to focus on the problems of a 'Western' method.

However, the application of Indigenous methodologies and epistemologies, such as those developed by scholars such as Wilson, Kovach and Potskin, in the interdisciplinary space of Indigenous Studies and research has produced knowledge that *re*-values Indigenous ways of understanding the world. This, now, sustained exploration and adoption of Indigenous methodologies has enabled Indigenous Studies and research to be 'a site of knowledge production' (Moreton-Robinson 2016: xvii). It has shifted Indigenous cultures from being an always lacking, object of research to something richer and more complex (Barry, Born and Weszkalnys 2008: 37).

The process of the development and application of Indigenous epistemologies methods has been uneven. Andrew Barry and his co-authors argue this shift represented a type of ontological interdisciplinarity (2008: 29): 'interdisciplinarity springs from a self-conscious dialogue with, criticism of or opposition to the intellectual, ethical or political limits of established discipline or the status of academic research in general'. Over time, these originally oppositional sets of ideas and practices come to have some 'value' and/or 'validity' in corners of the academy.

Tracing or mapping the 'topology of value' accorded to Indigenous research, in non-Indigenous studies, can demonstrate the ways in which Indigenous knowledge systems have been stretched and (de)formed in different circumstances (for example, disciplinary field, institutional location or professional status). Further, what counts as 'validity' or 'value' within Indigenous epistemologies and methodologies varies in different institutional, personal, cultural and political circumstances. For example, in the legal field the logic of Indigenous knowledge – as demonstrated through Dreaming stories – about land have been at different times vital to policy and research and at other times dismissed. In one significant domain, legal decisions that have recognized Indigenous land title have depended on Indigenous knowledge, while at other times similar knowledge has been classified as without sufficient value and instead Western knowledges have taken precedence (O'Brien and Elder 2014). For example, the Quandamoopah scholar Karen Martin (2007) remembers the process of providing Indigenous knowledge in an interdisciplinary environment – made up of historians, anthropologists, lawyers and scientists who were compiling knowledge on land ownership. She wrote: 'I became increasingly aware of how my knowledge and experience were measured against pre-determined categories of culture to which it was determined that I could provide no new or convincing examples' (2007: 30). In another case pertaining to land, Helen Verran (2002) explores the ways in which Indigenous knowledge about fire and land management shifted from being 'consistently dismiss[ed]' to a situation where scientists were funded to 'learn from members of an Aboriginal community' (Verran 2002: 731–732).

However, it needs to be remembered that the driving logic of Indigenous knowledge is not to achieve institutional validity. Again, we cite Karen Martin who states that the rejection of her Indigenous knowledge in a particular interdisciplinary space did not unravel her 'unwavering belief in my Ancestral relatedness to Quandamoopah and my sovereignty' (2007: 31). Other Indigenous Australian and Canadian research has demonstrated the ways in which 'validity' and 'value' are not designed to fit external worldviews, but instead reflect the internal or inherent value of Indigenous knowledge systems. For example, Shaun Wilson's work has developed the idea of 'research is ceremony' (2008). Explaining that preparing for and undertaking a research project is similar to preparing and undertaking a ceremony, Wilson makes it even clearer that involvement in Indigenous research, and this should be any research, is guided by protocols and responsibilities. In addition to Wilson's idea of ceremony, a group of Maori scholars introduced the notion of Kaupapa Maori (Pihama, Cram and Walker 2002), Karen Martin advanced the image of 'storywork' (2008), and I (Jonathon Potskin) developed the understanding of the Medicine Wheel. All of these epistemological logics firmly locate Indigenous research not only in terms of the aftermath of hundreds of years of colonial oppression and the affects this has had on our personal and communal living, but also within frameworks shaped by creative Indigenous knowledges that further sustain and reinvigorate Indigenous worlds.

## Conclusion

What connects us as Indigenous peoples today is our shared multigenerational colonial experiences. But beyond this we can use newer concepts to trace the recovery and the revaluing of the strong connections we have to our homelands through our languages and cultures. Indigenous methodologies sustain the links between community and the academy enabling a productive flow of ideas between these different locations. These are interdisciplinary practices that sustain the cultural and social purchase of Indigenous worldviews in relation to the 'lived subject positions [of Indigenous peoples] within modernity' (Moreton-Robinson 2016: xv). These practices enable us to do more than just re-enact what non-Indigenous people want us to be. However, for non-Indigenous scholars these practices and approach can also provide a metaphor or a frame for thinking about interdisciplinarity more broadly.

## Note

1 The term Indigeneity is used to describe the state of being Indigenous. Indigenous peoples are understood as pre-colonial communities residing in contemporary nation-states. Drawing on the United Nation's definitions these communities are often marginalized and/or subject to racism. In this chapter the term Indigenous is used as a generic term to describe a broad range of peoples. Similar terms could have been chosen (e.g. Aboriginal or First Nation). Indigenous people, of course, also have their own terms for themselves.

## References

Australian Human Rights Commission (2017). *Close the Gap: Progress and Priorities Report 2017*. Retrieved 16 March 2018 from: www.humanrights.gov.au/our-work/aboriginal-and-torres-strait-islander-social-justice/publications/close-gap-progress-0

Barry, A., Born, G. and Weszkalnys, G. (2008). Logics of interdisciplinarity. *Economy and Society*, 37(1): 20–49.

Behrendt, L. (2012). *Indigenous Australia for Dummies*. Milton, Queensland: Wiley.

Brennan, S., Davis, M., Edgeworth, B. and Terrill, L. (Eds.) (2015). *Native Title from Mabo to Akiba: A Vehicle for Change and Empowerment*. Sydney: Federation Press.

Burawoy, M. (2013). Sociology and interdisciplinarity the promise and the perils. *Philippine Sociological Review*, 61: 7–20.

Chilisa, B. (2009). Indigenous African-centred ethics: contesting and complementing dominant models. In D. Mertens and P. E. Ginsberg (Eds.) *The Handbook of Social Research Ethics* (pp. 407–425). London: Sage.

Chilisa, B. (2012). *Indigenous Research Methodologies*. London: Sage.

Elder, C. (2007). *Being Australian*. Sydney: Allen and Unwin.

Jacobs, J. A. and Frickel, S. (2009). Interdisciplinarity: a critical assessment. *Annual Review of Sociology*, 35: 43–65.

Kann, M. (1979). The political culture of interdisciplinary explanation. *Humanities in Society*, 2(3): 185–300.

Kovach, M. (2009). *Indigenous Methodologies: Characteristics, Conversation and Contexts*. Toronto: University of Toronto Press.

Kowal, E. (2015). *Trapped in the Gap: Doing Good in Indigenous Australia*. New York: Berghahn Books.

Langton, M. (1993). *'Well, I heard it on the radio and I saw it on the television': an essay for the Australian Film Commission on the politics and aesthetics of filmmaking by and about Aboriginal people and things.* Sydney: Australian Film Commission.

Martin, K. (2008). *Please Knock Before You Enter: Aboriginal Regulation of Outsiders and the Implications for Researchers.* Teneriffe, Queensland: Post Pressed.

McGregor, R. (1997). *Imagined Destinies: Aboriginal Australians and the Doomed Race Theory.* Melbourne: Melbourne University Press.

Moreton-Robinson, A. (2016). *The White Possessive: Property, Power and Indigenous Sovereignty.* Minneapolis: Minnesota University Press.

Morrissey, P. (2015). Bill Neidjie's *Story About Feeling*: notes on its themes and philosophy. *Journal of the Association for the Study of Australian Literature*, 15(2): 1–11.

O'Brien, R. and Elder, C. (2014). New ways for exploring who knows what in a native title case: a sociological approach. *Australian Aboriginal Studies*, 3: 30–42.

Parker, J. and Samantrai, R. (2010). *Interdisciplinarity and Social Justice: An Introduction.* Albany: State University of New York Press.

Pihama, L., Cram, F. and Walker, S. (2002). Creating methodological space: a literature review of Kaupapa Maori research. *Canadian Journal of Native Education*, 26(1): 30–43.

Rigney, L.-I. (1999). Internationalization of an Indigenous anticolonial cultural critique of research methodologies: a guide to Indigenist research methodology and its principles. *Wicazo Sa Review*, 14(2): 109–121.

Roe, P. (1983). *Gularabulu.* Fremantle: Fremantle Arts Centre Press.

Sillitoe, P. (2007). Anthropologists only need apply: challenges of applied anthropology. *Journal of the Royal Anthropological Institute*, 13: 147–165.

Strathern, M. (2004). *Commons and Borderlands: Working Papers on Interdisciplinarity, Accountability and the Flow of Knowledge.* London: Sean Kingston Publishers.

Strathern, M. and Rockhill, E. K. (2013). Unexpected consequences and unanticipated outcome. In A. Barry and G. Born (Eds.) *Interdisciplinarity: Reconfigurations of the Social and Natural Sciences* (pp. 119–140). Florence: Taylor & Francis.

Verran, H. (2002). A postcolonial moment in Science Studies: alternative firing regimes of environmental scientists and Aboriginal landowners. *Social Studies of Science*, 32(5): 729–762.

Wilson, S. (2008). *Research Is Ceremony: Indigenous Research Methods.* Nova Scotia: Fernwood Publishing.

# 10

# Troubling

*Anne Galloway*

The following entry can be found in the Encyclopædia of New Zealand History, published 6 March 2032, at Tāupo, New Zealand.

**National PermaLamb Programme★**

On 7 June 2019, in an unprecedented show of national cooperation, all of the country's political parties unanimously voted in favour of creating the NZ Ministry of Science and Heritage. For the first time in over 150 years, there were fewer sheep than people on our islands, and the new ministry's mandate was to ensure the protection of NZ's past and future. Within a few weeks, a meeting was convened in Oamaru, bringing together the nation's best scientists, historians, engineers, artists, economists and religious leaders to re-imagine our national icon – and so a new species was born.

Using well-established transgenics research and recombinant DNA techniques, select genes from prize-winning NZ huntaway dogs were introduced into the genetic sequence of the finest NZ merino sheep, and development halted at the juvenile stage. The resulting cross-bred animal embodied the behavioural traits of a dog and took on the physical appearance of a lamb for its entire life, making it the perfect companion species. Each *PermaLamb* would also be implanted with a full suite of networked identification, location and sensor technologies, enabling it to generate and collect petabytes of data over its lifetime.

The National *PermaLamb* Programme was launched on 12 August 2021 with government incubators and dispensaries set up in each region of the North and South Islands. To ensure the growth of the nation's new flock, every citizen and permanent resident of New Zealand over the age of 18 was required to adopt a *PermaLamb* and, in return, offered tax credits.

Ministry agents began meeting all international visitors at our ports in order to recruit *PermaLamb* foster-programme participants, and offer credit toward any future immigration applications. Almost immediately, people started camping outside ministry offices so that they could be among the first *PermaLamb* caregivers, and animals quickly became back-ordered from our national laboratories.

Nonetheless, within days the National *PermaLamb* Flock started to provide us with invaluable environmental data, and TVNZ began broadcasting each region's average temperature, humidity,

Figure 5.10.1   *PermaLamb*: 100 percent pure Kiwi transgenics (copyright Anne Galloway and Lauren Wickens)

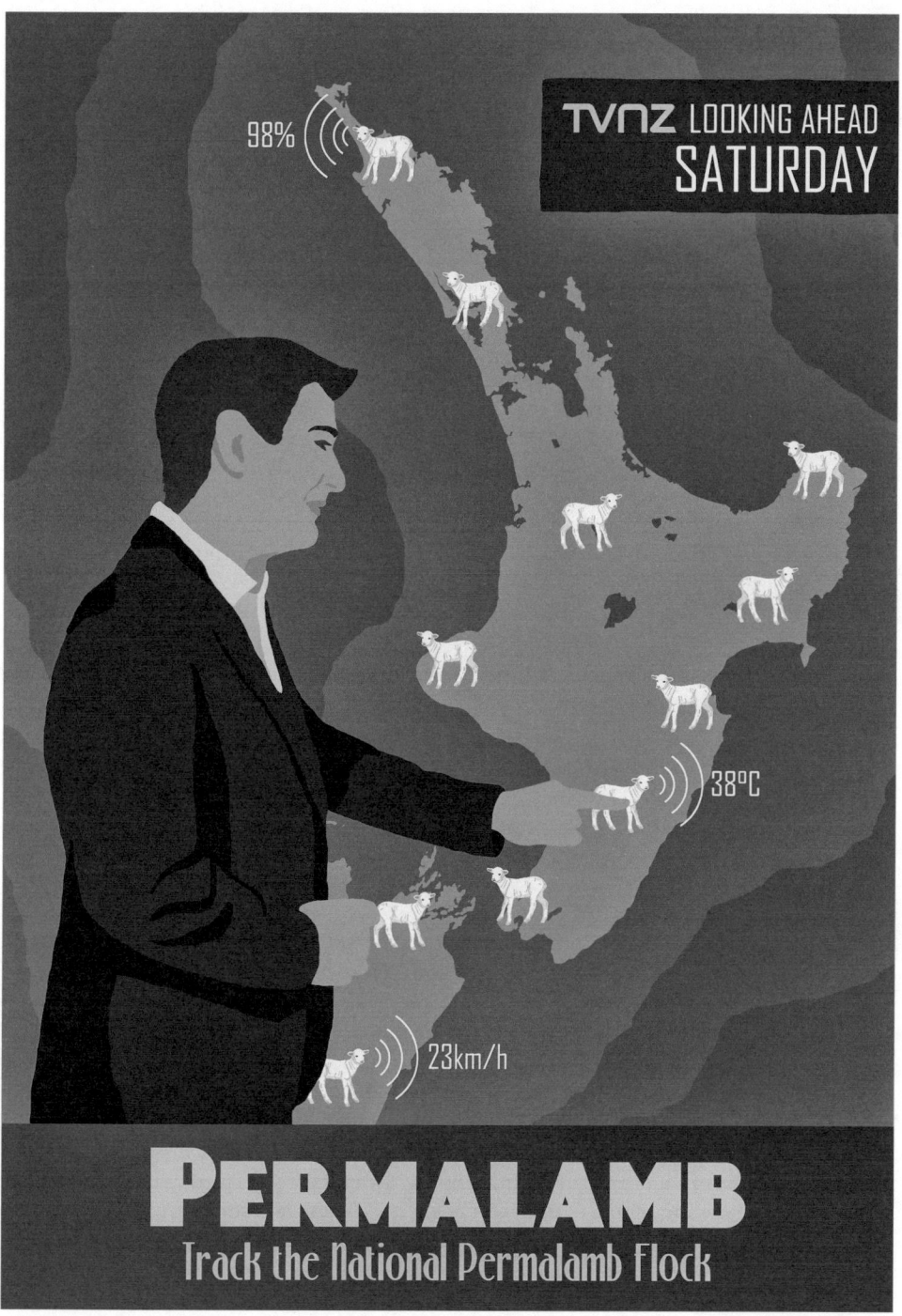

*Figure 5.10.2*    *PermaLamb*: track the National *PermaLamb* Flock (copyright Anne Galloway and Lauren Wickens)

wind speed, soil quality, air quality and sound quality as part of the national evening news, with real-time data tracking available via the Internet.

The *PermaLamb* Social Network offered caregivers the opportunity to share details of their animals' activities with friends and family, as well as to subscribe to the feeds of other *PermaLambs* across the country. By the end of the first year, #PermaLamb2393 and #PermaLamb38645 each had over 10 million followers from around the world. Within three years, both networks had over 4 billion combined global subscribers.

The ministry gave special attention to what it would mean to share our everyday lives with *PermaLambs* and still maintain our national sense of practicality and self-sufficiency. The merino's fine wool was re-established as a preferred textile fibre, and within two years every New Zealand home developed the capacity to clothe its members.

*PermaLambs* eliminated the need for shearing by naturally shedding their wool, and school-children often competed to see who could collect the most. Local *PermaLamb* parks brought communities together to play with their pets and sustainably produce their own clothing; knitting machines were activated if lambs registered as 'happy'.

The National *PermaLamb* Programme's ability to clothe the nation was accompanied by its capacity to feed us as well. Recognizing the need to identify a new source of protein for NZ's meat-eaters, without ever sacrificing the value our country places on animal welfare, each *PermaLamb* came with an optional slaughter kit. Designed to be as gentle and compassionate as possible, special grasses could be fed to a *PermaLamb* so that it would fall asleep and never wake again.

Maintaining such a close connection to a food source was new to many of us, and previously casual barbeques and Sunday roast dinners became treasured rituals among family and friends for years to come.

★The full set of NZ's Ministry of Science and Heritage National *PermaLamb* Programme posters is available from the Nation's Archives for personal, non-commercial download and printing. Visit http://countingsheep.info/permalamb.html/ to download hi-res, A3-size images.

## Speculation as method

I make things, and make things up. Things that do not yet exist, and things that might never exist. In my research, I conjure individual relations and entire worlds. I try things on, and see how they fit. I make adjustments, and I try them on again. I show them off. I ask people if they suit, if they are beautiful or ugly. Comfortable or constricting. I ask others to try them on. And I ask what they *could* do in them, if they *would* live in them. In other words, I play dress-up and invite others to join me.

This conceit of playing dress-up is useful, I think. I first learnt to do it as a child reading fantasy books, and spending a great deal of time in fictional and other enchanted places. Putting on special clothing was an integral part of troubling my sense of self and entering these worlds, and costumes still signify a transition to new ways or states of being. As Margaret Atwood (2011: 27) notes, when considering humanity's long history of god-kings and shamans, 'the man or woman and the costume and regalia were almost one and the same: you were the role, and the role was the garment and its embellishments. You inhabited it rather than just wearing it'. Carnivals still bring extraordinary, if temporary, powers to 'ordinary' people, and fancy dress allows us to discover who we are (and are not), as well as to explore who we want (and do not want) to be.

I believe that the ability to invoke, trouble and 'inhabit' other worlds and worldviews is instrumental to qualitative research's critical and creative role in knowledge-making. It is an

*Figure 5.10.3    PermaLamb*: designed for easy shedding and wool gathering (copyright Anne Galloway and Lauren Wickens)

*Figure 5.10.4* *PermaLamb*: designed for gentle and compassionate slaughter (copyright Anne Galloway and Lauren Wickens)

important part of the aesthetics and ethics of what we do. Speculative design – the more intellectual, word- and image-based version of playing dress-up – can disguise or rearrange the world as we know it, and let us see it differently. It can introduce new objects and relations, and allow us to explore their implications (see, for example, Dunne and Raby 2001, 2013; DiSalvo 2012; Malpass 2017). And, at its most fantastical, it can stretch the very boundaries of social imagination and possibility. The political stakes here are not small, as Ursula K. Le Guin (2009: 40–41) so eloquently explains:

> In reinventing the world of intense, unreproducible, local knowledge, seemingly by a denial or evasion of current reality, fantasists are perhaps trying to assert and explore a larger reality than we now allow ourselves. They are trying to restore the sense – to regain the knowledge – that there is somewhere else, anywhere else, where other people may live another kind of life. The literature of imagination, even when tragic, is reassuring, not necessarily in the sense of offering nostalgic comfort, but because it offers a world large enough to contain alternatives and therefore offers hope.

Speculative design ethnography is based on the promise and hope gifted by Le Guin's fantastic. As a research method it uses actual experiences and attitudes to elicit questions rather than provide answers – although unsolicited and unexpected answers have been known to appear. It values accumulation rather than reduction. It is open, performative and generative, excessive and uncertain. It strives for resonance rather than realism, and embodies what Patti Lather (1993: 686) calls 'voluptuous validity', or research that:

> goes too far toward disruptive excess, leaky, runaway, risky practice; embodies a situated, partial, positioned, explicit tentativeness; constructs authority via practices of engagement and self-reflexivity; creates a questioning text that is bounded and unbounded, closed and opened; brings ethics and epistemology together.

Ultimately, it is also a form of engagement that creates temporary, or mobile, publics gathered around a specific matter of concern, whose goal is to (be)come together (Galloway 2010) but not necessarily to become *one*. This is consistent with Dunne and Raby's (2001) original placement of sense-making in the hands of the audience, as they expected 'the user would become a protagonist and co-producer of narrative experience rather than a passive consumer of a product's meaning' (p. 46). However, this meaning may also be in partial or complete opposition to the intentions of the speculative design ethnographer.

## Valuing *PermaLamb*

*PermaLamb* – the speculative scenario that appeared at the beginning of this chapter – is part of *Counting Sheep: NZ Merino Internet of Things*, a three-year research project in which I conducted ethnographic research into merino wool and meat production, and how they might be affected by emerging technoscience. The second half of the project involved working with design students to 'translate' matters of interest and concern arising from my fieldwork into speculative design scenarios that might stimulate broader thinking around entanglements of technoscience and animals. As I was specifically interested in how speculative design ethnography might be used as a research method, matters of valuing and validating speculation were central to our work.

A total of four speculative scenarios were created and published online at http://countingsheep. info/, accompanied by a short survey. The decision to make the scenarios available online was primarily one of practicality. First, it allowed us to design objects that were not fully functional (they need only be photographed), and to focus on designing easily shared content. Second, it eliminated the need to secure physical exhibition space and manage access; however, it was also a deliberate decision to try to avoid a potentially exclusionary gallery context. The decision to work with a combination of written text and still images was also a calculated choice to experiment beyond the *de facto* standard of online video. Although the English-speaking Internet is neither inherently accessible nor inclusionary, the web provided stable locations for our creative content and questionnaire, and the project website continues to serve as an archive.

For potential comparison and contrast, each scenario was designed to occupy a place along a realist-fantasist spectrum: *PermaLamb* was the most fantastical narrative and *Kotahitanga Farm* (an urban exhibition farm) occupied the realist end-point. Building on a breeder's comment that 'Merino are a man-made animal', and concurrent efforts by the New Zealand Merino Company to breed the 'perfect sheep' (http://perfectsheep.co.nz), I tried to imagine how far those ideas could be taken. Knowing that sheep and working dogs are almost always tied together on New Zealand farms offered the opportunity to combine transgenics, or cross-species breeding, with Internet of Things technologies, and *PermaLamb* was conceived as a means to provide a cultural context for such technoscientific developments. The overall narrative was created to be a bit tongue-in-cheek, drawing on a number of well-trodden stereotypes of New Zealand culture and familiar discourses surrounding technology and neoliberal governance. The aesthetics of our 10-poster series also directly mimicked the visual style of early to mid-century New Zealand tourism advertising (see, for example: Alsop, Stewart and Bamford 2012) and high-resolution digital copies suitable for home printing were included online.

Ultimately, our audience was global, with respondents from New Zealand, Australia, the United States, Brazil, South Africa and the UK – representing university, industry, technology, farming and government workers. But it was also quite small and notably homogeneous – we received only 54 responses to all four scenarios, and a full 40% of the respondents identified as university-affiliated, with design as the most commonly selected personal interest. These data may do little to counter concerns that speculative design speaks best, if not also predominantly, to itself – but I remain optimistic that it can. The key, I believe, will be found in process rather than product: speculative designs need to be made *with* others, not just shown *to* others. But that is a discussion for another place and time!

Although the *PermaLamb* scenario garnered the least attention of the four scenarios, with only six respondents (two university-affiliated, two technology-affiliated and two farm-affiliated), it was the content of these responses that led me to choose it as the case study for this chapter. Not only was the narrative considered too fantastical for some, it was also considered the most troubling in terms of both social and research ethics.

One potential 'problem' with fantasy is that some people will assess it in terms of its plausibility or practical realism, and find it lacking:

Of course, sheep are not indigenous to NZ . . .

I don't know if is practical in the city [as] a lamb is not a pet, but if you have it in a special farm and you can track your lamb with a chip, could be plausible.

I'm not sure about the shedding. I think it might be better to have non-shedding animals and sheering [sic] stations (operated as lemonade stands to maintain the child engagement?). One of the complaints about dogs is shedding, so much that some of the most expensive breeds do not shed.

Using a huntaway [is a bad idea]. They are a naturally noisy breed of dog, and tend to rush things. Border collies or another suitable working breed which is easy for average people to train would make a far better end result.

Didn't find this scenario particularly realistic. While the other two [*sic*] scenarios I could see working in one sense or the other, this one just didn't sit well with me. I have never been good with, what I see as, unrealistic scenarios. My imagination doesn't stretch quite this far, not necessarily the huntaway gene into a sheep, but more that everyone has to have one etc. Sounds like a huge disaster to me.

Nonetheless, the responses above are reasonable and demonstrate valid knowledge held by different (including 'non-expert') publics.

On the other hand, an ability to treat a fantasy scenario as 'real' allowed other respondents to raise a number of ethical concerns about human–animal relations:

The genetic modification of the PermaLamb to remain forever in 'lamb' state is not a good idea. The name of the scenario might hint at how central this notion is to the animal, but I think it would be terrible for pets to be 'designed' in this way, as I think children in particular learn about the cycle of life and death, usually, through the ageing of their pets. I think that ageing process is important in recognizing that animals also have personality. My old dog got grumpier as she aged. A permanently young animal may not encourage a whole range of empathetic insights that pets can teach kids.

It blurs the lines between product, food and pet in ways that I hadn't thought of before. The convenience of the 'special' grass, the way it helps smooth over that 'seam' from pet to food was the most provocative. It encourages you to confront the relationship we have to animals in terms of a stark choice – is this animal cute enough to keep for a while longer? If not, no worries! Hand feed it, in a nurturing way, and then invite your friends over for the cook-up.

This scenario creates circumstances in which people can better consider the welfare of the animals that provide their clothing and food by being the relationship into the home.

Making every NZer have one [is a bad idea], not everyone is capable of taking care of animals. Creating a whole new species rather than working with, and promoting the ones that we already have [is also a bad idea].

Similarly, respondents raised ethical concerns about the roles of science and technology in society:

This scenario has linked consumerism, animal welfare and technology in new ways for me. It also highlights the moral and ethical dilemmas inherent in design, and shows how quickly the kind of utopian determinism around technology could lead to an 'advancement' that we, as a society, may not actually want.

I'm interested in the idea of quantifying pets and animals, and think this is the most likely to happen. 'Augmenting' our pets with an array of sensors and network technology is intriguing, probably from a health monitoring sense. I find this scenario most interesting when it comes to feeding data back to the animals themselves – not so much what kind of quantifiable data is possible to be captured, but whether it is possible to allow the animals themselves to act upon this data? The idea of an ad-hoc mesh network formed out of live-stock or 'companion animals' is also really interesting, like a kind of 'infrastructure fiction'. The data collection (IF truly private) offers an opportunity for understanding place better [but] there would need to be additional clarification about privacy protections.

Genetic engineering [is] not human's role on the planet. Information like weather can be gathered without implanting animals with technology. Social networking is for people-to-people contact not people-to-animal. No part of this scenario illustrates people at work, the people have no life purpose. The scenario of compulsory having an 'animal' is in biblical terms the mark of the devil without the human having the mark.

And finally, one Australian-based university-affiliated respondent used the *PermaLamb* scenario to express concerns about the research project itself:

Disturbing. I think people have done enough f*cked up stuff to animals, just look at all the problems that designer dogs have with their health. I cannot believe that you got funding to come up with any of these future scenarios. It just seems that you are trying to work on behalf of the wool/sheep industry to find even more ways to exploit sheep and other animals at their expense. There are so many ethical issues that you haven't even touched on. It is a very bad idea. This project is a classic example of the biotechnological turn in animal sciences which aims to increase the efficiency and profitability and output of animal production with little regard for ethics or animal suffering/welfare. This scenario has made me think that all New Zealand research must have to involve some sort of benefit for the sheep industry.

Created as tools for further questioning, our scenarios purposely leave open the kind of response they might inspire – and the comments immediately above suggest that research products stripped of any description of process run the risk of being interpreted (or mis-interpreted) any number of ways. Speculation has long risked being conflated with prediction, and I would add it equally risks being conflated with advocacy. At no point did we identify any of the scenarios as either desirable or undesirable, but the survey asked respondents to identify elements of each. The above comment was the only one in 54 responses that seemed to assume I and/or my team *wanted* our scenarios to one day come true.

## Validating speculative design ethnography

*PermaLamb* is an example of what Michael calls 'difficult objects' or things that:

warp the scientific and the social (as mediated by the designers) – they have implications that are good and bad, individual and collective, internal and external, biological and cultural, emancipatory and authoritarian, modest and arrogant, cruel and funny, academic and commercial, serious and playful, and of course, designerly and scientific.

*Michael 2012: 542*

Following Haraway (2008), our speculative design ethnography aimed 'to build attachment sites and tie sticky knots to bind intra-acting critters, including people, together in the kinds of response and regard that change the subject – and the object' (p. 287). But these responses and changes are not given; rather they are both more and less than what researchers may intend, expect, or hope.

Given the kind of respondents we had, we also find Michael's (2012: 541) description of speculative design's public to be useful:

> The public seems to be composed of more or less fully rounded persons, more or less able to confront with cognitive and emotional maturity (for want of a better phrase) [the] novel … designerly artifacts. What is particularly interesting is that this 'maturity' is characterized by a capacity to entertain, deal with, and explore the confusion, ambiguity, blurriness of the issues associated with these objects. This is a tacit model of the public where its members suffer neither from intellectual deficit nor citizenly shortcomings – rather, it is a constituency whose role is not to be 'citizenly' (whatever form that might take) within a context of policy making, but thoughtful within a context of complexity.

And just as our speculative designs did not seek to solve problems, the kind of public engagement that arose did not provide solutions – or even agreement! Rather than seeing this as a failure of either our research or its impact, we suggest that the respondents' thoughtful engagement indicates its own form of success. Recalling previously discussed approaches to evaluating speculative design, rather than defining success by whether or not the intentions of the designer were met – or if we were able to directly support citizen action – our case of public engagement might be best described as an experiment in cosmopolitics (cf. Stengers 2010), where again we might look for *resonance* in meaning among participants rather than an attempt to make a common present, or future. But mostly, I consider the methodological value and validity of this work to be its capacity for what Haraway (2016) calls 'staying with the trouble' – for creating and inhabiting spaces 'without guarantees or the expectation of harmony with those who are not oneself' (p. 98).

## Acknowledgements

The *Counting Sheep: NZ Merino in an Internet of Things* (2011–2014) research project was generously supported by the Royal Society of New Zealand Marsden Fund.

## References

Alsop, P., Stewart, G. and Bramford, D. (2012). *Selling the Dream: The Art of Early New Zealand Tourism.* Nelson: Craig Potton Publishing.

Atwood, M. (2011). *In Other Worlds: SF and the Human Imagination.* New York: Doubleday.

DiSalvo, C. (2012). *Adversarial Design.* Cambridge, MA: MIT Press.

Dunne, A. and Raby, F. (2001). *Design Noir: The Secret Life of Electronic Objects.* Berlin: Birkhauser.

Dunne, A. and Raby, F. (2013). *Speculative Everything: Design, Fiction, and Social Dreaming.* Cambridge, MA: MIT Press.

Galloway, A. (2010). Mobile publics and issues-based art and design. In B. Crow, M. Longford and K. Sawchuck (Eds.) *The Wireless Spectrum: The Politics, Practices, and Poetics of Mobile Media* (pp. 63–76). Toronto: University of Toronto Press.

Haraway, D. (2008). *When Species Meet.* Minneapolis: University of Minnesota Press.

Haraway, D. (2016). *Staying with the Trouble: Making Kin in the Chthulucene.* Durham, NC: Duke University Press.

Lather, P. (1993). Fertile obsession: validity after poststructuralism. *Sociological Quarterly*, 34(4): 673–693.

Le Guin, U. K. (2009). The critics, the monsters, and the fantasists. In *Cheek by Jowl: Talks & Essays on How & Why Fantasy Matters* (pp. 25–42). Seattle: Aqueduct Press.

Malpass, M. (2017). *Critical Design in Context: History, Theory and Practices*. London: Bloomsbury.

Michael, M. (2012). 'What are we busy doing?': engaging the idiot. *Science, Technology, & Human Values*, 37(5): 528–554. doi: 10.1177/0162243911428624

Stengers, I. (2010). *Cosmopolitics I & II*. Minneapolis: University of Minnesota Press.

# 11

# Problem-making

*Alan Irwin and Maja Horst*

All forms of research have their problems. But often it seems that this is especially true of interdisciplinary research (Fitzgerald *et al.* 2014). In this essay, we consider two dimensions of problem-making. Our approach is to introduce these and then to explore them with reference to one case of interdisciplinary research collaboration. This will then lead us to three lessons regarding problem-making in an interdisciplinary context which we think have larger significance beyond this specific case.

Before we get into this, however, it is important to observe that the notion of 'a problem' has a significant history within the human sciences and philosophy. As Thomas Osborne notes, the very definition of 'problem' has been greatly debated. As he inquires, 'What after all *is* a problem?' (Osborne 2003: 2). Alongside this definitional discussion, one can also find an extended series of reflections on 'problematization'. For Foucauldian scholars, 'problematization' generally refers to the social and cultural processes whereby certain phenomena become defined as problematic: madness, crime and delinquency, expressions of sexuality (see, for example, Foucault 1985). Osborne, drawing upon Foucault but also a selective comparison of Georges Canguilhem and Henri Bergson, has made the case for what he calls 'historical problematology' – an approach that involves attention to the diagnosing of problems, to the generation of solutions but also to a larger ethical commitment which presents the multiplication of problems 'as a sort of critical virtue in itself' (Osborne 2003: 15).

Clearly, notions such as 'problem', 'problematization' and 'problem-making' need to be seen in larger intellectual context but also used with considerable care. In what follows, we have opted for the term 'problem-making' as a means of acknowledging the connection to especially Foucauldian scholarship but also emphasizing the link to questions of method and methodology. Drawing upon the previous discussion, we present 'problem-making' as incorporating two significant dimensions.

The first dimension relates to what can be termed the 'social construction of problems', that is, the political, cultural and institutional processes through which parts of social life come to be seen as a problem (whether 'immigration' within recent European and US elections, or disagreements between professionals about how to tackle a particular question or concern). Considering this dimension in explicitly Foucauldian terms, one can think of it as the institutional and cultural movement between everyday difficulties, questions or challenges and the sense of a problem which needs a specific and practical solution.

The second dimension of problem-making relates to analytical strategy (Andersen 2003): usually, where the researcher finds a problem or dilemma which attracts her attention but which may be less apparent or interesting to the investigated subjects (why is 'immigration' considered to be a problem in areas where there has in fact been very little immigration? What are the challenges of interdisciplinary collaboration?). Problem-making here is closer to a research procedure or process with analytical attention drawn to often-unacknowledged or unrecognized paradoxes, points of tension or contradictions. To take an example from research around public engagement with science, this might mean exploring the relationship between governmental talk about 'open dialogue' and the actual organization of specific consultation processes (Irwin 2001). The problem here (put a little strongly) is how political rhetoric around democracy can clash with – or even contradict – the mundane realities of consultation. The point is that this 'problem' might be much more apparent – and interesting – to the science and technology studies (STS) researcher than for the consultation organizers themselves.

One is of course tempted to see the first as a 'social' process and the second as a 'method'. But what then is the connection between these two forms of problem-making? Is the researcher doing something fundamentally different from the research participants or else participating in the same sense-making process? For us, this question raises issues both concerning the valuing of knowledge claims within the social sciences but also of an ethical and political nature: why should the problems of social scientists be accorded greater value than those of others and what is the responsibility of the social scientist when constructing problems? If social scientists simply affirm the problem categories of social actors, then one can reasonably question what they add to the current situation. If they identify whole new problems, then why should these problems be given precedence over others and what is the relationship to actors' own problems? As we will discuss, these issues come especially to the fore when as social scientists we offer accounts that generate problems in a manner which can disconcert, provoke or annoy research participants and collaborators.

In our research, we have stumbled into these issues many times: including when giving academic accounts of the activities of policymakers or conducting qualitative research into public understandings of risk. But here we will focus on just one case, namely the interdisciplinary collaboration of one of the authors (Maja Horst) with a group of synthetic biologists. The intention behind this collaboration was to bring the perspectives of the social science and humanities (or SSH – in this case, including STS, science communication and philosophy) into a close working relationship with a cross-disciplinary group of scientists actively working in this field (for a rather different account of a similar collaboration see Rabinow and Bennett 2012).

As such collaborations go, this has been a mainly successful experience. In particular, the management of the project has been very open and respectful about the kinds of knowledges represented by Maja and her research group. The Synbio Management is keen to integrate the various disciplinary aspects and to share projects and work. However, this does not mean that the collaboration has been without challenges. In particular, Maja and her group encountered difficulties in connection with the writing of an academic paper. In our experience, this can often be a crucial point in any interdisciplinary collaboration: how do those who are the focus of research react to a formal and externally directed account of their activities, often written in an unfamiliar academic language and offering a form of 'valuing' which might seem rather alien?

The paper in question focused on a series of cross-disciplinary discussion groups that did not quite work as intended. Despite the stated wish of the synthetic biologists to engage more with SSH research, rather few showed up to the specially organized workshops introducing the area. It seemed that the overall commitment to interdisciplinarity was not reflected in the everyday operation of the collaboration. This 'problem' caught the attention of Maja's group which then began to consider why this was the case. In their paper, Maja and colleagues made an effort to investigate this interdisciplinary problem from different viewpoints, including that of the philosopher who had tried to organize the workshops, the young scientists who seemed less willing to collaborate than expected and the (scientific) project management team. These different viewpoints were then presented as three perspectives on 'the challenges of collaboration'. The broad conclusion was that, rather than primarily representing a matter of mutual disrespect, ignorance or embedded hostility to interdisciplinarity, various actors can have well-founded professional and career-related reasons for not engaging in this form of collaboration.

However, the Synbio Management was not happy with the paper. Putting it very simply, what the social scientists considered as measured analysis, they understood as criticism of the collaboration. What was seen as critical distance, they thought of as letting the collaboration down. What was intended as an honest admission of collective shortcomings, they had to struggle not to see as betrayal. They also expressed confusion as to the whole point of the paper. Why not stress the positives rather than being so critical? Surely the main issue is how we can develop legitimate forms of science to make the world a better place? Why are you writing about things that risk making collaboration even harder? Why, in short, are you making these issues into a problem?

The process of negotiations over the paper took more than half a year, and finished with a kind of truce. Not everyone agreed, but the discussion made certain things clearer. First of all, the whole interdisciplinary collective managed to maintain a collegial (if sometimes tough) dialogue right to the end. Second, however, the group simply had to give up explaining to the synbio scientists why a paper of this sort was important. The scientists accepted that Maja and her colleagues needed to write it and trusted their professional integrity, but they were never at ease with the paper. In that sense, they simply did not give the paper the kind of value that Maja and colleagues were hoping for. Nevertheless, and largely stimulated by these discussions, Maja, her group and the scientists are now working on a joint paper-writing initiative with a focus on making synthetic biology more socially robust. Third, we have to report, the paper-writing process has not up till now been successful. The paper in question has not yet been accepted for publication. Of course, Maja and her colleagues are left to wonder whether the compromises made during the internal discussion somehow took away the paper's edge. While we would prefer to view these interdisciplinary discussions in productive terms, it is undeniable that they can create demands beyond those of more conventional mono-disciplinary social science.

For the present authors, this whole episode provides a space for reflection about problem-making both as a social process and as an analytical strategy or method – and also about the value placed on certain forms of problems in specific settings. We could of course tell this as a story of heroic social scientists struggling against a powerful scientific establishment. But who says that the synbio scientists were not right? Perhaps this was their way of expressing vulnerability in the face of institutions (especially funders) that they saw as more powerful? Equally, there could be more than one form of truth at work here, suggesting also that dispute over the 'real' problems in any situation will reflect more fundamental differences, disagreements and tensions. For now, and in the spirit of a conclusion, we will draw the following lessons for problem-making.

First, and perhaps most obviously, the choice of 'problems' in any given context is not necessarily straightforward. Instead, there are likely to be multiple problem definitions available. This means that social scientists are inevitably engaged in a process of social construction and selection of which problems to prioritize and value. Seen in this way, the two dimensions of problem-making presented earlier do not seem so very far apart. In our opinion, it is time for a renewed discussion of the normative and intellectual bases for such prioritization and valuation decisions – particularly within interdisciplinary research. One example of this could relate to the treatment of power. Leaders of large cutting-edge science laboratories can be seen as fair game by SSH research (as the aforementioned Rabinow and Bennett analysis seems to demonstrate). Compared to the sense of insecurity and marginality often experienced by researchers in the social sciences and humanities, well-funded and high-profile scientific facilities can seem very robust indeed – and perhaps therefore deserving of challenge. However, might we not want to reflect a little more deeply about the assumption of innocence on the part of SSH, and cognitive and institutional power on the part of natural science? More generally, in contexts where multiple problems are in existence (or can be brought into existence), the intellectual and ethical implications of problem-making need to be carefully reflected upon and debated.

Second, problem-making is by no means the exclusive preserve of the academic social scientist (for a related argument regarding reflexivity, see Lynch 2000). Certainly, the interdisciplinary collaborators described above were highly productive problem generators: from the problems of building scientific careers in this pressured context to those of research managers in holding such a group together. SSH researchers are often rather good at identifying problems and this is a long and important tradition in our disciplines but, again, it might be time to reflect upon our own assumptions and valuations. We would suggest that it can also be useful to work with the problems of other actors rather than simply taking it for granted that one occupies an elevated place in this regard.

Third, far from being a disadvantage of interdisciplinarity, these contestations and problem-makings can be a major benefit. One of the strengths of interdisciplinarity is that it can open up the problem definitions and valuations of social science and humanities scholars to external scrutiny and challenge. One can choose to be offended by this, seeing it as an affront to one's professional standing and expertise. For us, this would have the general consequence of raising disciplinary boundaries – and perhaps decreasing the perceived value of social scientific research. Interdisciplinary collaboration might, at times, involve robust exchange, periods of self-doubt, discomfort and challenge to established traditions on all sides. But that might also be its very point.

As we have demonstrated, 'problem-making' from a social scientific perspective provokes many further questions, ambiguities and issues. We have explored some of these but also suggested that the notion possesses important strengths – and, more generally, productive possibilities. In the case explored here, and despite the acknowledged difficulties, the focus on problem-making allowed a cross-disciplinary dialogue, shared focus and sustained reflection that conventional research methods and procedures, developed primarily by and for researchers themselves, would simply not permit.

Problem-making can be viewed both as a process of social construction and as an analytical method. In the end, our argument is that these should not be seen as alternatives. Instead, an awareness of larger processes of social construction should inform the kinds of problem-making engaged in by researchers. And the problems identified by researchers should contribute to wider reflections on social and institutional challenges – including the identification of political and organizational problems. In that way, problem-making becomes central to the goals and methods of interdisciplinary research.

## References

Andersen, N. A. (2003). *Discursive Analytical Strategies: Understanding Foucault, Koselleck, Laclau, Luhmann*. Bristol: The Policy Press.

Fitzgerald, D., Littlefield, M. M., Knudsen, K. J., Tonks, J. and Dietz, M. J. (2014). Ambivalence, equivocation and the politics of experimental knowledge: a transdisciplinary neuroscience encounter. *Social Studies of Science*, 44(5): 701–721.

Foucault, M. (1985). *The Use of Pleasure*. New York: Penguin.

Irwin, A. (2001). Constructing the scientific citizen: science and democracy in the biosciences. *Public Understanding of Science*, 10(1): 1–18.

Lynch, M. (2000). Against reflexivity as an academic virtue and source of privileged knowledge. *Theory, Culture and Society*, 17(3): 26–54.

Osborne, T. (2003). What is a problem? *History of the Human Sciences*, 16(4): 1–17.

Rabinow, P. and Bennett, G. (2012). *Designing Human Practices: An Experiment with Synthetic Biology*. Chicago, IL: University of Chicago.

# 12

# Project-ing

## From differences to design

*Connor Graham*

This short essay considers the possibilities of three kinds of *differences* for studies of technology, particularly for productively engaging across disciplines and *project-ing* towards future technology design.[1] This distinct perspective confronts the sense of common humanness that is often created through human rights discourse (Montoya 2012) and the discourse of global threats (Miller 2004), discourses that are often produced and engaged in by elites. This essay considers instead that as contemporary, globalizing technologies such as the mobile phone travel across national borders they are productive less of a sense of 'common humanness', than situated at the edge of confronting and negotiating cultural *differences*. Studying such technologies through, for example, focusing on common experience, shared *being* and *action* in-the-world (Dourish 2001), might embrace a sense of 'common humaneness' but can also easily become a means of reducing and generalizing *differences* that persist across various kinds of borders. In this essay I consider the importance of understanding *differences within* and *between* studies of technology by reflecting on and re-examining an ethnographic study of the use of photographs in South West China.

The ethnomethodological work of Hughes, King, Rodden and Andersen (1994) suggests that the critical re-examination of ethnographic studies of ICTs can connect with the design of ICTs (in their case, 'systems') or, in the terms of this essay *project-ing*.[2] They characterize this approach as one of 'the different roles for ethnography in design' (Hughes *et al.* 1994: 436–437). They point out that this method is particularly valuable in the social science field where there is little sense of a 'cumulative corpus of findings to underpin any application of their knowledge' (ibid.: 436). The method of re-examination offers much potential value in interdisciplinary work, as long as the difficulties and hazards of deploying such studies across disciplinary boundaries are acknowledged, with regard to, for example, the particular kind of validity being appealed to, given the 'multi-paradigmatic character of social research' (ibid.).

I propose that taking *differences* in culture, in people, in places and in use is useful for interdisciplinary discussions of such studies' validity, use and usefulness. Specifically, I show how attention to *differences* within and between the results of technology studies can respond to reductive and generalizing moves in three distinct ways. *Differences* first support discussions of versions of validity across different disciplines and expose the relationship between 'social' and 'design' disciplines.[3] They can, second, reveal particular, discipline-specific analytical strategies.

*Differences* also support the re-examining and re-informing of technology studies through drawing attention to the process of imagining design possibilities or *project-ing*.

My exploration of the value of *differences* started in 2007 when I began studying ICTs across settings where the culture, people, places (and potentially) use were ostensibly quite different, despite a common focus on use of a thing, the thing in this case being photographs, both physical and digital.[4] The fieldwork I was conducting extended across China, the UK and Australia, although only the work in China has been written up and published. The work in China centred on middle-class city-dwellers in Chengdu city in Sichuan Province. As the study progressed it became clear that the limited number of people I was learning about were not only different from descriptions of other people in other studies but were also quite different from one another. These are two distinct types of *differences*: *differences without* and *differences within*. It also became abundantly clear that the status of my membership of society in China was quite different from my status in the UK and Australia, another category of *differences: reflexive differences*.

So *differences*, first, became important when establishing the focus and purpose of the studies of the then contemporary ICTs. It was important we understood the culture, people, places and use in past studies in quite a systematic way: *differences without*. This was powerful for contextualizing our own description, moving from comparison to analysis. In which countries did past studies take place? Who were the participants in terms of age, class, language group and ethnicity? In what spatial configuration and geographies were participants placed? What kinds of things were being used: physical or digital photographs and/or cameras? How was the description generated: through interview, observation or some other method? What disciplinary bias did the past studies have and how did this affect description? This initial framing involved tracing connections between the different contexts of past studies and the results. It also involved answering questions about how the context and methods of previous studies of ICTs were different from what we were attempting.

Focusing on the idea of *values* or what we termed 'beliefs about what people should do' (Graham and Rouncefield 2010: 81) was useful when understanding differences across study participants: *differences within*. Through how we defined values we allowed for the possibility of family resemblances – incorporating similarities and differences – across cultures, peoples, places as there are such resemblances for games: 'football, chess, patience and skipping are all games . . . it would be foolish to say that all these activities are part of one supergame, if only we were clever enough to learn how to play it' (Winch 2008: 18).[5] This move acknowledged both the commonness of practices and technologies such as photography and cameras. It also acknowledged heterogeneity from within as such technologies are shaped and 'made local' through members of a particular society. Thus, *values* provoked certain questions that related to primarily *differences within* but also *differences without*. We asked: what are the *similarities* and *differences* across current observations and between previous studies of ICTs and what we have observed? How might we account for these similarities *and* differences, given the difference in cultures, peoples, places and use? Answering this last question involved us examining differences across the informants in China and engaging the informants themselves in discussion.

Following and based on this, in a reflexive move, it was important for me to recognize not only our disciplinary context but also our understanding of our own membership: the culture, people, places and use of *our* 'photographic technologies', or, more simply 'where we were coming from': *reflexive differences*. This was complex enough without Mark and me being quite different, even if both of us were informed by ethnomethodology. Carrying out field-work in the UK helped with this somewhat, informing me about the familiar as I investigated the less familiar and independently learned about Chinese culture, history and religion as I have, in the past, learned about Irish culture, history and religion. One of the challenges in this process was not to exoticize, simplify or simply misread the unfamiliar and to disentangle perceived tradition-informed practices from contemporary practices when accounting for, for example, use. How were these observations *similar* and *different* from what is familiar? How could we account for these *similarities* and *differences* through values? This examination tuned us to factors concerning the people, places and other relevant uses under study that could account for *differences* and *similarities*.

Studies of technology are converged upon by, among other disciplines, different varieties of anthropologists, computer scientists, psychologists and sociologists, all with different commitments concerning what might be studied, why and how it might be studied, what such study might result in and how these results might be used.[6] Different disciplines privilege different views on what validity is or, in different terms, what they regard as valuable outcomes.[7] All these forms of validity approach the breadth and complexity of human *differences* in contrasting ways. Crabtree, Tolmie and Rouncefield (2013) point to some of the consequences of these differences in views. For example, time and extent of immersion or ecological validity can be regarded as the key measure of a study's worth. As with the study described here, seeing these engagements with different kinds of validity as established values in a discipline succeeds in both making these views visible and exposing certain techniques and uses. This 'cultural' way of seeing disciplinary *differences* is useful for inter-disciplinary discussions of (as opposed to feuds over) such studies' validity, use and usefulness. As illustrated above, questions about the status of the corpus become particularly important when studying technology in a society in which one is not a member, when *reflexive differences* are in play.

As Crabtree *et al.* (2013) argue, another, more ambitious kind of quite impossible validity is a common aspiration for studies of contemporary ICTs that extend beyond description: the tangible quest of reaching towards the initiation of new or different (that word again) ICTs of the future; to *project* beyond the possible general and the particular real towards new, as yet unconceived, unrealized situations; to infer what we have not understood, provide insights into an unrealized project or the consequences of an unrealized project or, more specifically to guide technology design.[8] I will term this kind of validity *project validity*[9] and the process as *project-ing*. Why disciplines might use the impossible as a norm is an important question. Some answers can surely be found in the study of the history of mass production and the contexts of technology design. The past, a place where future projects have already happened, is surely under-acknowledged in understanding such *project-ing*. To frame this observation as a question: what are the circumstances – culture, people, places, use – in which past *project-ing* has been done and what logics and rationale were deployed?

Returning to the study presented at the beginning of this essay, with all the care taken to acknowledge differences, and given the aim of *project-ing*, how did this diverse set of findings, this multi-faceted cognizance of *differences* help? After these studies we became more aware of not only the different contexts in which new ICTs might operate but also how these related to the contexts of other studies of photographic technology: *differences within* and *without*. Admittedly, we were less tuned to *reflexive differences* and valuing practices and different versions of validity than we might have been. What did concern us was considering how newly introduced ICTs operate in diverse contexts: past, present and future. These operations presented a spectrum of possibilities that supported ongoing discussion about future design. We also learned that ICTs, rather than being embedded in contexts where values lurked, were really permeated with values and allow, through their use, certain values to be made visible: Sacks' maxim that technology is 'made at home in the world that has whatever organization it already has' (Sacks 1992: 549) was borne out. Thus, the whole process of *project-ing* incorporated a view on past studies, present work and future possibilities.

So how might these kinds of *differences* be useful for interdisciplinary work when social scientists and computer scientists with different views on validity converge on description? This essay has already suggested their role in productively resolving interdisciplinary feuds. Another answer concerns how *differences* might be useful for *project-ing*, to serve the needs of computer scientists. The nature of the *project-ing* or analytical strategies is crucial to consider when answering this question. The process of *project-ing* towards engineering industrial, safety critical systems is quite different from *project-ing* towards creating playful domestic social technologies. In the former project, understanding the variability of *differences within* the people involved in the study and the acceptability of *differences without* in terms of established thresholds for risk is crucial. Both these kinds of *differences* operate in a context of *reflexive differences* across analysts. In the latter project, it is reasonable to interpret *differences* simply as being a kind of variety *within* and *without* studies which informs deeply *reflexive* imagining.[10] Carefully acknowledging not only these kinds of *differences* but also the analytical strategies deployed to confront these *differences* can potentially smooth over interdisciplinary conflict regarding how such studies *should* be used.

Yet there is also the possibility of the imagining work of project-ing re-informing description through particular disciplinary foci and supporting questions about what kinds of validity are important in the face of the 'description informing design' mantra. Imagining the future can inform the past and what is regarded as important about it. This operationalization of *differences* can support a more nuanced treatment of the results of studies of technology and a conversation about the nature and extent of knowledge, risk and creativity. It also positions past projects as first part of a corpus of studies that articulate a connection between provisional descriptions of orders in-the-world and instances of *project-ing* and, second, as provisional descriptions of *possible*, particular re-orderings of the world.

As common technologies such as ICTs circulate the globe, promoting a coarse sense of 'common humanness', so they are consumed and appropriated at different levels of local society and once intensely indigenous symbols and images become known. In this cycle of mass production and consumption and intense circulation, *differences* can be either forgotten or assumed not to exist. I hope that this short essay has shown how acknowledging differences, whether among disciplinary approaches or within and without past and present description is important (Graham and Rouncefield 2010). I also hope I have shown that understanding such *differences* as relational can inform *project-ing*, allowing *project-ing* to inform description.

## Notes

1  I deploy a dash to emphasize the work and activity involved in 'project-ing' and the two possible meanings of 'project'.
2  I acknowledge that ethnomethodology, with its focus on work and actors' constructions of their own life worlds, has a particular perspective on human experience.
3  I present three definitions of validity below.
4  This work was conducted with Mark Rouncefield from Lancaster University and was funded by his Microsoft European Fellowship 'Social Interaction and Mundane Technologies'.
5  This is a direct reference to Wittgensteins's idea of 'family resemblance': 'I can think of no better expression to characterize these similarities than "family resemblances"; for the various resemblances between members of a family: build, features, colour of eyes, gait, temperament, etc. etc. overlap and criss-cross in the same way. – And I shall say: "games" form a family' (Wittgenstein 1986[1953]: 33).
6  In this essay when I refer to 'computer science' or 'computer scientists' I mean academics and practitioners who are concerned with understanding humans as a secondary concern and developing computing technology as a primary concern.
7  Three instances of validity are familiar for most disciplines. These instances are less definitive categories or descriptors than place-holders supporting the discussion of key differences. One use of validity refers

to the internal workings of such studies being sound. Another is sometimes referred to as 'external validity' or a study's generalizability. This use of the word validity used in relation to observational studies of ICTs assumes the possibility of the generalizability of a study's results to other, actual and possible situations involving ICT use. The fixation on external validity becomes visible in the search for similarities, the counting of occurrences in observational data, the search for and extrapolation of themes and rules in how results are presented. Increased frequency in observations means it being more likely to be 'the way it is'. An extension of this use of validity is 'ecological validity' or the ability to generalize results to the real world. This view of validity centres on correspondence to a real situation involving ICT use that has happened. The fixation on ecological validity becomes visible through results based on the accumulation of evidence over time, the detailing of the particularity of the situational aspects of the study – the culture, the people, the place and the use – and even the role and positionality of the observational instrument (including the human observer). Increased authenticity of observation means it is closer to 'the real moment'.

8  This refers to Christine Halverson's ideas of 'inferential power' and 'application power' (Halverson 2002: 245).

9  Here I have deliberately avoided the equivocal term 'design' in favour of 'project' to communicate the provisional and speculative nature of such 'reaches' towards a possible future.

10  This distinction leans on Löwgren's (1995: 87–88) distinction between 'engineering design' and 'creative design'.

## References

Crabtree, A., Tolmie, P. and Rouncefield, M. (2013). 'How many bloody examples do you want?' Fieldwork and generalisation. In W. O. Bertelsen, L. Ciolfi, A. M. Grasso and A. G. Papadopoulos (Eds.) *ECSCW 2013: Proceedings of the 13th European Conference on Computer Supported Cooperative Work* (pp. 1–20). London: Springer.

Dourish, P. (2001). *Where the Action Is*. Cambridge, MA: MIT Press.

Graham, C. and Rouncefield, M. (2010). Acknowledging differences for design: tracing values and beliefs in photo use. In *Proceedings of the Ethnographic Praxis in Industry Conference* (pp. 79–99). American Anthropological Association.

Halverson, C. (2002). Activity theory and distributed cognition: or what does CSCW need to do with theories? *Computer Supported Cooperative Work*, 11(1): 243–267.

Hughes, J., King, V., Rodden, T. and Andersen, H. (1994). Moving out from the control room: ethnography in system design. In *Proceedings of the 1994 ACM Conference on Computer Supported Cooperative Work (CSCW '94)* (pp. 429–439). New York, NY: ACM.

Löwgren, J. (1995). Applying design methodology to software development. In G. M. Olson and S. Schuon (Eds.) *Proceedings of the 1st Conference on Designing Interactive Systems: Processes, Practices, Methods, & Techniques (DIS '95)* (pp. 87–95). New York, NY: ACM.

Miller, C. (2004). Climate science and the making of a global political order. In S. Jasanoff (Ed.) *States of Knowledge: The Co-production of Science and Social Order* (pp. 46–66). New York, NY: Routledge.

Montoya, A. (2012). From 'the people' to 'the human': HIV/AIDS, neoliberalism, and the economy of virtue in contemporary Vietnam. *Positions: East Asia Cultures Critique*, 20(2): 561–591.

Sacks, H. (1992). A single instance of a phone-call opening. In G. Jefferson (Ed.) *Lectures on Conversation Volume II* (pp. 542–553). Oxford: Blackwell.

Winch, P. (2008[1958]). *The Idea of a Social Science*. London: Routledge and Kegan Paul.

Wittgenstein, L. (1986[1953]). *Philosophical Investigations*. Oxford: Basil Blackwell.

# 13

# Qualifying

## *Gay Hawkins*

Qualification is a productive process whereby things and human identities are enacted as value is created.

*Calişkan and Callon 2009*

There are two important and related propositions in Calişkan and Callon's definition of qualification. First, that qualifying is a process that involves the production of qualities through various qualification methods and devices. And, following this, that qualifying is not a process of finding value but of provoking and shaping it. These two ideas highlight the complex relationship between qualification activities and the enactment or *doing* of value: the way in which 'qualities' do not so much refer to the fixed characteristics of something but, rather, how specific characteristics emerge and how they should be appraised, what sorts of judgement should be formed about them.

'Qualification' as a theoretical concept is usually found within Science and Technology Studies (STS) inflected economic sociology. It describes the multiplicity of devices and processes that actualize which particular qualities of a product will be considered valid in a market. This critical point – that qualities are actualized in particular settings for particular purposes – is very suggestive for thinking about interdisciplinary research methods; it foregrounds the way in which research could be considered as an ongoing qualification exercise. That is, a method for provoking and validating particular qualities in the objects under investigation. This might seem troubling for those who consider research an exercise in representing an unconfined raw reality or for those who assume that things have intrinsic, observable, objective characteristics. Qualification would seem to be a distortion rather than a representation of reality. But qualification does not work in this register, its intellectual lineage is the performativity paradigm and this means that it is concerned with the dynamics of *making things real* rather than capturing a fixed reality. This is a crucial distinction and it is at the heart of why qualification can be requalified from an economic practice to an interdisciplinary method.

What does it mean then to explore qualifying as method? How can we understand it as a technique that both produces objects of research and shapes how they should be considered and judged? And how do research objects participate in qualification strategies and suggest particular methods of valuing and validation? Finally, how might the dynamics of qualifying reveal the complexities and politics of research as an attempt to know a particular reality *and* also appraise it? These questions drive my argument. I explore them through some reflections on the methodological dilemmas I faced when examining the recent rise of bottled water as a new market and mundane object (Hawkins, Potter and Race 2015).

When we began this project it was apparent that the qualities of bottled water had already been determined and judged. It was a demonized object that had become the focus of much activism and opposition. This forcefield of critique was partially what attracted us to the project, what captured our interest and affect. But it rapidly became an impediment to the research process. Critique tended to frame the multiple qualities of bottled water – as product, as waste, as threat to water resources – in terms of fixed attributes that generated negative impacts. The effect of this was to render bottled water inert, to exclude it from the active realm of the social. Bottles' modes of acting in the world, their diverse relations and shifting realities, which we observed all around us, were obscured. Yet it seemed to us that bottles were far from stable and their impacts were difficult to generalize or determine in advance. They were constantly being qualified and requalified according to different situations and practices, just as they were also engaged in various qualifying exercises themselves – especially with the water they contained. The attributes of the bottle emerged always in relation with other things and in this sense they were not so much attributes as *qualities* – characteristics that were made present or validated according to particular associations and realities. The question was: what was influencing the emergence of these qualities? Were the intrinsic qualities of the bottle making distinctive social realities or were the realities the bottle found itself in provoking particular qualities – intrinsic and extrinsic? We found our answer in Callon, Méadel and Rabehariosa's (2002) account of the product as a series of reciprocal transformations: 'The product singles out the agents and binds them together and, reciprocally, it is the agents that, by adjustment, iteration and transformation, define its characteristics' (Callon *et al.* 2002: 198).

What a beautiful account of qualifying as method, of the dynamics whereby objects and methods mutually qualify and iterate each other. Callon *et al.* perfectly capture the way in which products are defined not by their objective characteristics but by the various metrological work or 'qualification trials' that seek to test, measure and singularize these characteristics in relation to other similar goods. In this way, the intrinsic features of products require work to become qualities, at the same time as these features participate in and shape the types of extrinsic devices that are used to measure and test them. The effects of these 'qualification trials' are to stabilize and objectify the qualities of things, albeit temporarily, so they can be fully exploited in market arrangements.

In the case of bottled water, attention to the dynamics of qualification was fundamental to making sense of how water became a market thing. We used one particular brand, Evian, to understand this. Rather than take the brand as a given we wanted to understand how it emerged historically and became a device for requalifying spring water in the town of Evian from a mundane liquid to a therapeutic liquid and product; something that was taken in doses, bottled and distributed far and wide, and capable of attracting bourgeois tourists from all over France. Explaining this significant transformation in the qualities of the water involved detailed historical research methods investigating everything from, for example: the earliest records of the 'discovery' of Evian water's miraculous therapeutic effects; the development of a company to bottle the water and distribute it; the growth of analytical chemistry and the ability to identify the mineral composition of the water; the rise of a regulatory regime within the French state to classify and protect various spring waters in the public interest; the development of the brand. The list of qualifying agents and devices went on and on. In all these investigations, our focus was on how the 'facticity of the commodity' (Atkins 2007) was established, how the distinct qualities of this water were both enacted and validated as real. This focus on qualification made it possible to see how the therapeutic qualities of the water were not socially constructed; they were not cultural or economic impositions on a passive material. Rather, they were actualized – called into being – as the water became a participant in new relations and market arrangements. In this way, the earliest forms of market assemblage for spring water from the town of Evian were not prompted by the intrinsic therapeutic properties of the water: rather these properties emerged as the markets were configured. As DeLanda (2006: 11) argues, the properties of elements involved in an assemblage – market or otherwise – do not cause relations. Properties only become expressive and potent with reference to their interactions and relations with the properties of other entities.

We can take this argument a step further by saying that qualification does not simply stabilize qualities – it also has the capacity to *provoke realities*. Drawing on Lezuan, Muniesa and Vikkelsø (2013), qualification could be considered as an experimental method for provoking realistic situations. In the case of Evian, the reality that was provoked was one in which the water did not simply become a product it also acquired therapeutic agency. While doctors and scientists in the nineteenth century expended much energy and effort measuring and attempting to explain the unique biochemical qualities of the water, seeking to confirm and represent its empirical reality, those drinking it under medical advice or taking spas in it were busy enacting this therapeutic quality. They were manipulating the water in new ways because it had practical relevance to them and, in this process, they were making those therapeutic qualities real, they were *provoking* or doing a new social reality.

Lezuan *et al.*'s argument about provoking reality is primarily focused on controversial social-scientific research experiments in the 1930s and 40s. These experiments choreographed artificial situations to induce spontaneous reactions and realities that were visibly made up but also compellingly real. These 'factitious' events were designed to both generate and challenge existing social reality within the contained space of the experiment. This notion of provocation or 'provocative containment' is helpful for investigating how qualification shapes judgement and value. If we accept that to study the dynamics of qualifying is to study how the qualities of things become real, how does this method determine which qualities to examine and why they matter? In the mess of researching bottled water and its rich social life it was quite obvious that beverage companies put an immense amount of effort into ensuring that only certain qualities were made present and valued in markets. Markets were a containment space. But containers can overflow and the activism that emerged to contest these markets revealed this with force.

By looking closely at the political campaigns that developed around bottled water it was possible to see how troubling new qualities and realities for the bottle emerged. These campaigns were often framed as exposés driven by the goal of revealing an empirical truth about bottled water's problematic effects that was hidden. However, the problem with this framing was that it fuelled the notion of reality as stable and robust. It also fuelled the idea that political critique operated at an analytical distance from this reality and was able to represent its truth to others who did not have such privileged access. In contrast to this understanding of politics, it seemed to us that what the methods and devices of activism did was *provoke* a political reality for bottled water. When activists documented plastic bottle waste or the depletion of water resources by beverage companies, they were not only exploring how other bottle realities were generated but also requalifying the bottle as a political object. They were identifying particular characteristics of the bottle *and* framing how they should be appraised.

Our method for understanding the impacts of activism on bottled water markets was to investigate the network of requalification practices that effectively enacted various controversies and political situations. Many of these practices were theatrical and playful. Standup comedy routines and confrontational images of people drinking oil situated the bottle in new relations and invited people to witness and validate the troubling qualities of the bottle that emerged. These practices could be considered as experiments in inciting public interest and their effect was to realize the externalities of bottled water markets. As Lezuan *et al.* say: 'realization should be understood here in the sense of making real, of bringing into existence, of generating a phenomenon whose existence might otherwise be doubted or resisted' (2013: 289).

To conclude: acknowledging or identifying qualities is to participate in the dynamics of qualification and the doing of value. While the intellectual development of this concept has largely occurred within economic sociology, its relevance resonates far beyond markets. Qualifying is a concept with significant traction as an interdisciplinary method. Using the example of a major study of bottled water markets, my aim has been to show how an investigation of qualifying practices was central to understanding the multiple realities bottles of water enacted: economic, social, environmental, political and more. But documenting qualification in action, framing it as the central object of research, is not the same as using it as a *means* to research. Perhaps this distinction is too stark? Can methods really be clearly separated from their objects as if they were mere instruments? No, they cannot: objects and methods are caught up in the dynamics of mutual qualification. The things we choose to research also choose us, also pose questions, also suggest suitable qualification devices and techniques. When we were investigating the dynamics of qualifying bottled water we were also, at the same time, engaged in myriad qualifying methods: identifying things or practices that struck us, that we wanted to consider, connecting them to other things, bringing archives and interviews and theories into play with them, selecting and appraising. In the process our research provoked a particular bottled water reality: entirely invented but entirely indebted to the multiple ontologies of bottles alive in the world.

## References

Atkins, P. (2007). Laboratories, laws and the career of a commodity. *Environment and Planning D: Society and Space*, 25: 967–989.

Calişkan, K. and Callon, M. (2009). Economization Part I: shifting attention from the economy towards processes of economization. *Economy and Society*, 38(3): 369–398.

Callon, M., Méadel, C. and Rabehariosa, V. (2002). The economy of qualities. *Economy and Society*, 31(2): 194–218.

DeLanda, M. (2006). *A New Philosophy of Society: Assemblage Theory and Social Complexity*. London: Continuum.

Hawkins, G., Potter, E. and Race, K. (2015). *Plastic Water*. Cambridge, MA: MIT Press.

Lezuan, J., Muniesa, F. and Vikkelsø, S. (2013). Provocative containment and the drift of social-scientific realism. *Journal of Cultural Economy*, 6(3): 278–293.

# 14

# Scaling

*Masato Fukushima*

One of artist René Magritte's favourite surrealistic tricks is to remove objects out of their proper context. An apple of extraordinary size that takes up an entire vacant room (in *The Listening Room* 1958) strikes us, rather paradoxically, with the impact of the image, despite the simplicity of the trick. Together with, say, Ron Mueck's photo-realistic sculptures of human figures of similar enormity, this painting offers a universal lesson about the meaning of scale in our everyday lives where things of ordinary size are peacefully immersed.

Scale does matter in many circumstances, including interdisciplinary collaborative work, if it is to be performed properly. Suppose, for example, that a country's president has been shot by international terrorists and is rushed to an emergency operation. The ER staff might dream of collecting a team of talented surgeons – the allure of doing so is undeniable – but such a team of 100 members of the brightest specialists from all over the world jostling one another in an overcrowded operating room – like Japan's world famous rush hour on public transportation – would create the kind of apocalyptic nightmare that Magritte might have been amused to paint.

In his classic work on software project management, Brooks (1975) argues that the idea of a 'man month' – that time and manpower are interchangeable – is a myth, especially in software development. Brooks convincingly maintains that even if time is short, it cannot always be complemented by the increase of manpower, most notably in software development. What might happen instead is that infusing additional manpower into an already tardy project further delays the process because a great deal of trouble is involved in providing instruction to the newcomers, setting up a more complex path of communication, and eventually, increasing the difficulty of managing such a project. Just remember the poor president who lost his life in an operation room packed with top-class doctors!

The lesson here is that scale is often essential in determining the success or failure of interdisciplinary collaborative work. These examples hint that to scale up the size of a group tends to result in decreased efficiency. This can be lethal if the needed operation is conducted in hazardous situations, such as the control tower of an airport or the command room for military operations, similar to the routines in an operation room. In such situations, the whole undertaking often requires concentration of mind and the tight collaboration that results from control by a single person (LaPorte and Consolini 1991).

One might wonder, however, if the cases thus far presented are adequate for considering the proper scale for interdisciplinary research practices, as the former require decisions within minutes or even seconds, not months or years, whereas research collaboration might not need to be as tightly organized as observed in these cases. Though the circumstances are not as dramatic, scale matters in research collaboration too. Collingridge (1996) enumerates a negative inventory of all the possible troubles that may be encountered by practitioners of any project on a large scale, ranging from the sheer difficulties of management to lack of flexibility, slow response, and inevitable bureaucratization.

Behind this view of the malaise inherent in scaling up projects lies a pivotal issue, generally referred to as the problem of 'transactional cost', which was originally conceived to prove that enterprises must exist to deal with market transactions so that all related costs are lower than those of individually handled transactions (Coase 1960; Williamson 1981). This concept can be extended to analyse required costs in organizations for interdisciplinary coordination, negotiation and collaboration when their scale grows. The man month myth also arose out of such analyses of 'economies of scale' for larger organizations.

Despite such insights, in interdisciplinary collaboration, the scale of number seems to continue to be important, as represented by so-called big science (Galison 1992) and, more specifically in recent years, big biology (Parker, Penders and Vermeulen 2010; Fukushima 2016). This bigness derives from the growing size of both the research organizations and the scientific instruments needed for more extensive research, with examples ranging from bigger accelerators and super-scale radio telescopes to the more advanced second-generation sequencers of DNA. Global issues, such as climate change, also demand such enlargement of size in research collaboration (Edwards 2010). The once lively argument about 'the end of science' stemmed partially from the pessimistic prediction that the growth of such a scale of instruments in terms of both cost and manpower would necessarily have limits.

There are at least two interesting corollaries to this issue of the rising size of interdisciplinary research. One is that coordination is needed not only for possible conflicts among the different instruments and even different epistemic cultures (Knorr-Cetina 1999) – such as, say, between the wet (traditional) and the dry (informational) approaches to biology (Fukushima 2016) – but also for managing the data that is produced from such research. Edwards, Mayernik, Batcheller, Bowker and Borgman (2011) refer to such conflicts between the different formats of data and metadata, as well as the difficulty of coordinating them, as 'science friction'. Their tentative but intriguing conclusion that such coordination is always ongoing and only temporary echoes Brooks' (1975) similar conclusion regarding the software project, as mentioned above. However, the somewhat pessimistic wisdom gained about restricting the size of the developing team of software – as well as that from realizing the impossibility of a dream team of 100 surgeons for operating on the president – apparently does not transfer well, as the growth in the number of scales (size, velocity, heterogeneity, speed) of data and metadata seems unstoppable.

Another issue, though having a slightly different focus, are the contradictory vectors of values used for evaluating scientific research at large: namely, the pressure to upgrade the scale(s) of interdisciplinary research and the intrinsically individualistic view of scientific feats, which becomes most evident in the reward system that is the essence of its valuation. Historical tragi-comedies related to the Nobel Prize – namely, who is in and who is out when the feat itself has been carried out collectively – exemplifies such contradictions. Intriguingly, this kind of contradiction demonstrates a striking kinship with the rising discussion about so-called 'organizational accidents' and issues relating to responsibility for them.

Researchers of organizational accidents have repeatedly emphasized that surface causes of large-scale accidents are, in fact, the outcome of an accumulation of chain reactions arising from smaller causes distributed throughout the organization (Reason 1997). The lethal mistake of a nurse who injected a patient with the wrong medication, for example, could have resulted from a series of blunders that went unchecked through various stages in the organization. Because of our individualistic assumptions concerning the cause of such accidents, those who are at the very end of this long chain of causal relations are often accused as the person most responsible.

The reward system in scientific endeavour seems to reflect this individualistic bias. The only difference is that scientists at the very end of such a large chain of causation are individually hailed, rather than put in jail. Alexander Fleming, who has been elevated to the status of a sort of scientific saint, is a case in point, as I realized in my observations of antibiotic labs: the impact of his antibiotic discovery is scaled out of proportion because of its ultimate global impact on various diseases, whereas the more substantial contributions by those who have made antibiotics available for medical purposes have never made the spotlight of the popular imagination (Macfarlane 1984).

In recent media coverage, Gregory Perelman also attracted my attention as the Russian mathematician who finally solved the Poincaré conjecture. Nonetheless, he rejected the Field's Prize, allegedly because of his distaste for the individualistic bias of the present academic award system (O'Shea 2008). A TV programme in Japan blatantly promoted this cult of individualism in science by not even mentioning Richard Hamilton, whose Ricci flow technique Perelman improved before applying it to prove the conjecture, in its dramatic presentation of Perelman as an extraordinary genius coming out of nowhere (NHK Enterprise 2010). These cases above are but small examples of the adamantly predominant individualistic image of scientific practice, which endures even against the growing scale(s) of accumulated contribution of past and present efforts that have culminated in success.

In sum, scales matter in various aspects of our everyday lives, but even more so in interdisciplinary research with the growing collaboration and historically accumulated efforts of the past. Such scaling up of research is Janus-faced: it is necessary for solving the more complex issues that now apply on a global 'scale', but it also confronts us with internal problems of disorganization and friction, along with the persisting individualism in the reward system. The imaginary case of the over-congested operating room easily demonstrates the internal constraints of proper organization in some situations. In other cases, however, the network of collaboration may be further extended, even if a re-examination is needed of the foundations that afford such expansion, either in the form of material infrastructure, or of our belief system that is intrinsically individualistic, if both accidents and scientific success are very often organizational.

Perhaps what Magritte wanted to convey to perplexed viewers with *The Listening Room* was exactly this: the dire consequences of large-scale apple growth in a vacant room – to whose creaking sound we should *listen* as a symptom of its acute need for remodelling – though, in fact, nobody seems to care in the typical manner of his paintings.

# References

Brooks, F. (1975). *The Mythical Man-Month: Essays on Software Engineering*. Reading, MA: Addison Wesley.

Coase, R. (1960). The problem of social cost. *Journal of Law and Economics*, 3: 1–44.

Collingridge, D. (1996). *The Management of Scale: Big Organizations, Big Decisions, Big Mistakes*. London: Routledge.

Edwards, P. (2010). *A Vast Machine: Computer Models, Climate Data, and the Politics of Global Warming*. Cambridge, MA: The MIT Press.

Edwards, P., Mayernik, M., Batcheller, A., Bowker G. and Borgman, C. (2011), Science friction: data, metadata, and collaboration. *Social Studies of Science*, 41: 667.

Fukushima, M. (2016). Constructing 'failure' in big biology: the socio-technical anatomy of the Protein 3000 Program in Japan. *Social Studies of Science*, 46(1): 7–33.

Galison, P. (1992). The many faces of big science. In P. Galison and B. Hevly (Eds.) *Big Science: The Growth of Large-Scale Research* (pp. 1–17). Stanford: Stanford University Press.

Knorr-Cetina, K. (1999). *Epistemic Cultures: How the Sciences Make Knowledge*. Cambridge, MA: Harvard University Press.

LaPorte, T. and Consolini, P. (1991). Working in practice but not in theory: theoretical challenges of 'high-reliability organizations'. *Journal of Public Administration Research and Theory*, 1: 19–47.

Macfarlane, G. (1984). *Alexander Fleming: The Man and the Myth*. Cambridge, MA: Harvard University Press.

NHK Enterprise (2010). *One Hundred Years' Struggle for Poincare Conjecture: Mathematicians Dream of Mushroom Hunting*. (DVD) Tokyo: NHK Enterprise (in Japanese).

O'Shea, D. (2008). *The Poincaré Conjecture: In Search of the Shape of the Universe*. London: Penguin Books.

Parker, J. N., Penders, B. and Vermeulen, N. (Eds.) (2010). *Collaboration in the New Life Sciences*. Farnham, Surrey: Ashgate.

Reason, J. (1997). *Managing the Risks of Organizational Accidents*. Aldershot: Ashgate Publishing Ltd.

Williamson, O. (1981). The economics of organization: the transaction cost approach. *The American Journal of Sociology*, 87(3): 548–577.

# 15

# Speculating

*Alex Wilkie*

For many social scientists, speculating encapsulates the very antithesis of sober empirical research. Where matters of urgent care and concern arise, such as climate change, speculation would appear the least appropriate mode of response-ability. Rather, and so the story goes, practices of speculation have gained notoriety in the contemporary imagination and the predominance of such practices presents a major challenge for those interested in the take-up of speculative thought as part of empirical research. On the one hand, the term connotes risky, irresponsible and opportunistic ventures, often in relation to the mercantile, such as the trading of financial derivatives, resource prospecting in the extractive industries or techniques employed in branding, product and service development to devise new attractions for and forms of consumption. On the other hand, speculation is ascribed to an outmoded branch of philosophy committed to a belief in, and the exercise of, pure imagination. In both cases, speculation relates to the absence, or dilution, of empirical modes of inquiry and thus connotes partiality, incompleteness and a high degree of uncertainty or unreliability. In the terms outlined above, speculation operates in two distinct ways, and often in combination, namely as a cognitive capacity or as an economic activity (Uncertain Commons 2013).

Common to such approaches to the speculative is also a preoccupation with temporality and futures. Social analysis has a long and rich tradition in examining the discursive practices and rationalities by which societal futures, often marked by developments in science and technology, are brought about and managed, perhaps most notably through the identification and management of risk in so-called 'reflexive modernity' (Beck 1992) as well as the construction of hopes and expectations associated with new scientific and technological developments (Brown, Rappert and Webster 2000). Speculation, however, requires a shift in approach from analysing how probabilistic futures are manifested, managed and contested in the present – how actors imagine, model, predict, coordinate and in turn configure the future to the present – to the construction of adequate concepts and devices for exploring possible latent futures that matter. A word of caution is in order here, however: speculation is both prospective and retrospective. It applies as much to the politics of explaining past events (what might have been) as it does to the capturing of future possibilities (what might be).

The notion of speculation has recently been reinvigorated by scholars interested in understanding, explaining and theorizing process, change, novelty, becoming and individuation as inherent features of existence and sociality. Arguably, two distinct realist approaches have emerged on this score. In brief, the first is a philosophical preoccupation with realism and ontology independent of thought and language that has been labelled 'Speculative Realism' (for an introduction see Bryant, Srnicek and Harman 2011) and variously motivated by Quentin Meillassoux's (2008) 'correlationist' argument where thinking and being are viewed as inseparable and thus render knowledge of an independent (ontological) reality beyond thought and language a matter of speculation. The second – one that holds more promise for scholars in the social sciences – draws on the work of A. N. Whitehead, Gilles Deleuze and Isabelle Stengers, in what can be understood as the *constructivist* approach. In this mode, speculative thought becomes a practice of designing and constructing adequate concepts and 'devices' that actively 'relate knowledge production to the question it tries to answer' (Stengers 2008: 92) and in so doing, the researcher, researched, research device and question become-with one another. As Bruno Latour (2003) points out, however, constructivism must not be confused with – or rather, must be rescued from – social construction and, for that matter, not mistaken for deconstructionism. That is to say, what counts as 'real' is a question of the strength and durability of human and non-human distributions (constructions), and not simply the 'meanings' ascribed by persons. Thus, to relate knowledge production to the question it tries to answer, and to Whitehead's ontological principle that states 'there is nothing which floats into the world from nowhere' (1978[1929]: 244) including the researcher, the research question and the instruments of research.

What, then, is a constructivist approach to method? One way to answer this lies in the contrast between constructivist and 'critical' approaches. For Latour (2004), constructivism aims to produce more adequate explanations rather than debunk the explanations of others. Here, debunking can be viewed as allied to the impetus to provide access to concrete truth hidden by explanation or abstraction. Constructivism, by contrast, does not aim to debunk, demystify or deflate but rather to strengthen the production of knowledge claims by actively acknowledging the practices and techniques by which knowledge comes into being. In this approach, constructivism concerns the fabrication of concepts and explanations as well as the devices, techniques, practices that partake in the milieu of the research event. For scholars interested in inventive methods (e.g. Lury and Wakeford 2012), it is precisely the question of devising adequate devices and instruments for research that matters. This involves a move from device-centred analysis (Marres 2012: 81) to a device-centred method. If, as Stengers argues, the speculative tools for philosophy are adequate concepts that act as 'lures' for thought, then for social researchers, arguably, research instruments are the tools for producing conceptions of the world that matter. In accounts of Whitehead's speculative philosophy as well as work that draws inspiration from Whitehead, it is precisely the practico-theoretical preoccupation with the experimental craft and care of the devices (conceptual, linguistic, material, machinic) that are made that comes to the fore.

In this constructivist light, speculative method can therefore be understood as a matter of situated becoming-with the researched. One way to think through this, or to approach the becoming-with nature of empirical research has been variously developed by Mike Michael. For example, Michael (2012: 535–536) draws on Mariam Motamedi Fraser's (2010) discussion of the notion of *event* in the work of Whitehead, Stengers and Deleuze. Briefly put, Fraser identifies two ways of understanding the event-as-process. On the one hand, the event denotes a process whereby elements combine and be together, retaining their individual properties. On the other hand, and in the Deleuzian sense, the event is the becoming-together of all the entities and phenomena involved in the process. As such, the event denotes a 'becoming together' of an event's components rather than a 'being together'. The implications of this for anti-reductionist research is threefold: first, to take seriously the historicity and specificity of phenomena under study; second, to include the becoming of the practices, concepts and technologies which give rise to that which participates in the research event; and, third, the upshot of the research event is a matter of practical investigation rather than the preserve of philosophical thought. On this last score, the act of speculating, understood as the engagement with the unfolding nature of 'social' or 'cultural' phenomena (which are necessarily heterogeneous), is also a practical matter of investigation. Crucially, this 'practical matter of investigation' can entail an interdisciplinary array of techniques designed to address processes of becoming.

Alongside an interdisciplinary team of designers and social scientists, I have taken up the challenge of speculating as an experimental approach to method in the context of energy-demand reduction and Twitter (Wilkie, Michael and Plummer-Fernandez 2015).[1] This involved the design, deployment and analysis of automated software robots (bots) to explore the existing as well as virtual, or immanent, dynamics of Twitter, not least in how this mediated energy-demand reduction practices. Most existing social media studies are preoccupied with numerical magnitude, value and force (e.g. frequency of tweets, quantity of followers, number of retweets) that appropriate measures of relevance and importance pre-built into the platform. By comparison, this project involved three different bots that were designed to intervene in Twitter exchanges (often nonsensically) to probe the proliferation and propagation of new political configurations, environmental concerns and practices (including Twitter practices).

Here, what was provisionally disclosed by our experiment into social media and environmental engagement was that energy-demand reduction communities are transient and are made and undone by indiscriminate accumulation (of followers, communications, etc.) as much as by explicitly shared content. This gave rise to the very concrete possibility that almost any activity or event can be subsumed into energy-demand reduction, and that, relatedly, ambivalence

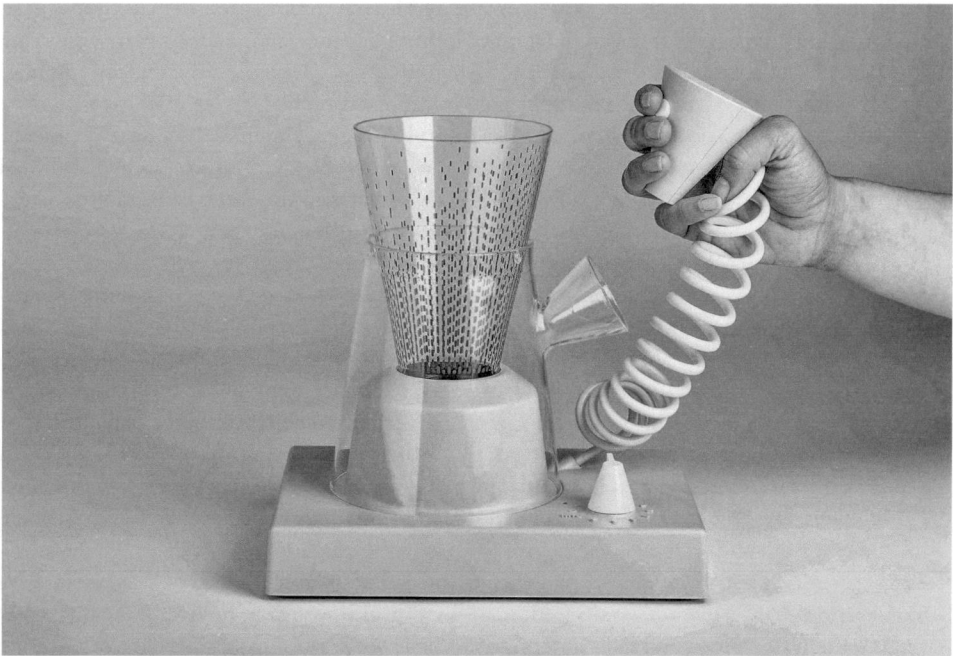

*Figure 5.15.1*   The Energy Babble (source © the author)

and disinterestedness do not necessarily prohibit engagement. What the above involved was reciprocal capture (Stengers 2010: 90) whereby the bots operated to disclose their own agency and capacities, and through their interventions (and the responses they precipitated) disclosed the emergence of other energy-related (human/non-human) actors on Twitter.

The bots, mentioned above, went on to resource the design and three-month deployment of 21 Energy Babbles (see Figure 5.15.1), automated talk-radio like research devices that were also informed by a plethora of other techniques, including workshops and ethnographic engagements. Developed by an interdisciplinary team of designers and social scientists, the Energy Babbles were distributed to members of seven local UK energy communities (in Cornwall, Devon, Kent, London, Norfolk and Nottingham). The babble device – using speech synthesis – drew on an aggregate of online environmental and energy-demand reduction content, such as Twitter feeds and scraped content, and live UK national grid updates to produce new, and often barely intelligible, spoken statements based on the corpus of content it collected.[2] Using the handset microphone, community members also had the ability to add to the cacophony of energy-related speech and were encouraged to share views, practices as well as respond to the device's output.

As such, and at base, the Babble sought to articulate (connect and express) the problem space of energy-demand reduction by scrambling, interjecting, provoking and inviting responses from both local community members and social media users. In effect, these articulations raised the possibility of a speculative engagement (for both researchers and community members) with energy-demand reduction and climate change. Consequently, during the three-month deployment of the devices to the energy communities, various unforeseen practices and sensibilities emerged with the Babble in its various eventuations across and within the communities. So, in some cases, the Babble was 'instrumentalized', used by community members in their efforts to create a formal alliance of local energy communities, or to promote community engagement and outreach events, or as part of experimental home set-ups. However, the Babble also opened up different modes by which communities enacted energy and energy-demand reduction; this included, on the one hand, a proliferation of instrumentalities, and, on the other, a querying of the very idea of instrumentality (Wilkie and Michael 2018). Along the way, the communities themselves also emerged and underwent change in various ways.

To summarize thus far: drawing on social science and design techniques, bots were developed to intervene and trace the unfolding of 'energy communities' on Twitter, which then informed the design and implementation of the Energy Babble which similarly intervened and traced the unfolding 'energy communities' in relation to seven local UK energy communities.

Now, given that these interdisciplinary speculative methods 'provoke' their objects of study one might query their value. Is not becoming (in this above case, of energy communities) simply the result of – the reaction to – a specific set of speculative interventions? In other words, can particular procedures be valued and formalized as a guide for the conduct of (speculative) research practices? There is no easy answer to this. But perhaps social and cultural researchers can draw on Martin Savransky (2016: 201–203) who favours a modest and practical approach to grounded speculative experimentation. This would seek to broaden the composition of empirically given situations through concepts and tools that are themselves actively involved in the very construction of possibilities that emerge from those situations. Put differently, but nevertheless in keeping with the ethos of constructivist speculation, speculating through interdisciplinary methods is a process that is itself not always transparent even to the practitioners involved (who are themselves becoming-together). Modesty would seem to be inescapable given that, in light of the preceding point, 'we don't consider ourselves authorized to believe we possess the meaning of what we know' (Stengers 2005: 995).

## Notes

1 This research was conducted as part of the project 'Sustainability Invention and Energy-demand Reduction: Co-Designing Communities and Practice' (ECDC for short) and was one of seven projects funded under the Research Councils United Kingdom's (RCUK) Energy Programme (project code ES/1007318/1).
2 See Gaver *et al.* (2015) for a more in-depth description of the Energy Babble.

## References

Beck, U. (1992). *Risk Society: Towards a New Modernity*. London: Sage.

Brown, N., Rappert, B. and Webster, A. (Eds.) (2000). *Contested Futures: A Sociology of Prospective Techno-Science*. Aldershot: Ashgate.

Bryant, L. R., Srnicek, N. and Harman, G. (2011). *The Speculative Turn: Continental Materialism and Realism*. Melbourne, Australia: re.press.

Fraser, M. (2010). Facts, ethics and event. In C. B. Jense and K. Rödje (Eds.) *Deleuzian Intersections: Science, Technology and Anthropology* (pp. 57–82). New York, NY and Oxford: Berghahn Books.

Gaver, W., Michael, M., Kerridge, T., Wilkie, A., Boucher, A., Ovalle, L. and Plummer-Fernandez, M. (2015). Energy Babble: mixing environmentally-oriented internet content to engage community groups. In *CHI 2015 Proceedings of the 33rd Annual Conference on Human Factors in Computing Systems* (pp. 2055–2064). New York: ACM Press.

Latour, B. (2003). The promises of constructivism. In D. Ihde (Ed.) *Chasing Technoscience: Matrix for Materiality* (pp. 27–46). Bloomington, IN: Indiana University Press.

Latour, B. (2004). Why has critique run out of steam? From matters of fact to matters of concern. *Critical Inquiry*, 30(2): 225–248.

Lury, C. and Wakeford, N. (Eds.) (2012). *Inventive Methods: The Happening of the Social*. London and New York, NY: Routledge.

Marres, N. (2012). *Material Participation: Technology, the Environment and Engaging Publics*. London and New York: Palgrave Macmillan.

Meillassoux, Q. (2008). *After Finitude: An Essay on the Necessity of Contingency*. London and New York, NY: Continuum.

Michael, M. (2012). 'What are we busy doing?': engaging the idiot. *Science, Technology & Human Values*, 37(5): 528–554.

Savransky, M. (2016). *The Adventure of Relevance: An Ethics of Social Inquiry*. London: Palgrave Macmillan.

Stengers, I. (2005). The cosmopolitical proposal. In B. Latour and P. Weibel (Eds.) *Making Things Public* (pp. 994–1003). Cambridge, MA: MIT Press.

Stengers, I. (2008). A constructivist reading of process and reality. *Theory, Culture & Society*, 25(4): 91–110.

Stengers, I. (2010). *Cosmopolitics I*. Minneapolis, MN: University of Minnesota Press.

Uncertain Commons (2013). *Speculate This!* Durham, NC and London: Duke University Press.

Whitehead, A. N. (1978[1929]). *Process and Reality: An Essay in Cosmology*. Gifford Lectures of 1927–8; corrected edition. New York, NY: The Free Press.

Wilkie, A. and Michael, M. (2018). Designing and doing: enacting energy-and-community. In N. Marres, M. Guggenheim and A. Wilkie (Eds.) *Inventing the Social*. Manchester: Mattering Press.

Wilkie, A., Michael, M. and Plummer-Fernandez, M. (2015). Speculative method and Twitter: bots, energy and three conceptual characters. *The Sociological Review*, 63(1): 79–101.

# 16
# Wedging

*Jane Calvert*

## The wedge in the door

It was during one of our weekly Skype calls that a synthetic biologist suggested what was to become the guiding metaphor for the Synthetic Aesthetics project: the wedge in the door. After three years of working together across disciplines and time zones, this conversation was one of the most significant moments. The metaphor worked for all of us on the project team: for the two synthetic biologists, for the critical designer, and for me and the other social scientific researcher. The brute physicality of a wedge was appealing. The door on synthetic biology was closing, and our project (and the book that we were writing together) would attempt to stop it from doing so.

The three 'disciplines' brought together by the Synthetic Aesthetics project (synthetic biology, critical design, and science and technology studies) are themselves hybrid, heterogeneous and relatively young. There were many things that we did not share, but the wedge became something that connected us. The closing that we were all resisting was the industrialization of biology. We wanted to prevent synthetic biology following unimaginative and entrenched paths – from becoming myopic and monolithic. The synthetic biologists did not want to see their field simply providing 'drop-in' biologically derived replacements for petrochemicals. Instead they aimed to harness the complexity of biology in a manner that was sympathetic to its capabilities (Elfick and Endy 2014). The critical designers wanted to make speculative artefacts that would challenge the futures that are imagined by dominant groups, by exploring ideas such as futility, frivolity and disgust. For me, resisting industrialization meant drawing attention to the fact that there are different ways of imagining how (or whether) we should make use of our increased powers to manipulate the biological world.

This might come across as an argument that participants in an interdisciplinary project should identify and adopt a guiding metaphor, but the wedge was more than a conveniently sharable metaphor, because it expressed a concern with keeping possibilities open, something that was central to our interdisciplinary endeavour. It is also significant that the idea emerged from our collaborative work. It was not a tool developed by one group (e.g. social scientists) to study another (e.g. synthetic biologists). This experience demonstrates the methodological importance of valuing the unplanned and emergent in interdisciplinary research.

## The Synthetic Aesthetics project

The Synthetic Aesthetics project itself was born three years earlier, at a week-long residential 'sandpit' event held in a remote hotel outside Washington DC, and funded by the UK's Engineering and Physical Sciences Research Council and the US's National Science Foundation. Sandpits involve real-time proposal writing and peer review, and grants are awarded on the final day. This rather unusual funding mechanism meant that many of the constraints of more conventionally funded projects (such as pre-defined outputs) were absent in Synthetic Aesthetics, giving us a considerable amount of freedom.

After the sandpit, the project team selected and then paired six scientists/engineers with six artists/designers. The pairs were tasked with investigating design in synthetic biology, with the freedom to take their work in any direction they chose. They were funded to spend time in each other's workspaces, and, somewhat unusually for projects of this sort, the scientists and engineers spent the same amount of time in the art/design studio as the artists spent in the science laboratory.

To give a snapshot of some of the resulting work: one pair looked at synthetic biology from the perspective of geological time (the sweep of which extends from the beginning to the end of the Earth), another made cheeses from the bacteria that grow on human skin, and another explored how plant cells 'compute' using shape and form. Significantly, all the pairs engaged in designing *with* biology rather than the design *of* biology. Their work highlighted the interconnectedness and evolution of living systems, the different dimensions of biological temporality, and our coexistence with our bacterial symbionts. The result of the project was a hardback, picture-filled book (Ginsberg, Calvert, Schyfter, Elfick and Endy 2014), substantial enough to be used as a doorstop.

## Opening up

One of the reasons why the wedge in the door was instantly appealing to me is because, by resisting closure, I saw it as a form of 'opening up' – an idea I have found myself returning to repeatedly in my work in the highly politicized field of synthetic biology. Stirling (2008) describes opening up in policy contexts as drawing attention to the often implicit framing conditions and assumptions that underlie discussions of a technology, which he argues can enable new questions to be asked, neglected issues to be addressed, and alternative technological pathways to be explored.

In my experience, interdisciplinary work can provide unique opportunities for opening up. For example, in the Synthetic Aesthetics project we all engaged in challenging and expanding each other's ideas about the future(s) of synthetic biology, as well as each other's ways of seeing the world. The latter is, for me, one of the thrills of interdisciplinary collaboration. It might be exciting, but opening up is not always comfortable, of course. If a crevasse opens up you can fall into it, or at the very least get vertigo (as experienced by Agre (1997) on transitioning from the technical to the social sciences). We all needed to move away from the edge and return to our (rather unstable) disciplinary homes with something to show for our joint efforts.

## Valuing interdisciplinary work

The value of both the Synthetic Aesthetics project and the book that we produced together has always been difficult to negotiate. The book itself was an unsatisfactory compromise for the artists and designers, who would have valued an exhibition far more. Although the academics on the project did receive institutional approval for engaging with professionals from outside the academy (ticking the 'impact' box), the project was simultaneously devalued by some for not being 'science' and, despite our best efforts, it has consistently been misinterpreted as PR or 'outreach' for the field of synthetic biology.

This is clearly problematic for me as a social scientist in respect to my own professional accountability, and also because it does not capture the value of what I found to be uniquely stimulating work, which I think embraced the social scientific virtue of critique. I have no interest in delivering artistically mediated outreach for synthetic biology projects in the future, but this is something that the synthetic biology community values highly, particularly in the context of perceived problems with public acceptance of the technology (see Marris 2015).

Outreach is the one-way dissemination of a particular worldview, but a wedge, in contrast, is something that stops, obstructs and resists closure. It is a metaphor that allows for dissonance, disagreement and ongoing challenge. But despite its crude force, as we all know from experience, attempts at wedging doors often fail, and spaces that were once open can close again (Callard and Fitzgerald 2015). As yet, it is too early to judge whether or not we were successful in our shared endeavour to stop synthetic biology from closing.

We did not write about the metaphor of the wedge in the door in the Synthetic Aesthetics book because the critical designer on the project team regarded it as 'too self-deprecating'. Admittedly no one wants their published book to be used for a mundane physical task of propping a door open (although as Latour (1992) has shown, the value of such mundane objects should not be taken for granted). The wedge might be a rather unsophisticated device, but I have decided to revive it here because its simplicity and force have stayed with us all as we have moved on to new work.

Six years after the project first started, one of the synthetic biologists emailed the project team. The email had no content apart from an attached image – a black foreground entitled 'industrialization of biology' and a door opening in the blackness, wedged open by our book, with blue skies beyond.

## Acknowledgements

This research was supported by the following grants: Synthetic Aesthetics (EP/H01912X/1), Engineering Life (ERC/616510) and SynthSys-Mammalian (BB/M018040/1).

## References

Agre, P. (1997). Toward a critical technical practice: lessons learned in trying to reform AI. In G. Bowker, L. Gasser, S. Star and W. Turner (Eds.) *Beyond the Great Divide: Social Science, Technical Systems, and Cooperative Work* (pp. 131–158). Mahwah, NJ: Erlbaum.

Callard, F. and Fitzgerald, D. (2015). *Rethinking Interdisciplinarity across the Social Sciences and Neurosciences.* Basingstoke: Palgrave Pivot.

Elfick, A. and Endy, D. (2014). Synthetic biology: what it is and why it matters. In D. Ginsberg, J. Calvert, P. Schyfter, A. Elfick and D. Endy (Eds.) *Synthetic Aesthetics: Investigating Synthetic Biology's Designs on Nature* (pp. 3–25). Cambridge, MA: MIT Press.

Ginsberg, D., Calvert, J., Schyfter, P., Elfick, A. and Endy, D. (2014). *Synthetic Aesthetics: Investigating Synthetic Biology's Designs on Nature.* Cambridge, MA: MIT Press.

Latour, B. (1992). Where are the missing masses? The sociology of a few mundane artifacts. In W. E. Bijker and J. Law (Eds.) *Shaping Technology/Building Society: Studies in Sociotechnical Change* (pp. 225–258). Cambridge, MA: MIT Press.

Marris, C. (2015). The construction of imaginaries of the public as a threat to synthetic biology. *Science as Culture*, 24(1): 83–98.

Stirling, A. (2008). 'Opening up' and 'closing down': power, participation and pluralism in the social appraisal of technology. *Science, Technology & Human Values*, 33: 262–294.

# Index